ELECTRONIC DESIGNER'S HANDBOOK

THIRD EDITION

ELECTRONIC DESIGNER'S HANDBOOK

a practical guide to transistor circuit design

T.K. HEMINGWAY, BSc (Hons)

Principal Circuit Consultant
British Aerospace Dynamics Group
Stevenage

BUSINESS BOOKS
COMMUNICA - EUROPA

First published by Business Publications, 1966
Second impression, 1967
Second edition published by Business Books, 1970
Third edition, 1979

ISBN 0 220 66339 4

Printed in England by T. J. Press (Padstow) Ltd., Cornwall
and bound by R. Hartnoll Ltd., Bodmin, Cornwall
for the publishers, Business Books Limited,
24 Highbury Crescent, London, N5.

Contents

Part Three USEFUL TECHNIQUES

Part Four APPENDICES

Preface

The newly qualified electronic engineer often finds difficulty in applying his technical knowledge to practical circuit design. Typical stumbling blocks are the choice of suitable operating currents and voltages, the fixing of these regardless of temperature variations which affect transistor behaviour, and conversion of the results of analysis into actual numbers when many of the transistor parameters included are not mentioned in the manufacturer's data sheets.

The object of this book is to explain circuit design methods which enable the engineer to overcome these obstacles and to design practical circuits. Part One describes these basic techniques, and emphasis is placed on designing circuits to operate correctly in spite of ambient temperature variations and spreads in transistor parameters. In the author's opinion it is more important that the engineer should understand the basic techniques of design than that he should acquire a superficial knowledge of a great number of circuits. Consequently only a few circuits are dealt with, but these are examined in very great detail. When the techniques are understood, the reader should have little difficulty in applying them to any circuit provided its mode of operation is known.

Part Two shows how novel designs can be synthesized and put into practical form; this section will also interest the experienced designer, as it contains several unusual circuits.

Practical difficulties in design and testing of circuits are examined in Part Three, some of the problems discussed rarely being mentioned in standard textbooks.

Approximations used in the text are justified where necessary by derivations given in the Appendixes.

I should like to thank Mr. P. Broderick who helped clarify many obscure points; the management of Marconi Instruments Limited for permission to publish this work; Mr. P. L. Burton, formerly of English Electric Company, who first taught me the principles of circuit design; and finally my wife who undertook the exacting task of preparing the typescript.

March, 1966 T.K.H.

Preface to the Third Edition

In this third edition, Chapters 1, 2, 3 and 6 have been expanded to include information on devices that have become commonly available since the publication of the first edition. Chapter 4 has been largely rewritten with the help of Peter Beatty (British Aerospace Dynamics Group). New Chapters 5 and 12 are devoted to integrated circuit operational amplifiers and their applications, in acknowledgement of the major role played by these devices in modern analogue circuit design. Chapter 17 is new and deals with the useful technique of Consequential Damage Analysis.

The opportunity has been taken to revise existing circuit drawings and to prepare the new drawings to conform with BS 3939, and to correct some errors present in earlier editions.

It is hoped that this considerably enlarged version of the Handbook preserves the style of the original edition with up-to-date format and contents.

November 1978 T.K.H.

Part One

BASIC ELEMENTS OF ANALOGUE CIRCUIT DESIGN

Chapter 1

Semiconductor diode properties

Like the thermionic diode, the semiconductor diode has a current–voltage relationship which changes with the polarity of the applied source.

FORWARD CONDUCTION

When the anode is made sufficiently positive with respect to the cathode, forward conduction begins, as shown in region (*a*) of Fig. 1.1. The voltage which has to be applied before appreciable forward current flows depends on the semiconductor. It cannot be given an absolute value unless the forward current which is considered 'appreciable' is more closely defined, but this 'turn-on' voltage, as it is often called, is approximately 0·1 V for germanium and 0·5 V for silicon at 25°C. As the temperature varies, this voltage (and, in fact, the whole voltage scale of the forward characteristic) changes at the rate of −2 to −2·5 mV/degC.

For higher forward voltages the current increases exponentially and is eventually limited only by the capability of the source and the bulk-resistance of the diode, although a lower limit must be adhered to in practice in order to avoid catastrophic failure due to overheating.

The forward incremental resistance of the diode is of importance in the design of circuits where a diode is subjected to signals of varying amplitude. It is defined as the rate of change of voltage with respect to current at a specified point on the characteristic. Because of the exponential law relating voltage and current, this quantity becomes smaller as the current or voltage at which it is specified increases; in fact, to a close approximation the incremental resistance is inversely proportional to the current at the specified operating point. Its actual value depends on the area and construction of the diode junction, being low for large junctions.

To summarize the properties of the forward characteristic, conduction begins at about 0·1 V for germanium and 0·5 V for silicon, at 25°C; incremental resistance at any current is inversely proportional to the current; the forward characteristic variation with temperature takes the form of a linear shift of −2 to −2·5 mV/degC in the forward voltage at any given current (*see* Fig. 1.1). Maximum permissible temperatures of operation are 90–100°C for germanium and 125–200°C for silicon, depending on the manufacturer. These are junction temperatures which are the sum of ambient temperature and temperature rise caused by the mean power dissipated in the junction.

REVERSE CHARACTERISTIC

When the anode becomes negative to the cathode, the current which flows is only a small proportion of the forward current which would flow if the polarity were reversed. This is shown by region (*b*) in Fig. 1.1, and the small current which does flow is called the *reverse leakage current*. This current is hardly dependent on applied voltage (until region (*c*) is reached) but depends greatly on junction temperature. Its variation with temperature is an exponential tending to infinity and is remembered most easily by the law that it doubles every 9 or 10°C rise.

The semiconductor material and junction area determine the order of leakage. For instance, a germanium diode of small area may have a reverse leakage of 2 μA compared with a silicon diode of the same area having about 20 nA, both at 25°C. Large-area power diodes might have values of 0·5 mA for germanium and 5 μA for silicon.

Because many circuits involve currents of less than 1 mA, leakage currents often cause changes of circuit operation when changes of ambient temperature occur. For this reason it is important when using diodes to know the maximum possible value of leakage current at the maximum expected junction temperature.

Often the required figures are not available from the manufacturer's data, since even if a high temperature value is quoted this rarely corresponds to the desired maximum temperature of operation. Fortunately, by applying the 'doubling every 9 or 10 degrees rise' rule it is simple to transform the data, e.g. a leakage of 100 μA at 100°C will be 50 μA at 90°C, 25 at 80°C, 12·5 at 70°C, and so on.

Some manufacturers give only a low temperature figure and although it might be thought just as easy to apply the rule to obtain values for higher temperatures, this will usually give a highly pessimistic figure since, strictly, not all the leakage current is subject to the 'doubling' law. Leakage caused by surface effects and not by semiconductor action remains more or less constant with temperature. The error caused by extrapolating the total leakage is usually negligible for germanium when using the 25°C value, but is often large for silicon. This is particularly true of silicon planar types, where semiconductor leakage is very low compared with surface effects at 25°C. There is no way of correcting for this, in the absence of further information from the manufacturer. The non-varying component of leakage values at 100°C or more can generally be assumed to be negligible; and, in general, the higher the temperature for which the leakage applies, the less will be the error in applying the 'doubling' law.

Breakdown, Zener, and Avalanche Phenomena

When the applied reverse voltage is increased in magnitude, the leakage current eventually rises and tends to a very high value limited (like the forward current) only by the capability of the source and the diode bulk-resistance (*see* region (*c*) in Fig. 1.1).

Depending on the material and construction of the junction, this effect may occur at voltages between about 2·5 V and several thousand volts. By appropriate doping of the semiconductor the voltage at which this current increase takes place can be controlled within close limits. Since the critical voltage is found to be constant throughout the life of the diode the effect, at first sight a defect of the semiconductor diode, has proved to be very useful in practical circuits.

The critical voltage is often referred to as the 'breakdown voltage'; in spite of this name the diode is not damaged, provided the product of this voltage and the current which is allowed to flow does not represent sufficient power to overheat the junction.

Zener diodes

Diodes deliberately designed to exhibit this effect at particular voltages are called Zener diodes. They are specified in several respects: the voltage corresponding to a certain current in the region (*c*) (*d*)—Fig. 1.1; the temperature coefficient of this voltage; the slope between (*c*) and (*d*) expressed as a resistance; and the permissible maximum current or power at various temperatures. At present they are invariably silicon diodes.

As will be seen later, the most common uses for the Zener diode are in obtaining stable supplies, in coupling between stages where

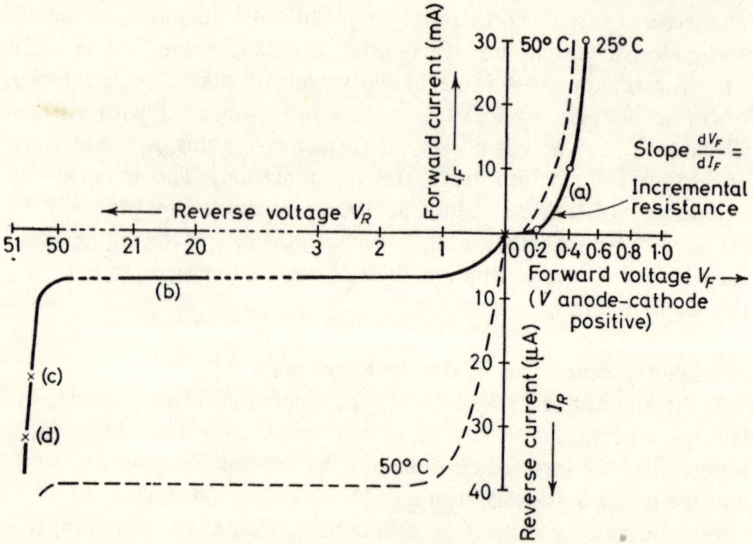

FIG. 1.1 Diode characteristic. Note differing forward and reverse scales

d.c. levels differ and in clipping circuits. Since the Zener diode is one of the most useful circuit elements, the orders of magnitude of the parameters should be remembered by the designer. This is fortunately easy, because all Zener diodes of a particular power rating have similar characteristics, regardless of the manufacturer.

The relevant facts are:

(1) The incremental resistance at a given current varies according to the nominal Zener voltage, the minimum resistance being obtained with a 7·5 V device. Typical figures are: 100 Ω for a 3·3 V or 16 V unit, 60 Ω for a 5·6 or 13 V unit and 15 Ω for a 7·5 V unit (all at 5 mA). These figures become proportionately lower as the current is increased.

(2) The nominal voltage is obtained at a particular operating

current which depends on junction size; for low-power Zener diodes (300 mW at 25°C) this current usually lies between 5 and 10 mA and is not critical.

(3) The temperature coefficient depends on the nominal Zener voltage. Below about 5·6 V the coefficient is negative, and above 5·6 it is positive. At 5·6 V the coefficient is roughly zero. Even if the tolerance on nominal Zener voltage is ignored, the critical voltage for zero coefficient is not exact, and for a nominal 5·6 V unit the usual coefficient is ±0·02 per cent per degree C. A 3·3 V unit is from −0·05 to −0·09 per cent per degree C and a 10 V unit is from +0·05 to +0·09 per cent per degree C. It is worth remembering that 3·3 and 6·8 V units have an actual voltage change with temperature of similar magnitude to that of a transistor or diode junction, namely from 2 to 2·5 mV/degC. This is often useful in circuits where the coefficients of Zener diodes and transistors can be made to add or subtract to give some degree of compensation.

Most manufacturers will supply specially selected units with a coefficient of ±0·002 or even ±0·0005 per cent per degree C. There are three methods by which selection is done. The simplest is to pick 'good' diodes of nominal voltage 5·6 V from a batch by measuring the coefficients at a given current. This is not usually satisfactory for ultra-stable circuits, since the coefficient of such a unit invariably changes with the actual temperature, e.g. ±0·005 per cent at 25°C, −0·01 per cent at −10°C, +0·01 per cent at 60°C, and changes in operating current alter the temperature for which the coefficient is zero. A second method is to carry out the above tests, but at two temperatures, e.g. 0 and 60°C, then choose units which when put in series yield ±0·005 per cent per degree C at both temperatures (one may be −0·01 at 0°C and +0·02 at 60°C, the other +0·01 and −0·025). This is a better method but is still not entirely satisfactory, because at other than the two selected temperatures the coefficient is uncontrolled (and tends to be >0·01 at −20°C) and the series addition naturally results in twice the incremental resistance, the nominal voltage being 11·2 V. The third method consists in adding normal diodes in series with a Zener diode so that the temperature coefficients cancel. Thus a 6·8 or 8·2 V Zener diode and one or two diodes in series can by careful selection yield a very small temperature coefficient. This gives the best results of the three methods when the units are to maintain a low coefficient over a very wide range of temperatures, because the coefficients which are being cancelled are more or less constant with temperature (only for units near 5·6 V nominal is this not so). Moreover, the incremental resistance is generally no worse than that of a single 5·6 V Zener diode. The resulting voltage of such a compound unit is in the 8–10 V region.

A similar idea is to cancel the Zener coefficient against the base-emitter coefficient of a transistor. This arrangement is particularly useful in stabilized power supplies where the transistor acts as a reference amplifier (*see* Chapter 9).

'Reference' and 'regulator' Zener diodes. The low-power Zener diode of about 300 mW maximum dissipation at 25°C is often called a 'reference diode' since one of its main uses is for producing a reference voltage in a stabilized supply. Higher-power Zener diodes of 1 W, 10 W, and even higher 25°C ratings are often used as simple direct regulators, the load being connected directly to the Zener diode. These high-power units are therefore known as regulator (Zener) diodes. Because of their large junction area, values of impedance are much lower than the typical reference diode values and currents are much higher, e.g. a 6·8 V, 10 W unit could be run at 200 mA giving an impedance of about 5 Ω.

High-frequency Effects

At high frequencies the junction capacitance of a semiconductor diode becomes significant. Its actual value depends on the construction used, on the area of the device, and also on the applied voltage if reverse-biased. Capacitance is always less at high reverse voltages and its law of variation is given by $C = K/V_R^n$, where n lies between 1/2 and 1/3.

This dependence of C on V_R is useful in enabling resonant circuits to be tuned by variation of applied voltage, the diode being used as a capacitive tuning element. Since the diode is reverse-biased, little current is taken from the voltage source and the arrangement is therefore suitable for remote control, there being negligible voltage drop even in long leads.

One snag is that the diode capacitance is lossy at high frequencies, so that circuit Q-values are reduced, but much improved results are being obtained with diodes specially designed for this application.

Hole storage

The phenomenon of hole storage occurs when a diode which has been conducting in the forward direction is rapidly reverse-biased.

Instead of cutting off and passing a normal reverse circuit, a semiconductor diode will remain conducting for a time τ_s (just as if its connections had been reversed at the same time as the current reversal) and then will suddenly cut off. The value of τ_s depends on the type of diode junction and its area (being greater for large-area devices), and is proportional to the forward current which had been flowing prior to the reversal and inversely proportional to the current flowing immediately after reversal.

Manufacturers' data gives a value 'Q' which means stored charge (no connection with circuit quality factor Q) for stated conditions; knowing the current I_R flowing after reversal, the time τ_s before conduction ceases is given simply by $\tau_s I_R = Q$, provided Q is the value for conditions just before reversal.

DIODE APPLICATIONS

Diodes are often divided into the two classes, signal diodes and rectifier (or power) diodes. This is an arbitrary division and is based

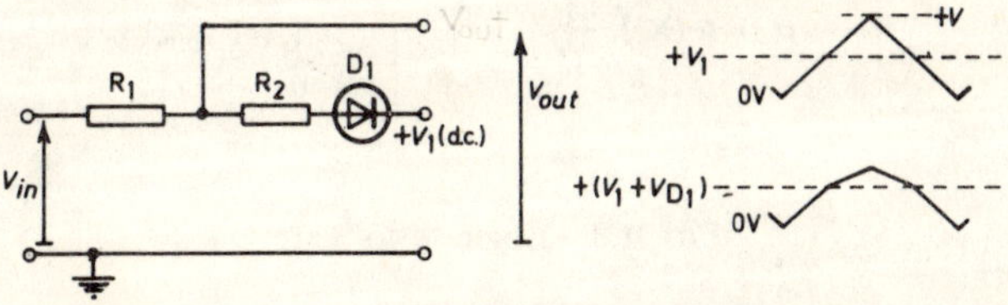

FIG. 1.2 Simple diode shaping circuit

on the application in which the manufacturer expects a particular device to be most useful. A large-area device intended for use as a power rectifier will normally be of little use for rectifying signals of 100 kHz, because its capacitance is likely to be high and its hole storage large and unspecified. Naturally, many borderline types exist which are useful for relatively low-power rectification and also for many audio-signal applications (e.g. Mullard OA202, Texas Instruments 1S121 and 1S922).

There are many applications of diodes which will be readily invented by the designer as required. For a simple example, suppose a triangular voltage waveform has been generated and it is required to lessen its rate of rise whenever its voltage exceeds a certain level V_1. Figure 1.2 shows a simple solution. V_{out} will follow V_{in} until its potential reaches V_1 plus the 'turn-on' level for D_1. Above this voltage alternation occurs, so that the waveshape given in the diagram is obtained.

This circuit is simple to understand, but note the slight error caused by V_{D1}, the forward drop before significant conduction begins, and note also that the reverse leakage current I_R for D_1 must be such that $R_1 I_R$ is negligible in comparison with V_{in}. At high frequencies this circuit will exhibit two faults: first, the capacitance of D_1 will load R_2, and, secondly, hole storage will cause the waveform to change slope at a higher point when rising than when falling. The first effect can be ignored if $C_D(R_1 + R_2)$ is small compared with the triangle period; the second is more complicated as it depends on the Q for the diode and on R_1 and R_2, since these determine the conditions prior to D_1 being reversed.

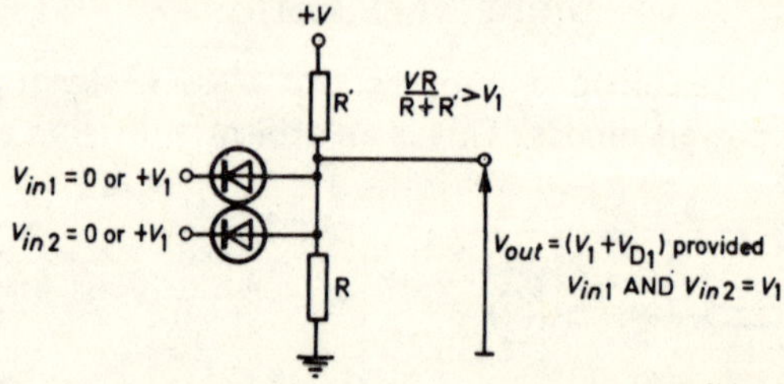

FIG. 1.3 Logic 'AND' gate

Difficulties will therefore be experienced if high-frequency operation is attempted; 'high-frequency' is a relative term and implies here a cyclic period so small that τ_s and $C_D R_2$ are a significant fraction of the period.

As indicated above, this type of circuit can be invented as required, and the same is true of the many diode logic gate circuits, of which one example is given in Fig. 1.3.

This is known as an 'AND' gate because if V_{in1} and V_{in2} can each have only the values 0 and $+V_1$, V_{out} remains at 0 unless both V_{in1} AND V_{in2} are at V_1. In this case V_{out} rises to

$$+V_1 + V_{D1} \quad \text{if} \quad \frac{VR}{R + R'} > V_1 + V_{D1}$$

This and many similar gates are often used in computer circuits, and their basic operation is obvious. The difficulty in designing this kind of circuit is mainly concerned with high-speed operation.

Rectifier Circuits

Much more subtle in operation are the rectifier and d.c. restorer circuits shown in Figs. 1.4 and 1.5. Although in very common use and apparently simple in function, the complete analysis of these circuits is difficult.

In the rectifier circuit (Fig. 1.4) diode D_1 conducts only when V_{in} is more positive than V_{out} so that with either pulse or sine-wave input

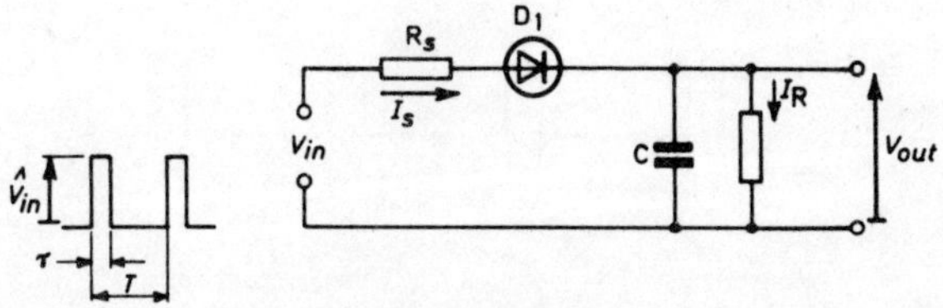

FIG. 1.4 Rectifier circuit

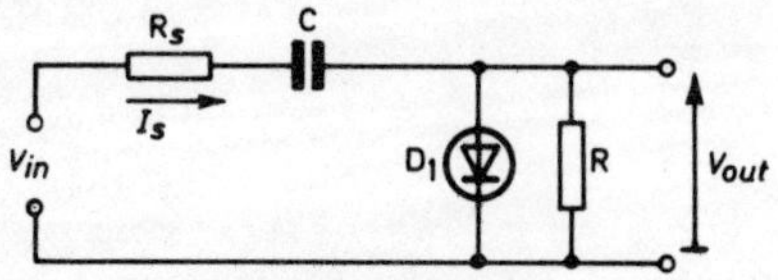

FIG. 1.5 Restorer circuit, d.c.

waveform conduction occurs as soon as V_{in} rises towards V_{in}. As a result, C charges through R_s and D_1 and may or may not reach $+\hat{V}_{in}$ before V_{in} descends towards zero. As soon as V_{in} has descended below V_{out}, C discharges through R towards zero until V_{in} rises again.

When used for power rectification, the object being to obtain d.c. from a.c. with the minimum of added ripple, the time constant CR is so large compared with the signal period that little discharge takes place. If R_s is small, then V_{out} (d.c.) is approximately equal to the positive input peak voltage for any input waveform; a circuit designed to operate in this way is known as a peak rectifier. In practice the source and diode possess some resistance, so that perfect peak rectification is impossible.

When R_s is so large that $CR_s \gg T$, and RC is also $\gg T$, V_{out} depends only on the positive average value of V_{in} and the circuit is then known as an average rectifier.

As illustrated, Fig. 1.4 is called a half-wave rectifier since D_1 completely ignores one half-cycle of the input signal; a full-wave rectifier uses two diodes so arranged that both half-cycles contribute to the d.c. output (Fig. 1.7). This has the advantage that for similar ripple performance to the half-wave circuit, C need only be half the value because it discharges through R for only half the signal period before being recharged. Also, the ripple frequency is twice the input frequency, which is advantageous since any additional ripple-

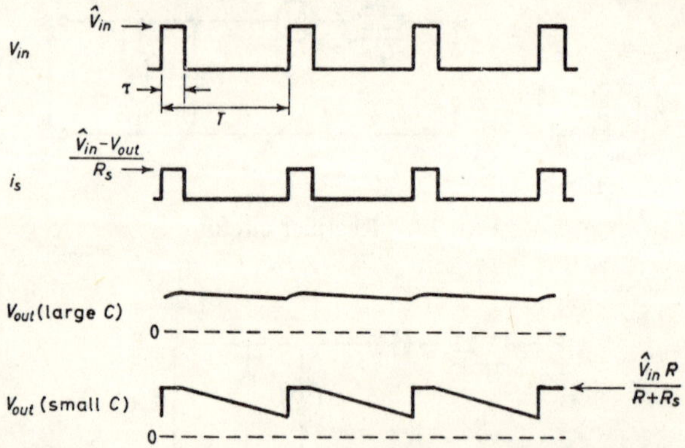

FIG. 1.6 Waveforms for rectifier circuit (Fig. 1.4)

removing circuits which may be required become less bulky as the ripple frequency increases. Practical snags are that a push–pull input is required (achieved by a transformer in Fig. 1.7) and that unequal values of R_{s1} and R_{s2}, or unequal input signals e_1 and e_2, cause the output contribution from two paths to be different, once more producing a ripple component at the input frequency.

Another full-wave circuit, the bridge rectifier (Fig. 1.8), uses four diodes, but no push–pull input is required. This arrangement possesses the advantages of full-wave rectification and avoids the problems of unequal input voltages and source resistance except for different diode characteristics. It also has a very definite advantage when the input has to be coupled through a transformer for reasons of isolation or voltage change. This is discussed more fully in Appendix 1, but the important result is that for a given V_{out} and V_{in}

the transformer core can be smaller if a bridge rectifier is used. The designer must therefore balance the cost of extra diodes against the size and cost of the input transformer before making his decision.

The above description of the principle of peak and average rectification is useful in showing the limit cases where R_s is in one case zero and in the other very large. For intermediate values, which are inevitable in real designs, a more complete analysis is required. The rough consequences of finite R_s and R are, first, that V_{out} could never reach $\hat{V}_{in}$, since during D_1 conduction C could only charge to $\hat{V}_{in}R/(R + R_s)$; and, secondly, that if $CR_sR/(R + R_s)$ (i.e. the charge time constant) is comparable with the time τ of D_1 conduction, C will charge even less than the above value. For the pulse input case

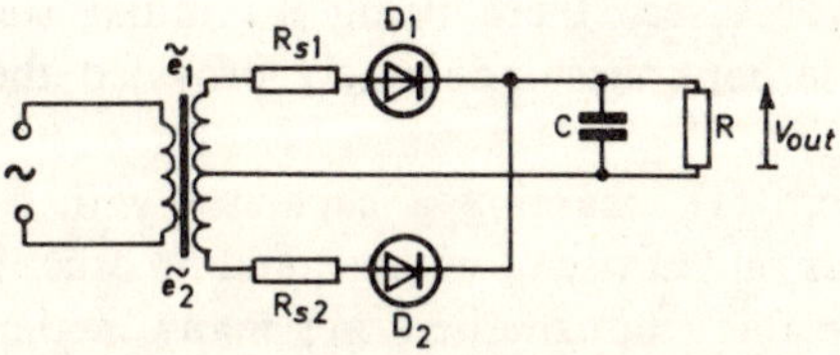

FIG. 1.7 Full-wave rectifier circuit

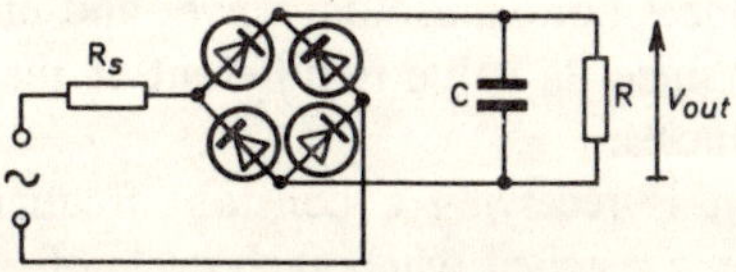

FIG. 1.8 Bridge rectifier circuit

τ is known and the analysis turns out to be simple once the procedure is known; for sine-wave input τ is initially unknown, so that the calculations are more involved (*see* Appendix 1).

The most straightforward method for analysing such circuits is to assume the capacitor carries no charge and then apply the input, noting instantaneous waveforms throughout the circuit. After many cycles an equilibrium state is eventually reached in which the charge gained and the charge lost by the capacitor in one cycle are equal. Knowing the magnitudes of charge lost and gained, equating the two leads to the steady (d.c.) component of voltage or charge which is present on the capacitor in this equilibrium state.

It is useful before demonstrating this technique to recall a few facts about charging capacitors.

(1) The current required to charge a capacitor at a rate of dV/dt V/sec is given by $i = CdV/dt$, where C is in farads and i is in amperes (equally correct and often more useful units are microfarads and microamps).

(2) From (1) it is clear that to charge a capacitor to any voltage in zero time would require infinite current, so that in analysis, if one plate of a capacitor is moved instantaneously, its other plate must move the same number of volts at the same time.

(3) From (1) it is also clear that a finite but instantaneous step of current into a capacitor causes the voltage to rise from zero at a finite rate of i/C V/sec, there being no initial step of voltage. The rate of rise continues indefinitely provided the current step remains.

(4) Again from (1), charging a capacitor with a fixed current produces a linear rate of rise of voltage, and no other form of charging will achieve this (though there are many methods which will produce this constant current).

(5) Regarding a capacitor as a reactance of $1/j\omega C$ is correct only in linear circuits with pure sine-wave input. Any non-linearity (such as a diode) produces non-sinusoidal waves and makes the reactance concept invalid, since its value is different at the fundamental and harmonic frequencies.

(6) If a circuit is receiving a constant frequency input and an equilibrium state is reached when each cycle of operation produces exactly the same waveforms (d.c. and a.c.) as the last, then the charge received and the charge lost by any capacitor in the circuit during one cycle must be equal. If not, the charge would change between successive cycles and waveforms would be different.

Consider now Fig. 1.4, where D_1 is assumed to be a perfect diode in that its turn-on voltage is zero, its forward resistance is zero, and its reverse leakage is zero. Assume also that the time constant CR is many times the input period.

The action of the circuit at the moment V_{in} is applied depends on which part of the V_{in} waveform is occurring at that instant. For the moment let V_{in} be at zero and be just beginning to rise.

Since C has zero charge, V_{out} is initially at 0 and D_1 is just beginning to conduct. As V_{in} rises, D_1 conducts and C charges through R_s, following but not equalling V_{in}. When V_{in} reaches the end of its positive peak, V_{out} is approaching its maximum possible value of $\hat{V}_{in}R/(R + R_s)$, but may not reach this since V_{in} now begins to descend.

As V_{in} descends, V_{out} tends to fall, not because of the fall in V_{in} which cannot itself pull V_{out} downwards (because reverse diode current cannot flow), but because of the discharge path through R. Under the assumed conditions where RC is large compared with a period of V_{in}, V_{out} now falls towards zero at a much slower rate than the descent of V_{in}.

This continues until V_{in} again rises to a level equal to V_{out}, when D_1 conducts and C again charges towards $\hat{V}_{in}$. The voltage waveform across C and the current, i_s, taken from the source are therefore as shown in Fig. 1.6.

Since R discharges C only slightly during one cycle, V_{out} and i_R are almost constant. Hence, the current always tending to discharge C is V_{out}/R and the charge lost by C in one cycle is $T(V_{out}/R)$.

The current which recharges C at the positive peak of each cycle is i_s which flows for a time τ. As the diode is assumed perfect (no forward drop), i_s is given by $(V_{in} - V_{out})/R_s$ so that the charge received by C each cycle is $(V_{in} - V_{out})\tau/R_s$. In the equilibrium state where each cycle is identical with the next, the net charge received by C must be zero, so that $(V_{in} - V_{out})\tau/R_s = TV_{out}/R$, i.e.

$$V_{out} = \frac{R}{R + (T/\tau)\,R_s}\hat{V}_{in}$$

A simple check on this result is given by making $\tau = T$, which represents a steady (d.c.) input of $+V_{in}$, resulting in an output of $R/(R + R_s)\hat{V}_{in}$. If, on the other hand, $\tau = 0$, V_{out} is zero, since C never receives charge. An alternative approach to the calculation is to equate the average power supplied by the source with that dissipated in the resistive elements in the circuit.

Now, the average power supplied by the source is

$$\frac{\tau}{T}\hat{V}_{in}(\hat{V}_{in} - V_{out})/R_s$$

The average power dissipation in R_s is

$$\frac{\tau}{T}\frac{(\hat{V}_{in} - V_{out})^2}{R_s}$$

and that in R is V^2_{out}/R. Hence

$$\frac{\tau}{T}\frac{\hat{V}_{in}(\hat{V}_{in} - V_{out})}{R_s} = \frac{\tau}{T}\frac{(\hat{V}_{in} - V_{out})^2}{R_s} + \frac{V^2_{out}}{R}$$

$$\therefore \quad \frac{\tau}{TR_s}(\hat{V}_{in} - V_{out})V_{out} = \frac{V^2_{out}}{R}$$

$$\therefore \quad V_{out} = \frac{R}{R + (T/\tau)R_s}\hat{V}_{in}$$

as before.

The trap to be avoided in this type of calculation is the assumption that the average power supplied by the source is the same thing as the average power dissipated in the source resistance R_s, and then to equate this to the power dissipated in R. This would lead to

$$\frac{\tau}{T}(\hat{V}_{in} - V_{out})^2/R_s = V^2_{out}/R$$

which is incorrect.

In the case where V_{in} is a sine wave $\hat{V}_{in}$ sin ωt, the above procedures are still correct but the analysis is much more complicated, since the time τ for which source current flows depends on V_{out} and V_{out} in turn depends on τ and R_s; moreover, the input current and voltage are not constant throughout the charging time τ. Average charge received by C, or, if the alternative approach is used, the average power supplied by the source and that absorbed by R_s, must be obtained by integration. This is given in Appendix 1, the results being interpreted graphically in Fig. 1.9. The graphs enable the output voltage and the conduction time τ to be deduced from the values of R_s/R, assuming negligible diode incremental resistance or forward voltage drop. It will be seen that a particular value of R_s, e.g. $R/20$ produces much more loss in V_{out} than a simple estimate would indicate: the drop in V_{out} from the case where R_s is zero is from $\hat{V}_{in}$ to about 0·7 $\hat{V}_{in}$, not, as might be anticipated, 0·95 $\hat{V}_{in}$. The reason is that the current which flows through R_s for a short time each cycle is much larger than the load current, since it has to replenish the charge on C (lost to the load) in this short time τ. The voltage drop in R_s is therefore many times larger than I_LR_s.

For the same reason the power dissipated in R_s is much higher than $I_L^2R_s$; its value is calculated in Appendix 1. Failure to appreciate this can result in unreliable power supplies in which R_s has been inserted to limit $\hat{i}_s$ to a safe level, but where the rating for R_s has been made much too low, so that this component eventually fails.

To take an example, the rectifier circuit for a radio or television receiver is often of this form, where $\hat{V}_{in}$ is 240 $\sqrt{2}$, $R_s = 70\ \Omega$ and $R = 1\ k\Omega$ (effective). Here $R_s/R = 0{\cdot}07$, giving $\tau/T = 0{\cdot}25$ and $V_{out}/\hat{V}_{in} = 0{\cdot}7$, i.e. $V_{out} = 240$ V (d.c.). The required power rating for R_s is *not* $70 \times i^2{}_L = 70 \times (0{\cdot}24)^2 = 4$ W, but must be $P_{Rs} = 30{\cdot}8$ W.

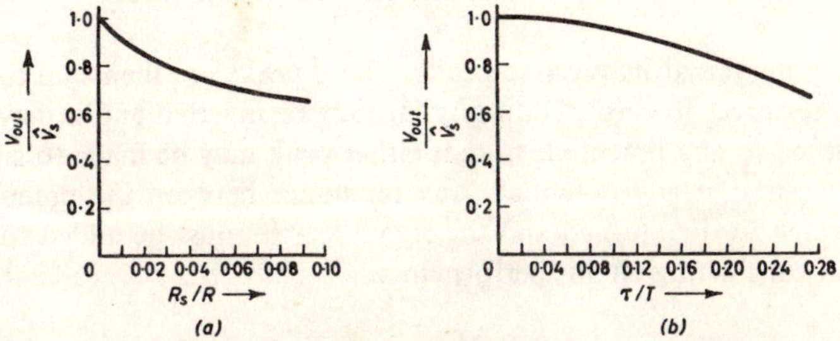

FIG. 1.9 V_{out} for sine-wave input in terms of (*a*) R_s/R and (*b*) τ/T

Choice of C

It has been assumed throughout the above analysis that C is so large that V_{out} has no a.c. component; this can be literally true only if C is infinite. However, the results are correct for practical purposes provided C is large enough to produce an alternating component of V_{out} which is negligible in the particular application. This means that the voltage by which C discharges through R in one cycle must be a small proportion of V_{out}. The discharge voltage is obtained from $i = C\mathrm{d}V/\mathrm{d}t$ and is $V_{out}T/CR$, since, for small discharge, the current is constant and equal to V_{out}/R. Hence, $V_{out}T/CR \ll V_{out}$, giving $CR \gg T$. The time constant of the load circuit must therefore be much greater than one period of the input if output ripple is to be small, and the analysis correct. The relationship between ripple and time constant is such that a time constant 1 per cent of the period produces ripple which is 1 per cent of the d.c. output.

The capacitor waveform is shown in Fig. 1.10.

d.c. Restorer

When the positions of diode and capacitor are interchanged, as shown in Fig. 1.5, the circuit operation is basically unchanged, but the output is now the difference between the capacitor voltage (i.e. V_{out} for the rectifier circuit) and V_{in}. The resulting waveform (Fig. 1.11) has ideally an alternating component which is identical to V_{in} and a d.c. component of $\hat{V}_{in}$, resulting in the waveform just reaching

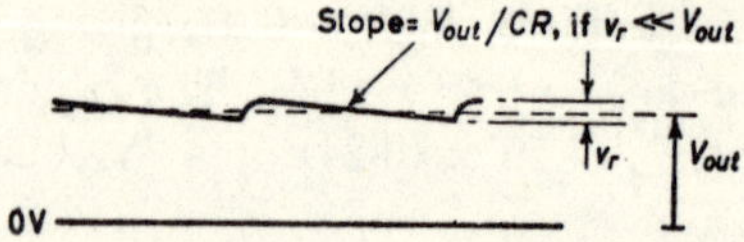

FIG. 1.10 Capacitor waveform in rectifier circuit

zero potential at its positive peaks. These peaks are then said to be 'd.c. restored' to earth. Naturally D_1 may be inverted and it may be returned to any potential, so that either peak may be made to sit at any desired steady potential. Any resistance between the potential to which D_1 is returned and the signal 'earth' must be added to R_s when considering circuit performance.

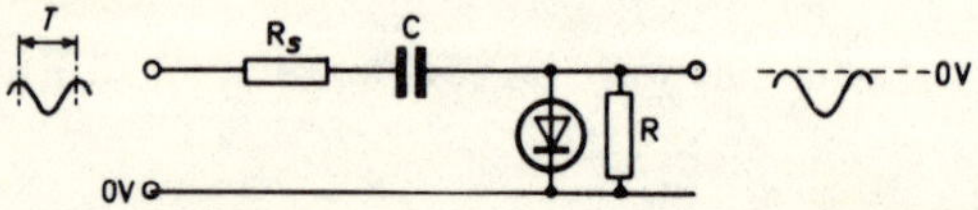

FIG. 1.11 Waveforms for d.c. restorer ($CR \gg T$)

This circuit can again be analysed by the methods used for Fig. 1.4, and circuit performance may in fact be derived directly from that analysis. However, the usual requirement for d.c. restoration is that it should be as near the equivalent of peak rectification as possible, giving the restoration level equal to the diode-return poten-

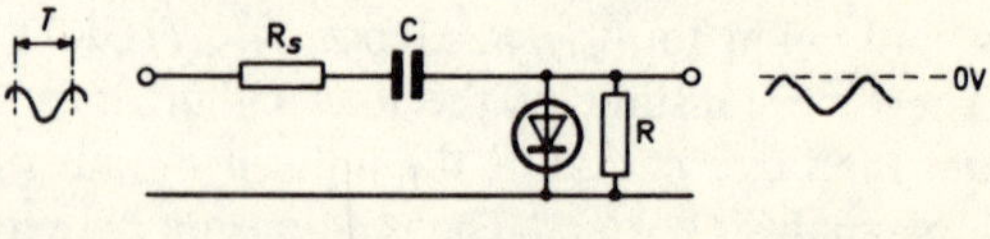

FIG. 1.12 Waveforms for d.c. restorer ($CR \not\gg T$)

tial. It will therefore normally be designed with $R_s \ll R$, $CR \gg T$ and $\hat{V}_{in} \gg$ diode forward drop.

Practical problems are that if C is too small, the ripple waveform across C will noticeably affect the waveform V_{out}, as shown in Fig. 1.12; on the other hand, if C is very large many cycles of V_{in} will pass before restoration level at the output is reached. R_s should be as low as possible, but, if very low, peak charging current requirement for C will be high and the source may be unable to supply this. As the source temporarily fails, the effective value of R_s becomes high and the restoration level wrong.

Finally it should be noted that any configuration equivalent to that of Fig. 1.5 produces d.c. restoration whether or not intended by the designer.

Zener Diode Applications

The Zener diode has the property of sustaining a voltage drop which is virtually constant over a wide range of currents applied in the direction which corresponds to reverse current in a normal diode. Its most obvious use is therefore to produce a stable voltage for supplying other circuit elements when the main supply is subject to wide variations; these variations may be very slow fluctuations or low-frequency or high-frequency ripple, the Zener diode acting as a low impedance over a wide band of frequencies.

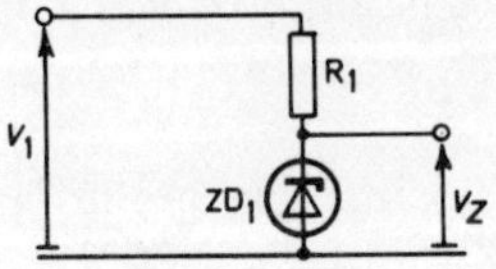

FIG. 1.13 Simple Zener diode stabilizer

The circuit of Fig. 1.13 shows a typical application in which a Zener diode is used to obtain a stable voltage source from a power supply V_1 which is itself subject to large variations. In order to illustrate the significance of the various quoted parameters, the circuit performance is assessed below assuming that

(i) $V_Z = 6{\cdot}8$ V $\pm$ 10 per cent for $I_Z = 5$ mA, i.e. ($V_Z(I_Z = 5$ mA) $= 6{\cdot}8 \pm 10$ per cent).

(ii) Zener incremental resistance for $I_Z = 5$ mA is 30 Ω maximum, i.e.

$$\frac{dV_Z}{dI_Z}(I_Z = 5\text{ mA}) = 30\ \Omega \text{ max.}$$

(iii) Temperature coefficient of V_Z at 5 mA is $\pm 0{\cdot}02$ per cent per degree C, i.e.

$$\frac{dV_Z}{dT}(I_Z = 5\text{ mA}) = \pm 0{\cdot}02 \text{ per cent.}$$

(iv) V_1 is 20 ± 5 V.
(v) R_1 is 2·7 kΩ ± 5 per cent.

Since the above values for the Zener diode apply for a nominal Zener current of 5 mA, R_1 has been given a value which gives approximately this Zener current under nominal conditions. Thus the nominal output V_Z will be 6·8 V. The effect on the output of the above specification figures is detailed below under the relevant number.

(i) The manufacturer's tolerance of ±10 per cent at 5 mA gives an output variation of ±10 per cent, i.e. ±0·68 V. The effect when I_Z is not 5 mA is dealt with in (ii).

(ii) The value of R_Z of 30 Ω at 5 mA implies that any departure of I_Z from 5 mA by an amount δI_Z will produce an output voltage change of $R_Z \delta I_Z$; since R_Z is positive, the direction of output change is the same as that of δI_Z. Since the value of δI_Z depends on factors (iv) and (v), and slightly on factor (iii) also, the effect of R_Z cannot at this stage be calculated.

(iii) Temperature coefficient produces a direct effect on the output of the same proportion, namely ±0·02 per cent per degree C, i.e. ±1·36 mV/degC.

(iv) Variations in V_1 cause changes in I_Z; this is the only reason why V_Z is affected by V_1. Since $V_1 = 20$ V gives $I_Z \approx 5$ mA with $V_Z \approx$ 6·8 V (it is generally unnecessary to be more precise except in highly critical cases), then a change of ±5 V in V_{in} will produce a change $\delta I_Z = \pm 5/R_1 = \pm 5/2{\cdot}7$ mA $= \pm 1{\cdot}85$ mA.

This gives a change in V_Z of $\pm 1{\cdot}85 \times 10^{-3} R_Z = \pm 1{\cdot}85 \times 30$ mV $= \pm 55{\cdot}5$ mV.

(v) Variation in R_1 again changes I_Z, and because of R_Z this changes V_Z. A tolerance of ±5 per cent changes I_Z by 5 per cent (since $I_Z = (V_1 - V_Z)/R_1$), i.e. $\delta I_Z = \pm 1/4$ mA and V_Z therefore changes ±(1/4)30 mV = ±7·5 mV.

Summarizing the above results, the value of V_Z is given by

$$V_Z = 6{\cdot}8 \pm \underset{\substack{\text{initial}\\ \text{tolerance}}}{0{\cdot}68} \pm \underset{\substack{V_1\\ \text{changes}}}{0{\cdot}055} \pm \underset{\substack{R_1\\ \text{tolerance}}}{0{\cdot}075} \pm \underset{\substack{\text{temperature}\\ \text{coefficient}}}{0{\cdot}001\,\theta}$$

where θ is the number of degrees Centigrade temperature variation. There are several points of interest in the above result.

First, the major item causing output uncertainty is initial tolerance;

this, however, does not change with use and is therefore of no importance if the only requirement of V_Z is stability rather than its absolute value, which is often true in practice. When the absolute value is important, it is necessary to specify a tighter initial tolerance and this is expensive, since it involves selection by the manufacturer (it is usually much more expensive for the customer to make the selection).

Secondly, all the causes of change are to a small extent self-compensating. For instance, when V_1 increases, I_Z also increases, causing a change of $\delta I_Z R_Z$ in V_Z. Because V_Z has now increased, the value of I_Z is not so high as was assumed at the high value for V_1. It is in fact given by $I_{ZH} = (V_{1H} - V_Z - \delta I_Z R_Z)/R_1$, where I_{ZH} and V_{1H} are the high values of I_Z and V_1; the calculation used earlier in the chapter assumed that $I_{ZH} = (V_{1H} - V_Z)/R_1$.

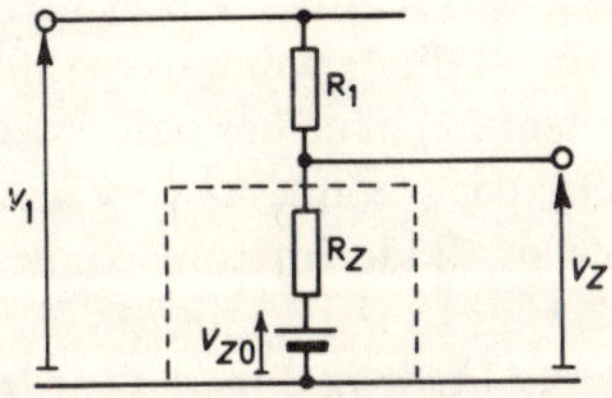

FIG. 1.14 Equivalent circuit for simple Zener diode stabilizer (Fig. 1.13)

The result is that the limits given for V_Z were slightly pessimistic. How slight this effect is can be shown by calculating the case for the initial tolerance of V_Z, namely $\pm 0{\cdot}68$ V. Taking the positive limit, I_Z is less than the expected value by $0{\cdot}68/R_1$, i.e. $0{\cdot}68/2{\cdot}7 = 0{\cdot}25$ mA, giving a drop in V_Z of $30 \times 0{\cdot}25$ mV $= 7{\cdot}5$ mV from its apparent value of $6{\cdot}8 + 0{\cdot}68 = 7{\cdot}48$ V. This is completely negligible in this and almost every other case.

Thirdly, it is easy to see that instead of calculating the changes of current caused by each effect and then using R_Z to put this in terms of V_Z, the operation can be simplified by regarding the diode as a perfect battery in series with R_Z (Fig. 1.14).

The value of the 'battery' voltage V_{ZO} is the Zener voltage at a given current I_Z minus the product of R_Z and I_Z. In the above example V_{ZO} would be ($6{\cdot}8 \pm 10$ per cent) $- 5 \times 10^{-3} \times 30 = 6{\cdot}8 \pm 0{\cdot}68 - 0{\cdot}15 = 6{\cdot}65 \pm 0{\cdot}68$ V. If R_Z were constant for all I_Z this equivalent circuit would apply for any condition; in fact, R_Z is a function of I_Z and the quoted R_Z is true for only a small region; e.g. at

$I_Z = 6$ mA R_Z may be 25 Ω. Since any calculations where accuracy is important would normally be used only for cases where I_Z has small variations, this equivalent circuit is a useful design aid.

Assuming in Fig. 1.14 that circuit values give I_Z in the region corresponding to the value of V_{ZO} and R_Z, the effect of variations is readily calculated; in particular, it is clear that any change δV_1 in V_1 will cause an output change of $\delta V_Z = \delta V_1 R_Z/(R_Z + R_1) = \delta V_1(30/2{\cdot}73)10^{-3}$; if $\delta V_1 = 5$ V, then $\delta V_Z = \dfrac{5 \times 30 \times 10^{-3}}{2{\cdot}73} =$ 0·055 V, as obtained earlier by direct calculation.

If R_Z is truly resistive (rather than reactive), then the above calculation applies to variations in V_1 at any frequency. V_1 can have a mean value in our example of +20 V and a ripple content which causes V_1 to oscillate by ±5 V about this value. The output V_Z will have a steady value of about +6·8 V and its ripple content will be only ±0·055 V. An actual diode will have parallel capacitance which causes this ripple content to be further reduced at frequencies above a few hundred kilocycles per second. At very much higher frequencies the series inductance of diode internal connections causes output ripple in the above circuit to rise again.

From the equivalent circuit of Fig. 1.14 it is evident that for the most constant V_Z the value of R_1 should be as high as possible and that of R_Z as low as possible. Changing R_1 in our example to, say, 27 kΩ would give no improvement, however, because I_Z then falls to about 0·5 mA and R_Z would be found to rise by a factor of at least 10. Performance would therefore be unchanged.

The only practical improvement is to raise the mean value of V_1 so that R_1 may be increased while maintaining the same value of I_Z and, hence, R_Z. The available supply voltage will naturally be limited, but the aim should be to use as high a value as possible. The use of a 'constant-current device', to be dealt with later, is sometimes appropriate and involves the use of a transistor (*see* Chapter 6).

An additional complication which often affects the calculation is the presence of a load current I_L, assumed negligible in the above example. If I_L is large, e.g. 4 mA, the value of R_1 must be changed. To maintain I_Z at 5 mA, R_1 must pass 9 mA and will have a value of about (20 − 6·8)/9 kΩ, namely 1·5 kΩ (*see* Fig. 1.15).

By causing R_1 to be thus reduced, the presence of I_L clearly reduces the stability of the circuit. Changes in V_1 are now reduced by a factor of about 1500/30, i.e. 50/1, instead of 2700/30, or 90/1.

The effect of load current changes are accounted for by noting that the apparent source resistance of the Zener circuit is 30//1500 Ω, i.e. about 30 Ω. (It is futile in these calculations to consider small effects such as the influence of 1500 Ω when placed in parallel with 30 Ω, since the figure of 30 Ω given by the manufacturer is merely typical and is subject to considerable variation.) Thus a change in load current of ±10 per cent, or 0·4 mA will produce an output change of $\mp$(0·4 × 30) mV, i.e. $\mp$12 mV.

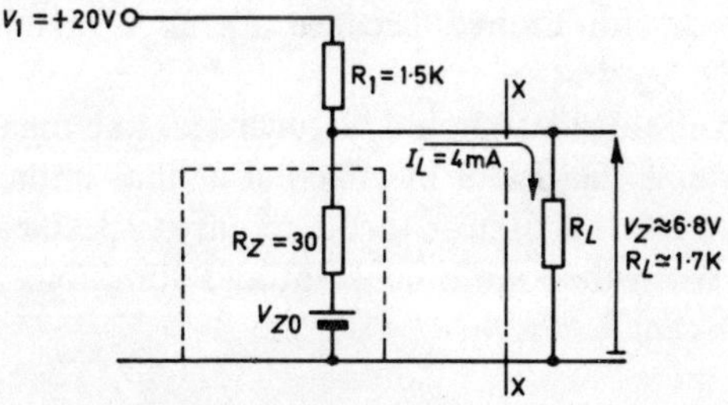

FIG. 1.15 Practical example with load R_L

The design of the simple stabilizing circuit merely requires care in adding all the effects which cause output variation. Improvements to the circuit can be made by using two or even more stages in cascade, i.e. V_1 is itself the Zener voltage of another diode.

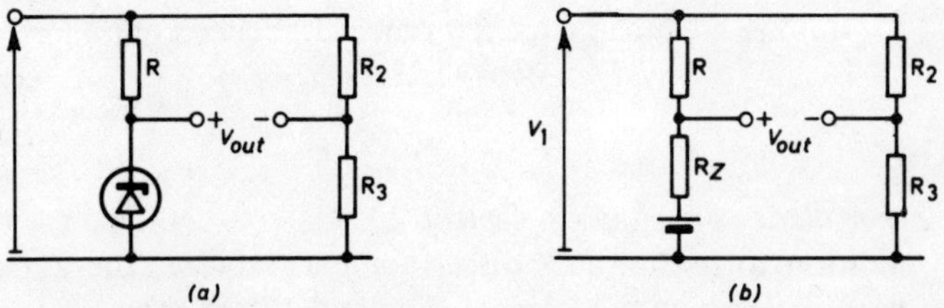

FIG. 1.16 (*a*) Improved stabilizer, (*b*) equivalent circuit of improved stabilizer

A more subtle improvement is shown in Fig. 1.16 (*a*), where the 'earthy' side of the load is returned to a potential divider R_2R_3. The idea is that when V_1 rises, causing the positive output terminal to rise, the negative terminal also rises. If both terminals rise the same amount, V_{out} remains constant, giving perfect stabilization against changes in V_1. This condition is achieved if $R_3/R_2 = R_Z/R$, as is obvious from Fig. 1.16 (*b*).

This arrangement is limited in its use because the output impedance (i.e. the source impedance seen by the load) is now $R_Z//R + R_3//R_2$, assuming V_1 itself has zero source impedance. This rise in impedance can be a big disadvantage if the load resistance is variable; if, on the other hand, this effect is reduced by making $R_2//R_3$ comparable with R_Z, the extra load on V_1 may be an embarrassment.

Another point is that the predicted infinite improvement against V_1 changes is not achieved in practice: R_2, R_3, R and R_Z are not known exactly, and if R_2 or R_3 is made adjustable for initial setting up, the improvement is still limited because R_Z is a function of Zener current.

In conclusion, this idea should be regarded, like many other compensating systems, as useful in effecting a final improvement to a circuit which performs almost to the required specification. It is of special value when the load is constant and variations in V_1 are less than ± 10 per cent.

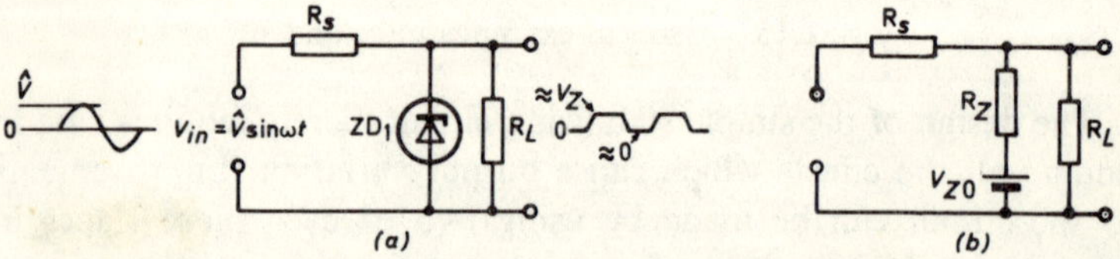

FIG. 1.17 (*a*) Zener clipping circuit, (*b*) equivalent circuit of (*a*) for Zener conduction

The Zener diode as a clipping element

In addition to its use as a direct voltage stabilizer the Zener is often used in clipping a.c. signals. Figure 1.17 (*a*) shows a typical circuit with the input and output waveforms; Fig. 1.17 (*b*) shows the equivalent circuit when ZD_1 conducts in the 'Zener' direction. In the opposite direction of conduction ZD_1 naturally behaves like a forward-biased silicon diode giving the normal voltage drop of such a diode.

The action of this circuit is obvious and the only design point is the choice of R_s, which must be much larger than R_Z if a flat-topped waveform is required, but not so large that attenuation due to the load prevents the level V_Z being reached. If the output is intended to be a sharp-edged square wave, then V must be much larger than V_Z; the actual rate of rise achieved is

$$\frac{R_L}{R_s + R_L}\frac{\mathrm{d}}{\mathrm{d}t}(\hat{V}\sin\omega t)$$

at any time when clipping is not taking place.

Temperature affects the mean level and the peak-to-peak value of the output according to the temperature coefficients of V_Z and V_F. If the positive clipping level is to be constant, then V_Z should be a low coefficient diode (e.g. 5·6 V); if it is more important that the peak-to-peak be constant, then V_Z should have a positive coefficient of magnitude equivalent to 2–2·5 mV/degC in order to cancel the V_F coefficients, e.g. a Zener diode of 8·2 V. In this case the output waveform moves positive by 2–2·5 mV/degC but the amplitude remains relatively constant.

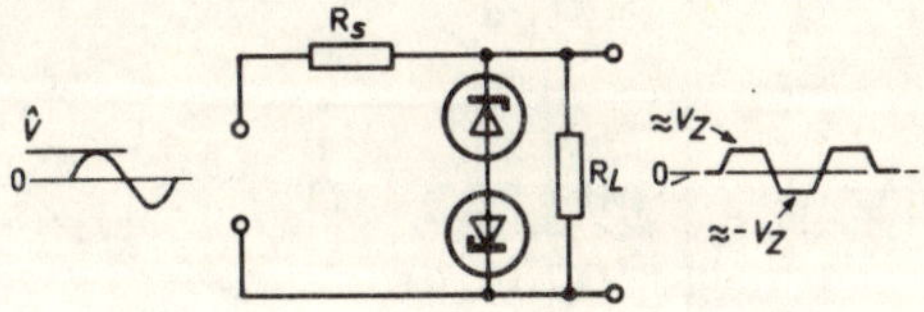

FIG. 1.18 Double Zener clipper

When the output is to be symmetrically disposed about zero level, two Zener diodes can be used 'back-to-back' in series—either the two anodes or the two cathodes may be joined (*see* Fig. 1.18). The output peak value is now ($V_Z + V_F$), where V_F is the drop of the forward-biased diode. Asymmetry is caused mainly by inequality between the two values of V_Z and to a lesser extent from inequality between the two values of V_F.

Temperature effects are slightly less in this circuit because, if the Zener coefficient is chosen as before to represent +2 to 2·5 mV/degC voltage change, each clipping level is stable, and no mean level shift occurs.

It must be emphasized that in either circuit such compensation is by no means exact; all that can be said is that an attempt to match coefficients helps, since linear quantities are involved.

Other uses of the Zener diode

As will be seen in later chapters, Zener diodes can often replace resistors, and sometimes capacitors, as coupling devices. By regarding the Zener diode as a battery and series resistor, such circuits are readily designed by the methods described earlier.

Precautions in Zener diode usage

When a voltage is applied to a Zener diode at a level below the Zener voltage, the current which flows (ideally zero) is often unspecified. This current, if of a significant level, will cause clipping circuits to distort waveforms before the correct clipping level is reached. Where this is important, the circuit should be redesigned so that the Zener diodes are operating in the conducting state, using normal diodes to perform the clipping function. Fig. 1.19 shows how a

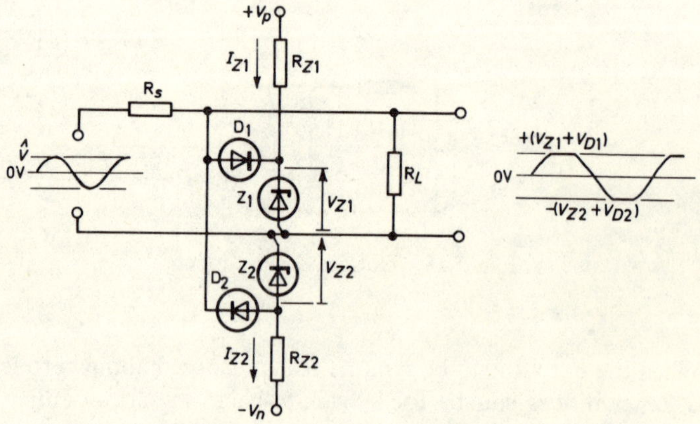

FIG. 1.19 Modified clipper

modification to the circuit of Fig. 1.18 can be devised to avoid curvature of the transfer characteristic before clipping occurs, using this principle. The penalty is that supply voltages are now required in addition to the extra components. An alternative is to specify in Fig. 1.18 Zener diodes with guaranteed leakage figures such as the 1N4099–1N4135 series.

Operation of a Zener diode, for coupling or non-critical reference purposes, at currents below the standard quoted level (usually 5 mA) is bad practice. Such operation often causes large low frequency variations in Zener voltage (called $1/f$ noise) and in some circuits saw-tooth waveforms will be generated if a capacitor is connected in parallel with the diode. If low current usage is required, then a diode should be chosen with an appropriate specification. Again, the 1N4099–1N4135 series is useful in such cases.

LIGHT-EMITTING DIODES AND LIGHT-DETECTING DIODES

Junction diodes are available with the property of emitting light when forward-biased. The most efficient diodes, using gallium arsenide, emit in the near-infra-red region. Various related semiconductors, e.g. gallium arsenide phosphide, can be used to produce light in the visible region.

Light-detecting diodes, usually of silicon, are small signal diodes so constructed that external light can reach the junction; often a simple lens is incorporated, focused on the junction. When reverse-biased the leakage current depends on the received light, the sensitivity being greatest in the near-infra-red region.

Applications

The simplest application for the light-emitting diode (known as the LED) is as an indicating lamp using a visible emitting device. To operate, a forward current of between 10 and 50 mA is commonly required, the junction drop being 1·5 to 2·5 V. It is essential to limit the current to the specified maximum (50 or 100 mA) and also to ensure that the supply voltage is adequate even for a diode sample needing 2·5 V. Protection must also be provided against reverse-biasing the LED; its maximum rating is only 2 to 3 V. A safe method is always to include a normal diode either in series or (as shown in Fig. 1.20) in back-to-back connection when any possibility of reverse drive exists. When deciding whether such a precaution is necessary the momentary conditions occurring as supplies are switched on and off (inevitably not quite simultaneously) should be examined.

As illustrated in Fig. 1.20 the drive circuit for an LED is simple and logic driving from TTL is achieved with the aid of a single transistor buffer stage.

The attractions of using an LED for panel indication are the low current and voltage combined with long life and immunity from vibration compared with a filament lamp. Moreover the construction allows an integral lamp holder to be incorporated so that the overall

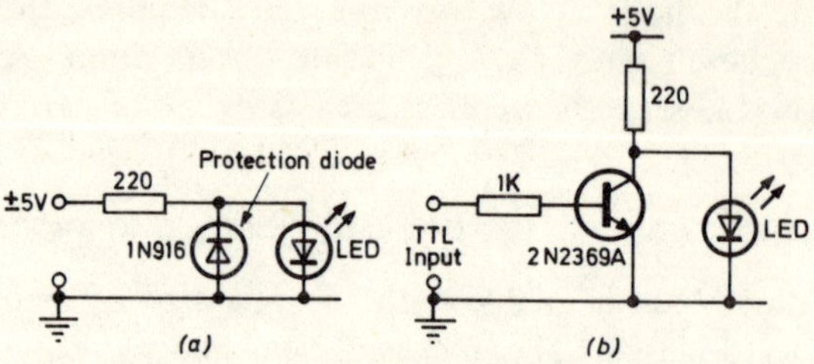

FIG. 1.20 Driving the LED

cost is no higher than the conventional lamp plus holder.

Apart from the simple lamp, there are many varieties of multi-section display as used in calculators, digital voltmeters and other panel instruments. Each segment of these devices is a simple LED and is driven in the manner described, some of the circuitry often being included in the integrated circuit package.

By using a silicon photodetector diode in conjunction with an LED, signals may be transmitted through the intervening space. Devices assembled as a pair for this purpose are known as optical couplers and can be used for counting objects or reading coded cards when these are allowed to interrupt the light. Couplers for this usage are provided with a slit between the two components. Sealed units, often called opto-isolators, are used to transmit signals between parts of a system which are at different potentials or where earth loops (see Chapter 16) must be avoided. Although in principle optical couplers can be used to transmit analogue signals by modulating the LED current, the resulting transfer law is linear only for small excursions about the standing current. For accurate transfer of such signals, they should be converted first into either serial logic or to one or other form of pulse-width modulation.

Although these devices are individually simple, the combinations available and the possible modes of operation are so numerous that the field of opto-electronics has become a separate branch of electronic technology.

Chapter 2

The transistor: d.c. characteristics

No attempt is made here to describe the physics of transistor operation. Instead its characteristics as a circuit element are discussed and orders of magnitude of parameters are also given, since these must be known in order to understand how the approximations are made when deriving usable formulae.

D.C. OPERATING CONDITIONS

Figure 2.1 shows both *p–n–p* and *n–p–n* transistors, and indicates how in some respects the transistor can be represented as two diodes,

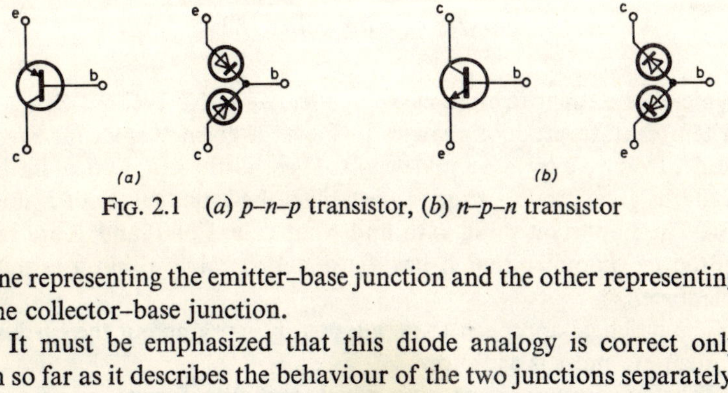

FIG. 2.1 (*a*) *p–n–p* transistor, (*b*) *n–p–n* transistor

one representing the emitter–base junction and the other representing the collector–base junction.

It must be emphasized that this diode analogy is correct only in so far as it describes the behaviour of the two junctions separately, but nevertheless the idea is helpful as an aid to visualizing the directions of forward currents and voltages, and also leakage currents.

The Emitter Circuit

From the above representation, and bearing in mind that it is correct only for junctions energized separately, it is readily seen that if the base is returned to a more positive potential than the emitter on an *n–p–n* transistor, current will flow and a potential difference will appear across the junction as shown in Fig. 2.2. The magnitude of the drop, named V_{be} in Fig. 2.2, clearly depends upon the current flowing into the emitter–base diode, i.e. it depends on V and R; the drop also depends on the forward characteristic of the emitter–base diode.

If a plot of the forward characteristic were available, it would therefore be possible to calculate the current I_e and the voltage V_{be} in Fig. 2.2, either by trial and error (until $(V - V_{be})/R$ gave a current which produced that value of V_{be}) or more scientifically by drawing a load line on the diode characteristic.

Such a calculation would, however, be subject to considerable error because of variations between the actual transistor used and the

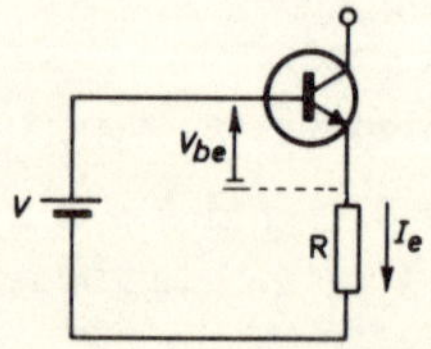

FIG. 2.2 Emitter current flow

'typical' one for which the curve applies, and also because changes in temperature produce changes in V_{be} at a given current (as for a diode, from −2 to −2·5 mV/degC). There is therefore no point in going to great trouble to try to establish the exact values of I_e and V_{be}. The important thing is to find what values of V and R are required to guarantee that I_e has the intended value within a certain tolerance.

The practical approach to be adopted in working out the current in such examples is as follows.

The temperature range over which the circuit must operate is e.g. 0–50°C. If the transistor is silicon, its V_{be} at the current it is intended to operate will lie between 0·5 and 0·9 V at room temperature (25°C), will be larger by another $(25 \times 2{\cdot}5)$ mV $\approx$ 60 mV at 0°C, and smaller by $(25 \times 2{\cdot}5)$ mV $\approx$ 60 mV at 50°C. Therefore, the limits of V_{be} are 0·44 and 0·96 V for a temperature range of 0–50°C. It is required that the emitter current be constant to, e.g. ± 10 per cent. If the simple arrangement of Fig. 2.1 is to be used, V must be at least so large that the two limits of $(V - V_{be})$ produce less than ± 10 per cent change in I_e. This choice of V leaves no permissible tolerance on R and on V itself, and since precision resistors and stabilized lines are expensive, a practical solution would be to make R a ± 5 per cent type (including temperature effects) and make the $(V - V_{be})$ tolerance less than ± 5 per cent. To achieve this V must be at least 10(0·96 – 0·44), i.e. 5·2 V, and it is convenient to use here, say, 10 V which is already available. R is now given by the nominal voltage across it, i.e. 10 – [(0·96 + 0·44)/2] divided by the intended value of I_e, i.e. 1 mA, for example. Therefore R is 9·3 kΩ, or to the nearest standard value 9·1 kΩ.

Note that the only transistor information required for the above (other than the desirable operating current, which will be dealt with in Chapter 7) is the value of V_{be}, its change with temperature and its variation from one unit to another. The figures used above are typical for a silicon transistor, i.e. 0·5–0·9 V at 25°C for any operating current which is likely to be reasonable for the device. The corresponding figures for germanium are 0·15–0·5 V. Temperature drift of about –2·5 mV/degC applies to both types. When dealing with power transistors carrying more than 1 A it is advisable to check on the maximum values given above.

The emitter circuit of a *p–n–p* transistor is designed in exactly the same manner, the only difference being the polarity; values and drifts of V_{eb} remain the same.

No difficulty need be experienced in remembering the correct emitter circuit polarities for the two types, since the arrow on the circuit symbol points in the direction of conventional current flow (as in the diode symbol) and the external voltage source has to be connected so that current will be supplied by this source in the direction of the arrow.

To establish I_e within the required tolerance, one therefore needs to know that tolerance, the temperature range, and whether the transistor is germanium or silicon. Knowing these facts, V is made sufficiently large and constant, and R is made sufficiently accurate. Apart from the use of temperature-compensating elements to counteract V_{be} temperature changes (to be dealt with later in this chapter) this is all that can be done to be sure of operating at the intended value of I_e — excepting the use of d.c. negative feedback shown in Fig. 2.9.

The Collector Circuit

If a voltage is applied between the collector and base of an *n–p–n* transistor so that the collector is positive, then the collector–base diode is reverse-biased. Provided no emitter current is flowing, because this would cause the two-diode analogy to fail, then the collector current will be merely the leakage of the diode (Fig. 2.3).

This current is known as the collector–base leakage current and is designated I_{cbo}.

As in a normal diode, I_{cbo} is the sum of two components, one of which is invariant with temperature; the other doubles itself every 9–10°C. The value of applied voltage has little effect on I_{cbo} until the 'breakdown' voltage is reached, when I_{cbo} rises rapidly.

Typical values for small silicon types range from about 10 nA

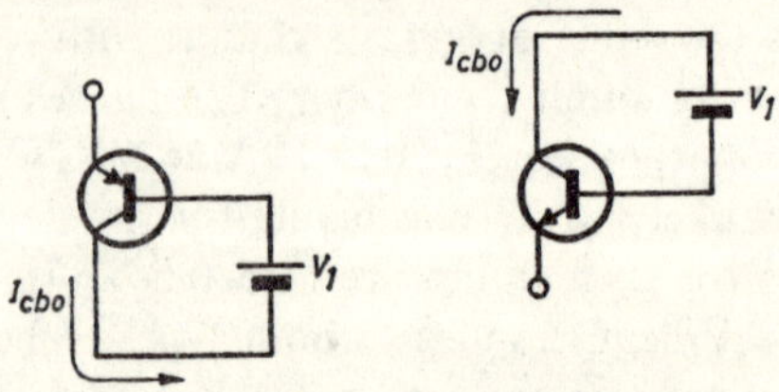

FIG. 2.3 Collector leakage

at 25°C to 1 μA at 50°C; since I_{cbo} depends directly on junction area, power transistors often have values of from 2 or 3 mA (25°C) to 20 mA (50°C).

Germanium values are much higher, e.g. small signal types 2 μA (25°C)–100 μA (50°C). Again, power types may be many orders of magnitude higher than the above.

It is evident that I_{cbo} is subject to much more uncertainty than V_{be} and that although the ratio of maximum to minimum I_{cbo} for silicon is similar to that for germanium, the absolute values for germanium are much larger.

Because of the exponential way in which I_{cbo} varies with temperature, it is almost impossible to predict with any confidence what value I_{cbo} will have at, for example, 65°C when its value is known only at 20°C. If its value at 20°C were, for instance, 10 μA ± 10 per cent, the 65°C figure assuming a law where I_{cbo} doubles every 9 or 10 degC rise, would have a lower limit of $9 \times 2^{4 \cdot 5}$, i.e. 200μ A and an upper limit of 11×2^5, i.e. 352 μA.

This uncertainty in I_{cbo} leads to many difficulties in circuit design and all the designer can do is to assume the worst case, i.e. he must know the maximum possible value of I_{cbo} for the device *at the maximum temperature at which the transistor will operate*. The 25°C figure is no guide to the performance at, for example, 50°C, as pointed out above.

Effect of Collector Load

The addition of a load R_L in series with the collector supply causes the applied collector voltage to be reduced by $I_{cbo}R_L$, and this voltage drop will be present whenever the collector supply is connected, adding to any other drop caused by transistor action.

As mentioned earlier, increasing the collector supply voltage has little effect on I_{cbo} until breakdown of the collector–base diode occurs. This effect is again analogous to the same action in a normal semiconductor diode, so that if R_L is present and V_1 is very large, the collector–base voltage rises to the breakdown value BV_{cbo} and the collector current is given by $(V_1 - BV_{cbo})/R_L$. The collector current can be very large compared with normal values of I_{cbo}, but damage will not occur provided the power rating of the junction is not exceeded, i.e. provided $BV_{cbo}(V_1 - BV_{cbo})/R_L < P_{cmax}$. Although this mode of operation is not generally useful owing to uncertainty in the actual value of BV_{cbo} (often much higher than the guaranteed minimum), such action often occurs transiently under overload conditions, or immediately after switching power supplies on and off. In these cases the power must be calculated and if necessary reduced to a safe level.

Typical breakdown voltages range from 10 to 80 V; less common but obtainable are ratings up to 500 V.

Comparison between Base–emitter and Base–collector Diodes

Although, as indicated above, the normal mode of operation is to forward-bias the emitter diode and reverse-bias the collector diode, each can also be used in the opposite connection.

Usually, but not always, the breakdown voltage of the reversed emitter–base diode is less than that of the collector–base diode, especially in diffused transistors (often 1 V only), and the reverse leakage of this diode before breakdown, which is usually less than I_{cbo}, is called I_{ebo}.

In other respects the two diodes are similar, and indeed the collector and emitter loads can be interchanged and provided the changed ratings are not exceeded, no damage will be done. However, performance as a transistor will be poor, except for a 'symmetrical' type, as will be seen from the following section.

Transistor Action

If the emitter–base diode is biased forward with a current I_e and the collector–base diode is simultaneously reverse-biased, the two-diode analogy fails because of transistor action. The resulting currents are shown in Fig. 2.4, which indicates that the collector

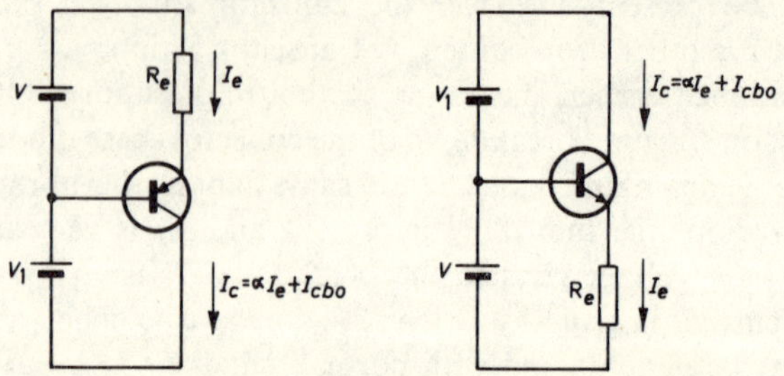

FIG. 2.4 Transistor currents

current has increased from I_{cbo} to $(I_{cbo} + \alpha I_e)$, where α is a transistor parameter whose value is close to but less than unity (generally within 5 per cent). Most of the emitter current therefore flows out of the collector, and since α is found not to vary appreciably with collector–base voltage (provided this is at least a few hundred millivolts) the total collector current is independent of V_1.

The addition of R_L into the collector circuit, as in Fig. 2.5, does not therefore change the currents, provided that $(\alpha I_e + I_{cbo})R_L$ is less than V_1 by a few hundred millivolts, i.e. provided the collector junction is still reverse-biased.

In designing a practical circuit it will always be necessary to know the collector potential and so the designer must be aware of parameter variations which cause drift in this voltage.

Naturally, variations in α and in I_e (discussed earlier) result directly in variations in collector current I_c and, hence, in V_{cb}. Variations in I_{cbo} again directly affect V_{cb}. The rate of change of α with temperature is ill-defined but usually lies between 0 and $+1/25$ per cent per degree C; its value also varies from one unit to another of the same type by a spread of approximately ± 1 per cent.

The variations of α have usually little direct effect on the collector circuit in comparison with the effect of V_{be} (on I_e).

In the practical design previously discussed I_e was established to within ± 10 per cent by using $V = 10$ V, $R_e = 10$ kΩ, giving $I_e =$ 1 mA. Suppose now that $V_1 = 10$ V and $R_L = 5{\cdot}6$ kΩ, and that it is required to calculate the value and expected variation in V_{cb}. The designer proceeds as follows.

I_e is known to be nominally 1 mA so that V_{cb} is 10 V less the drop in R_L, which is nominally 5·6 V, giving $V_{cb} = 4{\cdot}4$ V.

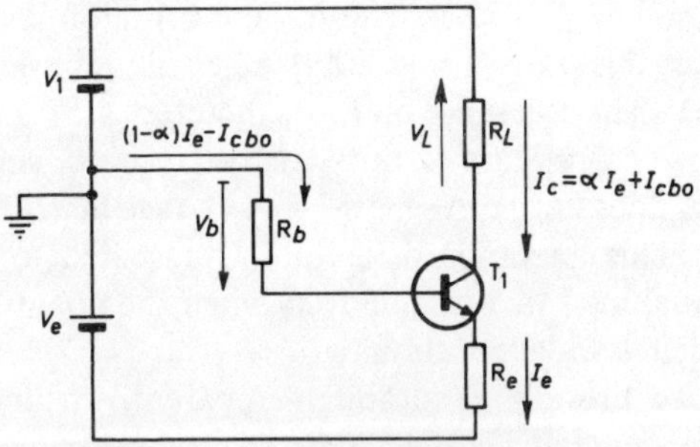

FIG. 2.5 Complete bias circuit

Note the assumptions made that $\alpha = 1$ and $I_{cbo} = 0$. The correct allowance for these quantities is not made at this stage, since if the very simple nominal calculation gave $V_{cb} = 0$ V, which could easily be the case if, e.g. $R_L = 10$ kΩ or $V_1 = 5{\cdot}6$ V, the circuit would clearly not operate and a new value of R_L or V_1 would have to be allocated. The argument that the correct allowances for α and I_{cbo} might influence the answer and then give a reasonable value for V_{cb} is invalid, because if these side-effects give a marked change in I_c, the design is bad (spurious variable effects should not predominate); if the effects have only slight influence, a slight change in R_L or V_1 would again lead to circuit failure. This example where the failure criterion is $V_{cb} \leq 0$ is naturally only a particular case; for many applications it could be that $V_{cb} \gtrless V^*$ is a failure condition, where V^* is a voltage limit determined by signal levels or by a following circuit. Returning to the design procedure, it has been established that nominally $V_{cb} = 4{\cdot}4$ V.

The actual value of V_{cb} is affected by I_e, which causes ± 10 per cent variation in that part of I_c which does not include I_{cbo}. This is a change of about 0·1 mA, which produces 0·56 V change across R_L and therefore in V_{cb}. Change in α from 0 to 50°C will be about 2 per cent at worst and changes from one transistor to another will be another 2 per cent. Since the minimum α for any transistor of the type to be used is, e.g., 0·95 at 0°C, the maximum could be 0·99, so that a nominal 3 per cent must be taken from the assumed value of 1 mA and a spread due to α of ± 2 per cent taken as tolerance. Hence, the drop in R_L (assumed 5·6 V) should be

$$[5{\cdot}6 - (3/100)5{\cdot}6] = 5{\cdot}43 \pm 0{\cdot}56 \pm (3/100)5{\cdot}43 = 5{\cdot}43 \pm 0{\cdot}75 \text{ V}$$

giving $V_{cb} = 4{\cdot}57 \pm 0{\cdot}75$ V. The I_{cbo} contribution is highest at 50°C, when I_{cbo} for this transistor is 100 μA, giving an additional drop in R_L of 5·6 mV. The tolerance on V_{cb} is therefore $4{\cdot}57 \pm 0{\cdot}75 - 0{\cdot}0056$ V.

The above calculation is not strictly accurate, since 10 per cent change in I_e does not represent 0·1 mA change in αI_e but more nearly 0·097 mA. This 'error' is deliberately presented to emphasize the futility of making exact calculations when the quantities being dealt with experience wide variations.

The above procedure, although apparently tedious, takes little time in practice and leads to a better understanding of which are the worst contributors in a particular case to variations in I_c or V_{cb}. It is clear, for example, that the value of I_c is uncertain by the proportion that I_{cbo} represents in I_e, quite apart from any other cause. If, therefore, $I_{cbo,max.} = I_e/10$, then I_c has at least 10 per cent uncertainty from I_{cbo} alone, because the I_{cbo} of some transistors of the specified type may be almost zero, whereas others will be $I_{cbo,max}$. Similarly, if the expected V_{be} variation is equal to $V_e/10$, the uncertainty in I_e is again about 10 per cent from this cause alone.

These points, if borne in mind at the beginning of a design, will enable at least a reasonably good first attempt to be made. Corrections can then be made after calculating errors. If, instead, equations such as

$$V_{cb} = V_1 - (V_e - V_{be})/R_e + I_{cbo}R_L$$

are solved and prove that V_{cb} is in error, the necessary corrective measures are not obvious.

Using the methods described above, the designer can now calculate the emitter and collector currents, and collector voltage for the circuit of Fig. 2.4. By using the successive approximation method presented he can equally well calculate the values of R_e and R_L necessary for a specified I_e and V_L to be obtained.

Effect of Base Circuit Resistance

The inclusion of a base resistor R_b as shown in Fig. 2.5 can have a considerable effect on the values of I_e, I_c, and V_{cb}. The circuit equations, as shown in Appendix 2, now appear very complicated and are of little use in practical design. The reason for all these changes in operating conditions is, however, very simple: the base is no longer at zero potential, because the difference current between I_e and I_c, which must flow out of the base, causes a voltage drop in R_b. The magnitude of the base current is clearly $I_b = (I_e - I_c)$, i.e. $I_e - (\alpha I_e + I_{cbo})$, or $(1 - \alpha)I_e - I_{cbo}$, and therefore the base voltage differs from zero by $V_b = R_b[(1 - \alpha)I_e - I_{cbo}]$. The complication caused by this is evident, because the calculation of I_e is immediately changed to $(V_e - V_{be} - V_b)/R_e$, which itself is dependent on I_e. Note that I_b can have either polarity and may be zero

The way in which the designer minimizes the problems of the calculation is very simple. He merely assumes that V_b is so small that it may be disregarded, and then, knowing I_b from the simply calculated I_e, chooses R_b so that V_b really is negligible. An objection which may be raised here is that perhaps it would be advantageous to allow V_b to be large so that large values of R_b could be used (even though it puts the designer to more trouble to work out his sums). This is rarely valid in serious design, however, because of the nature of I_b. Referring to the equation $I_b = (1 - \alpha)I_e - I_{cbo}$ it will be noted that the first part of the expression has a factor $(1 - \alpha)$ and that the second term is I_{cbo}. Now, α varies in transistors of any type by a few per cent and since α is close to unity, $(1 - \alpha)$ varies widely for these changes in α. A typical spread in $(1 - \alpha)$ is 3 or 4 to 1. The I_{cbo} component varies by about 100 to 1 in most transistors, and is often comparable in maximum value to the first term (e.g. in germanium transistors and in silicon transistors run at low current). It is clear therefore that the value of I_b is uncertain by a factor of at least 3 or 4 to 1 and, hence, so is V_b. If V_b is allowed to have a maximum value which is enough to affect I_e significantly, the spread in I_e, and therefore in I_c and V_{cb} will be very large. Since it is rare that large spreads (of more than about 20 per cent) can be tolerated, the whole idea of allowing V_b to become significant must be discarded in good design practice.

The procedure then is to ignore R_b initially, calculate I_e, I_c, and V_c as if R_b were zero, then assess I_b. If R_b is already fixed (i.e. it is part of a previous drive circuit), I_e is now recalculated knowing that V_b is not zero but is really I_bR_b. This gives a new value of I_e; if this is different from the original value by more than the tolerable uncertainty, the design is unsatisfactory. If acceptable, the new, slightly different, I_e is now used to recalculate I_b and V_b. This again affects I_e but considerably less even than before. V_b can yet again be recalculated, but convergence will have been obtained in most circuits by this stage.

If in the above design R_b is not known but is to be specified by the designer, a reasonable value of V_b is decided (a certain percentage of V_e), I_b is calculated, and R_b chosen to give not more than that value of V_b for the highest limit of I_b. As above, the value of V_b changes I_e: hence, I_b, V_b, and again I_e, but the sequence converges rapidly.

This iterative process is similar to that used for most practical circuit design. Although the description is long the actual process is not, and the designer soon becomes aware of which circuit values need respecifying and in which direction.

Taking the values previously used in illustrating the calculation of V_{cb}, the permissible value of R_b will now be worked out using the (arbitrary) condition that the presence of R_b should not cause I_e to change by more than 5 per cent.

The first step is to calculate I_e approximately (in view of the corrections which will follow as a result of R_b there is no point in exact calculation). This is easily done and yields, as before, $I_e \approx 1$ mA. Now, $I_b = (1 - \alpha)I_e - I_{cbo}$ and α was assumed to vary from 0·95 to 0·99, giving a variation in $(1 - \alpha)$ of 1/20–1/100. The value of I_{cbo} can be anywhere in the range 0–1 μA. With nominal I_e (1 mA) the limits of I_b are therefore from (1/20 mA − 0) to (1/100 mA − 1 μA), i.e. from +50 to 9 μA. The effect of I_{cbo} is therefore less than that of $(1 - \alpha)$ in this example and leads to a maximum I_b of 50 μA. If I_e is to be upset by only ±5 per cent, then V_b must be only 1/20 V_e, i.e. 0·5 V. Therefore R_b should not exceed 0·5/50 MΩ, or 10 kΩ. A value of 4·7 kΩ is therefore satisfactory. The value of I_e will now be different and the limits of I_b can be used to calculate the limits of I_e. These will clearly be within the 5 per cent limits set (from the effects of R_b only), so that the design is successful.

Incompatible design requirements

It may occur (and very often does, in the first attempt to meet a given specification) that the permissible maximum R_b obtained from the indicated design procedure is too small for other circuit requirements. In such a case the designer must resist the temptation to play down some of the possible variations which led to the low value of R_b. The possible solution may be to change the transistor for a higher α type, or a device with lower I_{cbo}, e.g. silicon instead of germanium. Similarly, a reduction in I_e, if allowable from output considerations, also reduces I_b.

Failure of these simple remedies means that too much is being asked of one transistor and either the specification must be changed or an additional transistor used as described in Chapter 7.

Temperature Drift

So many complicated equations are presented for the determination of the effects of temperature on operating conditions that the student will be pleased to know that the procedure outlined in the previous sections already includes all necessary figures.

The claim that a circuit is stabilized against temperature drift merely implies that the designer has been able to produce a circuit meeting the required specification while taking into account (as in the above examples) the parameter spreads with temperature as well as spreads from one unit to another.

It may be difficult to imagine a circuit capable of operating correctly with any transistor of a specified type and yet failing because of temperature change. This can, however, come about for the following reasons. First, if the operating temperature rises by, for example, 40 degC, the value of I_{cbo} is likely to rise by a factor of 16, which would usually represent a much greater current change than the possible values of I_{cbo} for all transistors of that type at normal room temperature. Secondly, most transistor parameters, although differing widely from one transistor to another of the same type, remain stable for any particular transistor at constant temperature. For example, a high-α transistor will maintain its high value indefinitely; V_{be} does not normally change during the life of a transistor; I_{cbo} is not so predictable but generally falls slightly as the transistor ages (rising I_{cbo} with operating time indicates an unreliable transistor owing to contamination, often caused by imperfect sealing from the atmosphere). This means that if factory pre-adjustment of one of the circuit values which determine operating conditions is allowed, enabling for example V_{cb} to be set to a particular voltage by selecion of R_e, the circuit will remain stable thereafter provided the temperature conditions are also stable. This procedure is often used for 'entertainment' circuits, where the use of the smallest number of components is more important than great temperature stability.

In most industrial and military designs, however, wide temperature variations are encountered and it is rarely practicable to construct a constant-temperature enclosure to house the electronic circuits. It should nevertheless be borne in mind that small temperature-controlled ovens are often used to enclose certain critical components such as quartz crystals, reference Zener diodes, and sometimes transistors.

It has already been established that temperature drift is caused by three transistor parameters. The list below explains how each of these affects the value of I_c and, hence, V_{cb}.

(*a*) V_{be}. For a given value of I_e the corresponding value of V_{be} varies with temperature at a rate which usually lies between − 2 and − 2·5 mV/degC. Note that this is a linear relationship and that although the spread quoted is wide, this covers all junction types and includes both germanium and silicon transistors. Any one type of transistor (i.e. alloy, planar, diffused, etc.) has a much narrower variation of the order of ± 0·1 mV/degC. The value of I_e also affects the V_{be} coefficient to about the same extent over the useful emitter current range of the transistor.

This V_{be} variation, as has already been shown, has a direct effect on the value of I_e in Figs. 2.4 and 2.5. The magnitude of the change in I_e is simply calculated, as has been seen, and *its direction is such that I_e increases when temperature rises.*

(*b*) I_{cbo}. I_{cbo} rises with temperature exponentially, doubling itself every 9–10 degC rise. Since, strictly, only part of its low-temperature value obeys this law (the other part remaining constant), the value at high temperature cannot be predicted from a low-temperature figure. By assuming that all of thelow-temperature I_{cbo} obeys the law, a pessimistic estimate for high temperatures can be made; in practice this is not far wrong for germanium but is often greatly pessimistic for silicon. It is usually necessary for design therefore to know absolute maximum I_{cbo} for the transistor at the maximum junction temperature to be used in the circuit (i.e. the maximum ambient temperature plus the rise due to dissipation within the transistor).

Its effect on the circuit has been seen to be twofold. First, I_{cbo} adds directly to the collector current. Secondly, the same I_{cbo} is flowing into the base and therefore causes a base voltage drop of R_bI_{cbo} *in such a direction as to increase I_e as the temperature rises.*

The latter effect usually predominates and its magnitude is simple to calculate, since it is equivalent to an increase in V_e of R_bI_{cbo}.

(*c*) α. As stated previously, the variation in α is approximately 0 to +1/25 per cent per degree C. Although for a particular transistor the law is more or less consistent over a wide temperature range, different transistors of one type often have α temperature coefficients at the two extreme limits.

Like I_{cbo}, α variation has two effects on operating conditions. The first, direct, effect is caused by the relationship $I_c = \alpha I_e + I_{cbo}$. The maximum magnitude of this effect is therefore approximately +1/25 per cent change in I_c per degree C, generally negligible in comparison with other drifts. The second effect is usually much more significant and occurs by virtue of the dependence of I_b and, hence, V_b (when R_b is present) upon $(1 - \alpha)$, which in turn depends critically on α. For most transistors the α temperature coefficient can therefore more conveniently be described in terms of $(1 - \alpha)$, which normally has a maximum temperature coefficient of +2 per cent per degree C. The quantity $\alpha/(1 - \alpha)$ is often called* β; and since α is close to unity, $1/(1 - \alpha)$ is approximately β.

As in the case of I_{cbo}, the effect of β temperature coefficient on the operating conditions of the circuit of Fig. 2.5 is simply calculated by working out the change in I_b and, hence, V_b when β changes by the expected maximum amount. This should be calculated using the value of β which is the minimum for the transistor type at the lowest operating temperature to be considered, and then assuming +2 per cent per degree C rise, since lowest initial β gives the worst case.

Note that, as with I_{cbo}, β variations *change V_b in such a way that I_e rises as temperature increases.*

Summary of temperature effects

Three parameters drift with temperature: V_{be}, which varies linearly at a rate of −2 to −2·5 mV/degC; I_{cbo}, which doubles some or all of its low temperature value for every 9 or 10 degC rise; and α, which drifts in such a way as to cause β to vary 0–2 per cent per degree C.

The main effects on the circuit are caused by the direct influence

*Also known as h_{FE}

of V_{be} on I_e and by the changing base current caused by both I_{cbo} and α, which cause V_b to differ from its intended value, and thus cause I_e to change.

Smaller effects are caused directly by the change in I_c, even at constant I_e, brought about when α and I_{cbo} vary.

All these effects cause the collector current to increase as temperature rises.

Design for stability

As indicated, the procedure described earlier for ensuring that circuit conditions are as required includes temperature-stability criteria. The changes in procedure for exceptionally wide tempera-

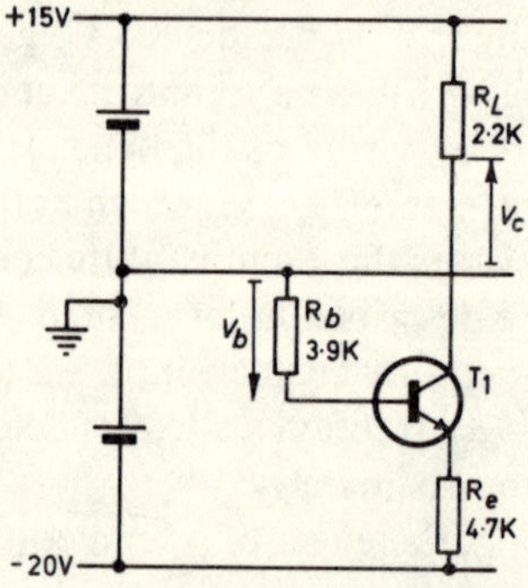

FIG. 2.6 Practical example of bias circuit

ture ranges (such as -55 to $+100$°C) in some military applications are changes in parameter magnitudes only. For I_{cbo} the 100°C $I_{cbo,max.}$ figure would be used (or more if transistor dissipation raises the junction temperature appreciably); for β a temperature rise of 155°C would be used and the worst case obtained by using the -55°C minimum β for the transistor; for V_{be} would be assumed a drift of $-(2{\cdot}5 \times 155)$ mV. Otherwise the procedure is unchanged.

Design example

As a practical example, the circuit of Fig. 2.6 will be examined and its temperature drift from 0 to 50°C predicted, assuming knowledge of the relevant transistor parameters.

It will be assumed that T_1 is a small signal (300 mW) silicon transistor having a minimum value for β of 20.

First, low-temperature operating conditions are found by assuming R_b to be zero and V_{be} its maximum value, e.g. 0·9 V.

Then,
$$I_e = (20 - 0{\cdot}9)/4{\cdot}7 \text{ mA}$$
$$= 4{\cdot}06 \text{ mA}$$

since
$$\beta_{min.} = 20$$
$$\alpha_{min.} = 0{\cdot}95$$

and
$$I_c = \alpha I_e = 0{\cdot}95 \times 4{\cdot}06$$
$$= 3{\cdot}86 \text{ mA}$$

Hence,
$$V_{cb} = 15 - 3{\cdot}86(2{\cdot}2)$$
$$= 6{\cdot}5 \text{ V}$$

Using the principle of calculating changes from nominal rather than recalculating every value for each extreme, the effect of R_b = 3·9 kΩ can now be accounted for.

I_b is known to be I_c/β, i.e. 0·2 mA, giving

$$V_b = -0{\cdot}2 \times 3{\cdot}86$$
$$= -0{\cdot}77 \text{ V}$$

The value for I_e should therefore be

$$I_e = (20 - 0{\cdot}9 - 0{\cdot}77)/4{\cdot}7$$
$$= 3{\cdot}9 \text{ mA}$$
$$I_c = 0{\cdot}95 \times 3{\cdot}9$$
$$= 3{\cdot}7 \text{ mA}$$

and
$$V_{cb} = 15 - 3{\cdot}7(2{\cdot}2)$$
$$= 6{\cdot}86 \text{ V}$$

Note, then, $I_b = I_c/\beta$ is still about 0·2 mA, so no further correction is required.

So far, temperature effects have been ignored, and since the minimum value of β and the maximum value of V_{be} have been used, the above values of I_b, I_c, and V_c represent the low-temperature limit. The changes which occur when the ambient changes by 50 degC are as follows.

(*a*) V_{be}. V_{be} falls by 125 mV. Since V_b is assumed constant (for the moment base voltage changes are to be ignored), the change in I_e must be $125/R_e$ mA, i.e. 125/4·7 μA, or 26·6 μA. This causes a change of 0·95 × 26·6 μA in I_c, i.e. 25 μA, and, hence, a change in V_{cb} of 25 × R_L μV = 25 × 2·2 mV, or 55 mV. The direction of the V_{be} change increases I_e and I_c and therefore reduces V_c.

(*b*) I_{cbo}. In the absence of data a maximum value of 10 μA at 50°C for almost any small silicon transistor is a slightly pessimistic figure and this will be assumed here.

There are two effects, as described earlier. The first is that I_c increases by 10 μA because I_{cbo} adds directly to the collector current. This gives a V_{cb} change of 10 × 2·2 mV, i.e. 22 mV, and this is again a reduction in V_{cb}.

The second effect is to change V_b by 10 × 3·9 mV, i.e. 39 mV. This movement of V_b will itself cause I_e to change by $39/R_e$ mA, i.e. 39/4·7 μA, or 8 μA, giving a change in αI_e of 0·95 × 8, i.e. 7·6 μA. This causes V_{cb} to change by 7·6 × 2·2 mV, i.e. 17 mV, and again V_{cb} is reduced.

Hence, the total effect of I_{cbo} is a V_{cb} reduction of 39 mV.

(*c*) β. The minimum value for β, namely 20, was used to derive the 0°C conditions. Change in β totals 2 × 50 = 100 per cent and β therefore changes from 20 at 0°C to 40 at 50°C.

The direct α change therefore alters I_c by 2·5 per cent, giving a change of 3·7 × 2·5/100 = 92·5 μA, reducing V_{cb} by 92·5 × 2·2 mV = 0·2 V.

When β changes from 20 to 40, I_b changes by I_e [(1/20) – (1/40)], i.e. by 0·025 I_e, or 3·9 × 0·025 = 97·5 μA, giving a V_b change of 97·5 × 3·9 mV = 380 mV. I_e therefore changes by $380/R_e$ = 380/4·7 = 81 μA, I_c by 0·95 × 81 = 77 μA, and V_{cb} by 77 × 2·2 mV = 0·17 V.

Hence, the total effect of β is a V_{cb} reduction of 0·2 + 0·17 = 0·37 V.

The total drift of V_{cb} from all transistor changes with temperature is therefore 0·055 + 0·039 + 0·37 = 0·46 V for silicon.

This method of calculating drift is not exact, since each parameter drift affects the contribution caused by another; for instance, when β changes from 20 to 40 the effect of a change in I_e due to V_b change is different in the collector circuit, since I_e drifts will be multiplied by 0·975 instead of 0·95.

In practice, however, the method is quite satisfactory, because these interacting effects are important only when one of them represents a large percentage change from the nominal state. If this proves to be true the design will require modification, and, using this method of calculation, it will be immediately obvious which remedy to take.

Temperature compensation

In view of the severe effects of drift in transistor circuits, attempts are often made to obtain stability by adding another temperature-sensitive element in such a way as to oppose drift caused by the transistor.

Since, as described in this chapter, transistor drifts are predictable only in their maximum values, it is generally impossible to provide an equal and opposite drift unless the compensating device can be adjusted at two or more different temperatures until its law is correct. This is normally out of the question, representing as it does a prolonged setting-up procedure, but may occasionally be accepted if the alternative is a more expensive system (e.g. chopper amplifier, Chapter 9).

The exception is V_{be} temperature variation which, first, has a limited range of values (from -2 to $-2{\cdot}5$ mV/degC) and, secondly, is linear with regard to temperature. This means that if it can be arranged to 'back off' V_{be} changes, and the backing-off device and the transistor have the opposite extreme values, then at least the drift has been reduced to $\pm 0{\cdot}5$ mV/degC. With a little more care, by using for the compensating device a similar junction type (e.g. alloy, diffused, planar, etc.) and running the device and transistor at similar currents, the spread is greatly reduced and is typically $\pm 0{\cdot}1$ mV/degC, an improvement of about 10:1 on the uncompensated transistor. The compensating element may be a diode or, for better matching, an extra transistor preferably mounted within the same can as the amplifier transistor.

The actual arrangements for compensation may be the circuit of Fig. 2.7, or any configuration in which the added device does not seriously affect circuit operation and in which its drift and the transistor's act in opposite directions and with equal significance.

Other Bias Arrangements

Although the bias circuit discussed is the most usual arrangement, there are a few other possibilities. The emitter may be held at a fixed potential (e.g. 'earth') and a base current supplied as shown in Fig. 2.8. In this circuit the base current is accurately known provided $V_1 \gg V_{be}$ but the resulting collector current depends greatly on I_{cbo} and β (*see* Appendix 2, standard bias circuit where $R_e = 0$). The practical implication is that in order to be certain that the transistor collector voltage will not reach zero even with a high-β transistor, R_L must be much smaller than the value it would be given in the standard bias circuit. The output voltage swing has to be restricted

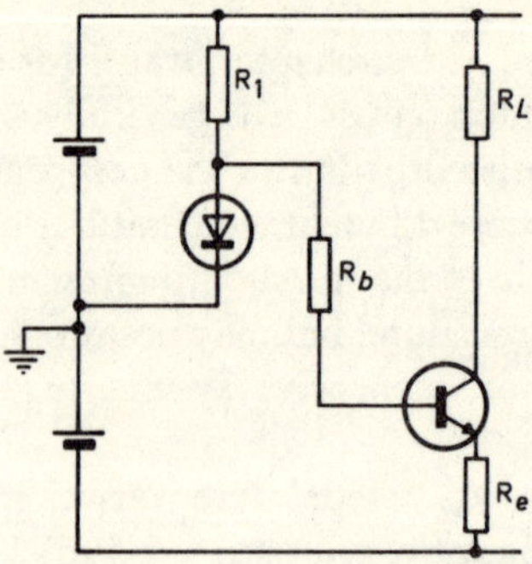

FIG. 2.7 V_{be} compensation circuit

to a small fraction of V_1 to avoid possible cut-off or saturation, and so the circuit is useful only in simple applications where temperature drift is unimportant and signal levels are low, e.g. domestic radio equipment.

An even worse version of Fig. 2.8 is obtained by omitting the base

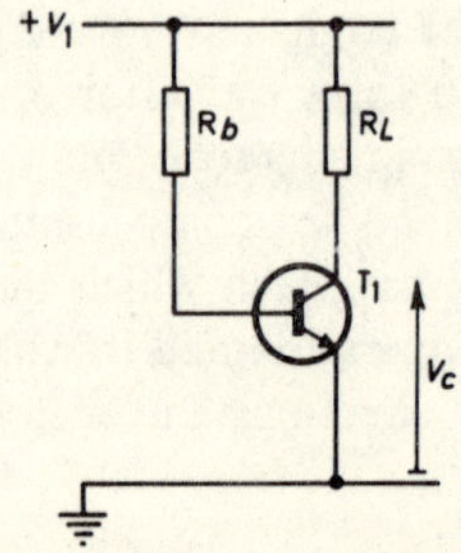

FIG. 2.8 Base current bias

resistor, and capacitor coupling the signal to the base. The defects of this method are even greater than those of the last, since the collector current is now given by βI_{cbo} (often called I_{ceo}), which is an amplified version of the extremely variable I_{cbo}.

A much better attempt is shown in Fig. 2.9, which uses d.c. negative feedback to define the collector voltage. Note that in the intended operating condition T_1 base will be just positive and, if the base current is assumed negligible, then the current in R_{b1}, approximately V_n/R_{b1}, must equal that in R_{b2}, namely V_c/R_{b2}. Therefore $V_c = V_n R_{b2}/R_{b1}$. This suggests that R_L has no influence on V_c and this is substantially true provided T_1 passes current and that I_b remains $\ll V_n/R_{b1}$. To design this bias circuit, just assume that conditions are

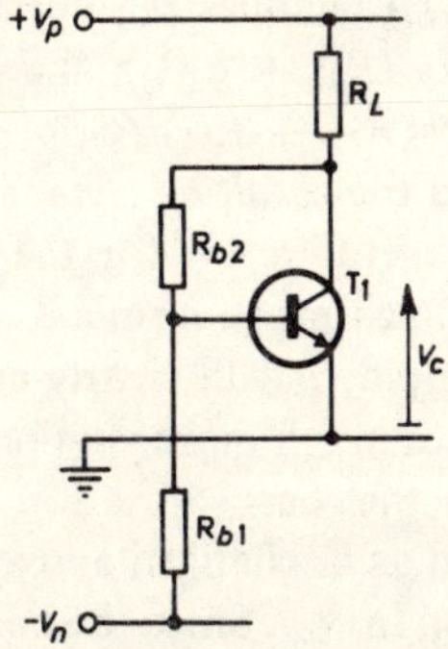

FIG. 2.9 Feedback bias

as desired and use Ohm's law to deduce circuit values. For example, suppose that $V_p = V_n = 10$, that V_c is to be approximately $+5$ V, and that T_1 collector current is to be 3 mA and its β at least 30.

T_1 base current is at most $3/30 = 0{\cdot}1$ mA. Let $V_n/R_{b1} = 1$ mA, giving $R_{b1} = 10$ kΩ. $V_c = +5$, so that $5/R_{b2} = 1$ mA, giving $R_{b2} = 5$ kΩ. The current in R_L is the sum of 1 mA from R_{b2} and 3 mA from T_1, so that $R_L = (V_p - V_c)/4 = 1{\cdot}2$ kΩ.

These figures are all approximate and no allowance for I_b was made in calculating R_{b2}.

To understand fully the operation of this bias circuit, note the effect of variations in V_p, V_n, R_{b1}, R_{b2}, R_L, V_{be}, β, and I_{cbo}. An increase in V_p increases $(V_p - V_c)/R_L$ thus increasing I_c and I_b, but until I_b becomes significant V_c remains constant. An increase in V_n increases V_n/R_{b1}, so that R_{b2} drops more voltage, thus increasing V_c in direct proportion to V_n. Decrease in V_p similarly has no effect on V_c until $(V_p - V_c)/R_L$ is equal to the current in R_{b2}, implying $I_c = 0$. The transistor is now cut off, its base voltage is no longer just below zero, and the stabilizing loop fails. Changing R_{b1} or R_{b2} is similar to changing V_n and a proportionate change in V_c results.

The temperature variations of V_{be}, β, and I_{cbo} are of special interest in that the method for assessing their effect is applicable to all d.c. negative feedback systems.

V_{be}, which was ignored in the previous calculations, affects the conditions by determining the potential at the junction of R_{b1} and R_{b2} and the correct equation for V_c (still ignoring I_b) is

$$(V_n + V_{be})/R_{b1} = (V_c - V_{be})/R_{b2}$$

When V_{be} varies by δV_{be}, it changes R_{b1} current by $\delta V_{be}/R_{b1}$ and so changes V_c by $(\delta V_{be}R_{b2}/R_{b1} + \delta V_{be})$, i.e. $\delta V_{be}(R_{b2} + R_{b1})/R_{b1}$. Therefore V_c increases by $(R_{b2} + R_{b1})/R_{b1} \times$ (2 to 2·5) mV/degC.

β and I_{cbo} change I_b and the *resulting change in V_c is approximately* $\delta I_b R_{b2}$. This is a most useful result for the assessment of current drifts arising at the feedback input terminal of a negative d.c. loop. The argument is very simple: I_b will clearly affect the base potential but the collector will move much more, so that in calculating voltage drops in R_{b1} and R_{b2} the base variation is negligible. Current V_n/R_{b1} can then be taken as unchanged and any I_b change flows into R_{b2}, thus varying the drop in R_{b2}. Since the base is fixed, the collector moves by this amount, and if I_b is inwards and then increases by δI_b, V_c will beco me more positive by $\delta I_b R_{b2}$ V.

Sometimes the circuit of Fig. 2.9 is used with R_{b1} omitted. Although this is advantageous in removing the need for V_n, the current in R_{b2} is now I_b, so that V_c depends greatly on β and I_{cbo}, which in Fig. 2.9 could be made only a small fraction of V_n/R_{b1}. Collector voltage is therefore badly defined, but not so badly as if R_{b2} were returned direct to V_p.

Finally, it should be pointed out that the commonly used bias circuit of Fig. 2.10(*a*), which requires only a single supply, is basically the same as that of Fig. 2.5. This is easily proved by Thévenin's theorem and its equivalent circuit is given in Fig. 2.10(*b*). For simplicity of description, most circuits in Part I use the form of Fig. 2.5.

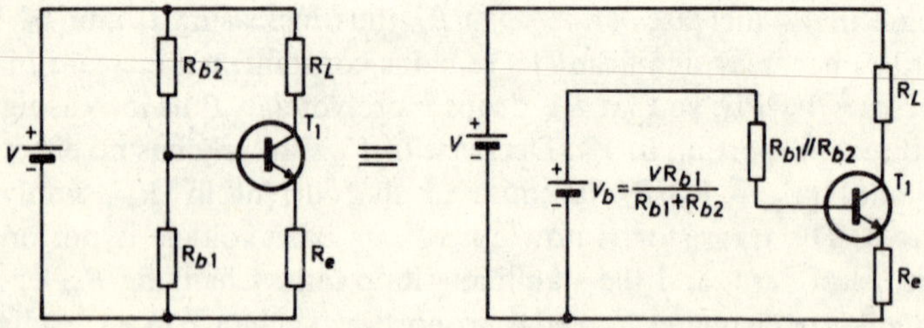

FIG. 2.10 Single supply bias circuit: (*a*) bias circuit, (*b*) equivalent circuit

SUMMARY

To summarize, designing a biasing circuit for a single transistor consists in ensuring that the operating currents and voltage remain within limits set by various external circuit requirements. This is achieved by making an initial nominal choice of values and assessing by calculation the effect of each varying parameter on the operating conditions. In the light of these effects, values are modified until the operating conditions are within the required tolerance limits.

Temperature stabilizing is not a separate problem and the calculation of 'stability factor', although indicating how bad stability is, does not point to the particular cause of drift. Instead the recommended procedure is to include temperature effects when assessing the effect of parameter variations.

Temperature-compensation techniques are of doubtful value except to make an already good or marginal design better. Reliance on these methods for correction of large drifts is unsound because the compensating device and the transistor may not always be at the same temperature; even if they are, a small imperfection in matching can give large drifts. In particular, I_{cbo} cannot be satisfactorily compensated for because of its great variability; the same applies to β, but to a very much smaller extent. V_{be} can be balanced out provided an improvement of no better than 10 : 1 is anticipated; special pairs of transistors ('dual' transistors) housed in one encapsulation do, however, give excellent balance of better than 100 : 1. The inherent V_{be} and β matching which exists between transistors on the same chip is exploited by the integrated-circuit designer in circuit configurations which would be unsound using discrete devices.

D.C. CHARACTERISTICS OF THE FIELD EFFECT TRANSISTOR (FET)

The FET is sufficiently similar to the bipolar transistor in its properties to enable the foregoing principles for bias arrangements to be used. However, there are some important differences in terminology and in magnitudes of parameters which need to be clarified.

n-channel FET

There are several varieties of FET, the most common of which in discrete circuitry is the *n*-channel junction FET (or *n*-channel J-FET). Although physically different from the bipolar transistor described

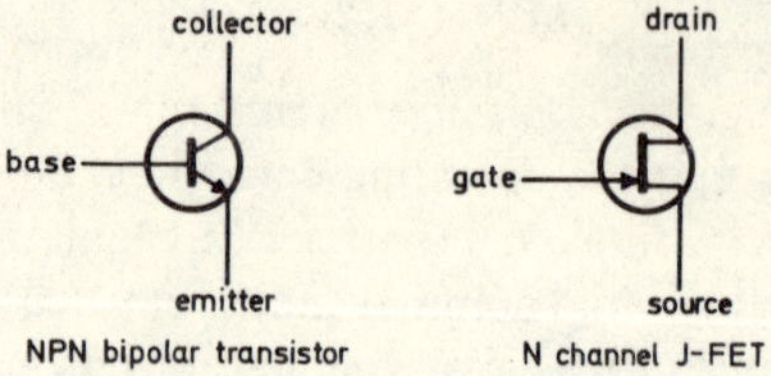

FIG. 2.11 Bipolar/FET comparison

earlier, it resembles in many respects an *npn* device in which the *collector*, *emitter* and *base* electrodes are renamed the *drain*, *source* and *gate*, respectively (see Fig. 2.11).

With the drain positive with respect to the source, the drain-source current approaches zero when the gate is highly negative and increases as the gate becomes more positive. This roughly corresponds to the bipolar device, but there are two major differences. Firstly the gate-source potential for passing an intermediate value of current such as would be used in a practical amplifier, is *negative* by a few volts, not positive by 0·5–0·9 V as in the bipolar case. Secondly the current in the gate electrode remains at a very low level (leakage only) over the whole normal range of operation unless the gate potential exceeds that of the drain or source by 0·5 V. The FET then resembles the two-diode model of Fig. 2.1(*b*).

The gate-source bias V_{GS} for normal conduction is, as stated above, a few volts negative and is highly variable for a given drain-source current I_{DS}. Consequently the negative supply voltage in the conventional arrangement of Fig. 2.10 needs to be larger than for a bipolar device to achieve a given tolerance in the defined value of operating current.

Base current bias (Fig. 2.8) has no equivalent for the J-FET; on the other hand the so-called auto-bias circuit used with thermionic valves (Fig. 2.12) and which cannot be used with bipolars, is fre-

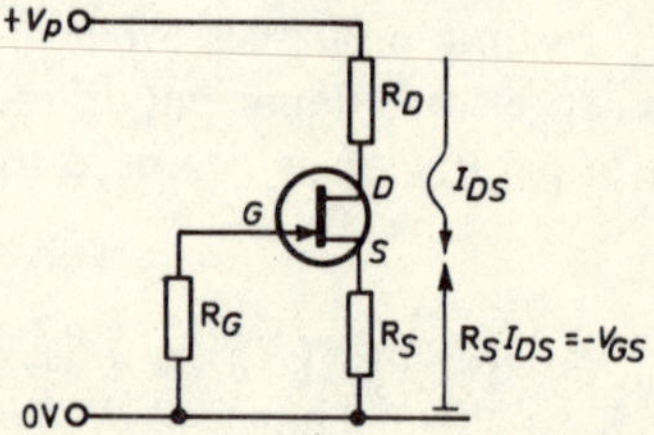

FIG. 2.12 Auto-bias for the J-FET

quently used with the J-FET when accuracy of bias current is unimportant. This configuration works correctly because the standing gate-source voltage is negative and the drop across R_s is in the correct direction to provide it.

Compensating for the inconvenience of the large gate-source voltage is the negligible gate current, implying that gate resistors may have values up to several megohms without seriously affecting the gate potential. Since the input resistance of the J-FET itself is in the hundreds of megohms region, this enables very high input resistance amplifiers to be made very simply—the commonest reason for using an FET as an amplifier.

Symmetry

Although specific reference has been made to drain and source these electrodes are interchangeable in function and, if the electrode areas are equal, identical performance will result. This is the only inference to be drawn from the statement that these devices are 'symmetrical': in particular, it does not mean that shaping circuits involving the non-linear features of the J-FET will treat positive and negative drain-source voltages similarly.

Other FET Varieties

The *p*-channel J-FET differs from the *n*-channel variety in much the same way as a *pnp* bipolar transistor differs from an *npn*. For normal biasing the drain-source voltage is negative and the gate course voltage is positive.

The *p*- and *n*-channel J-FETs so far described are 'depletion' types and their state of drain-source conduction is high when the gate is at source potential.

The metal oxide silicon transistor (MOST), commonly used in both *n*- and *p*-channel forms in modern logic integrated circuits has a metal oxide insulating layer between the gate and other electrodes. The gate does not form junctions with either source or drain so that only leakage current flows in the gate whatever its potential is, until breakdown occurs. This apparent advantage is a mixed blessing in discrete circuits since the charge picked up from clothing or finger contact can damage the device which lacks the inherent diode protection of the J-FET. Special handling is required, such as the use of tin foil or tweezers to short-circuit the electrodes until equipment assembly is complete. Some types have diode protection built in to avoid these difficulties.

As well as being available in 'depletion' form, MOST devices are made in the 'enhancement' form, in which for an *n*-channel type I_{DS} is substantially zero when V_{GS} is zero and increases as V_{GS} becomes more positive. Similarly the *p*-channel enhancement type also has I_{DS} near zero when V_{GS} is zero, but I_{DS} increases as V_{GS} becomes more negative.

Leakage Currents

Only passing mention has been made of leakage currents because their magnitudes are very low in the small signal devices being discussed, being typically 0·1 μA at a junction temperature of 150°C. These become significant only when circuit resistances of several megohms are present in very high ambient temperatures.

SUMMARY

Biasing the FET presents similar problems to those associated with the bipolar transistor. The main differences are the polarity and magnitude of the gate-source voltage compared with the base-emitter voltage, and the negligible gate current compared with the base current.

Chapter 3

The transistor as a switch

One of the assumptions made in the last chapter on the biasing of a transistor was that the collector–emitter voltage would never be allowed to reach zero. This chapter deals with the use of values deliberately chosen to make V_{ce} as near zero as possible.

Figure 3.1 shows a simple circuit which in terms of normal bias arrangements is unsatisfactory because the only current directly determined by the circuit appears to be I_b, which must be V_p/R_b provided $V \gg V_{be}$. Hence, I_c must be $\beta I_b = \beta V_p/R_b$ and $V_L = \beta V_p R_L/R_b$,

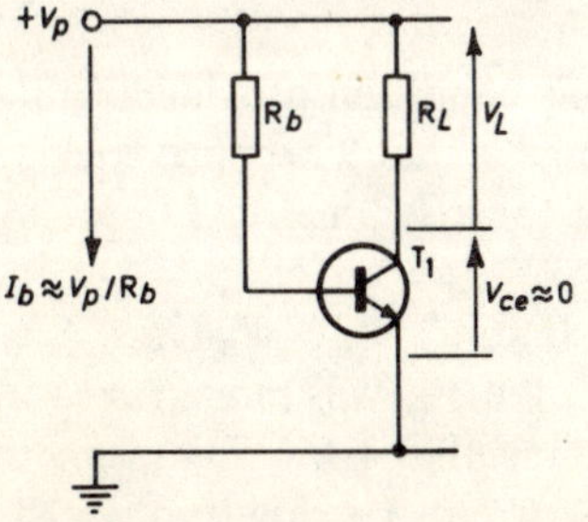

FIG. 3.1 Saturated transistor

which is greatly dependent upon β. But suppose that, even with the lowest possible value of β, $\beta R_L/R_b > 1$; then V_L is apparently $> V$. This is clearly impossible, since the collector voltage would then become negative with respect to earth and no negative supply exists which could cause this to happen.

In fact, as V_{ce} approaches zero β falls until, with V_{ce} almost zero, $\beta = 0$. Since β is the ratio of a change in I_c to the change in I_b causing it, this implies that further increase of I_b has no further effect on I_c and I_c remains equal to V_p/R_L.

In this condition the transistor is said to be 'saturated' or 'bottomed', and to ensure this state the designer merely makes $I_b > I_c/\beta_L$, where β_L is the large signal current gain* from 0 to the desired I_c. When the same supply rail is used for R_b and R_L and its voltage is much greater than V_{be}, then for saturation $R_b < R_L\beta_L$.

A similar situation is shown in Fig. 3.2, where the collector is fixed and two supplies are used. To ensure saturation the same criterion must be obeyed, $I_b > I_c/\beta_L$ or in this case $V_p/R_b > V_n/(R_e\beta_L)$.

In the saturation condition the transistor has some especially useful properties. Consider the circuit of Fig. 3.2 to be changed by reversing the polarity of V_n; the circuit is now identical in form to Fig. 3.1 except that the emitter and collector are interchanged. It has already been pointed out in Chapter 2 that a transistor operates

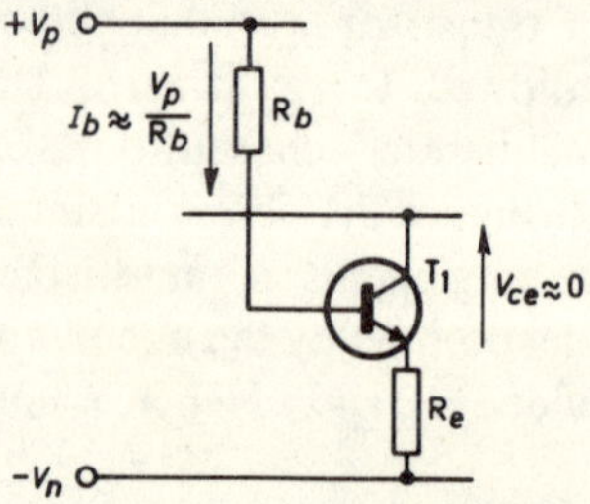

FIG. 3.2 Alternative saturation circuit

quite normally in this condition provided ratings are observed. The only significant change is that β becomes much lower (e.g. 3 or 5) in this direction except for a symmetrical transistor, where β is similar in both connections.

Hence, provided that $V_p/R_b > V_n/(R_e\beta_{Lr})$, where β_{Lr} is the β_L in reversed connection, the V_{ce} will remain zero in the Fig. 3.2 circuit even when $-V_n$ is changed to $+V_n$. This means that the transistor acts as a short-circuit to currents passing through R_e *in either direction.*

This concept is most important in understanding transistor switching circuits (especially d.c. choppers). Although it is possible to design the circuit to saturate only with $-V_n$ negative (by $V_p/R_b > V_n/R_e\beta_L$ but $< V_n/(R_e\beta_{Lr})$), it is equaliy possibly to design for bidirectional saturation by making $V_p/R_b > V_n/(R_e\beta_{Lr})$.

SATURATION POTENTIALS

The collector–emitter voltage at saturation, known as $V_{ce(sat.)}$, depends on several factors: the type of transistor construction, the

* β_L is the ratio of collector current to base current at a given V_{ce} and β is the ratio of a small change in collector current to the change in base current which causes it, at a specified value of I_c and V_{ce}.

value of load current I_L ($=I_e$ or I_c) and its direction, the base current, and, to a slight extent, the junction temperature.

In general, alloy transistors, germanium and silicon, *p–n–p*, and *n–p–n* have low values of $V_{ce(sat.)}$. Planar epitaxial transistors have values almost as low. Alloy diffused, 'straight' planar, and diffused types have much higher values.

The load current naturally influences $V_{ce(sat.)}$, because, when it increases beyond $\beta_L I_b$ or $\beta_{Lr} I_b$, the transistor begins to leave saturation and so V_{ce} rises. If this is offset by arranging an increase in I_b when I_L rises, V_{ce} may still rise or alternatively may change sign, passing through zero at some value of I_L. This is puzzling at first sight but careful examination of Figs. 3.1 and 3.2 reveals how this comes about. In Fig. 3.1 it is clear that V_{ce} is positive, i.e. collector positive to emitter. An increase in I_c accompanied by an increase in I_b would maintain saturation (assuming β_L constant), but I_c still has to flow through some bulk semiconductor not under the influence of transistor action and the voltage drop of this part of V_{ce} must increase. In Fig. 3.2, if the special case is considered where $I_e = 0$, V_{ce} will clearly be negative, i.e. emitter positive to collector since no supply exists which could cause the open-circuit emitter to fall below earth potential. On reconnecting I_e, the emitter potential can be negative or positive depending on I_b and the bulk resistance through which I_e flows. As I_e rises from zero, V_{ce} therefore changes sign, becoming positive, and then continues to increase in this direction.

The effect of increasing the base current depends on its initial value. If it is assumed to be zero and is increased, V_{ce} will approach zero as I_b approaches $I_L/(\beta_L$ or $\beta_{Lr})$. As I_b further increases, V_{ce} becomes even closer to zero, and then either increases again with the same sign as with low I_b, or passes through zero, changes sign and then increases. The reasoning for similar behaviour for changing I_L is again applicable: the circuit of Fig. 3.1 always has V_{ce} positive, and that of Fig. 3.2 will change sign.

Temperature rise has two main effects: first, β increases with the same results as a change in I_b, and, secondly, the bulk resistance changes, causing a direct change in $V_{ce(sat.)}$. If the circuit is designed to be well in saturation at all temperatures of operation, a 50 degC rise typically causes $V_{ce(sat.)}$ to change 100 per cent, which is usually tolerable, since the circuit is normally designed to make $V_{ce(sat.)}$ itself negligible.

Significance of the Above Results

Where low V_{ce} is of great importance and base current of I_L/β_{Lr} (and β_{Lr} cannot be assumed >2) is available, the circuit of Fig. 3.2 is better than that of Fig. 3.1 whether the emitter load is returned to + or $- V_n$.

When, however, a slightly higher figure for V_{ce} can be tolerated, Fig. 3.1 requires much less I_b; this is particularly important when I_L is really large, such that $I_L/\beta_{Lr} = I_L/2$ would represent a large drain from the base supply circuit.

In practice this generally means that Fig. 3.2 is used when $I_L <$ 5 mA and Fig. 3.1 for higher currents: it is fortunate that higher V_{ce} can usually be tolerated in high-current circuits.

When $I_L < 100$ μA, then, by using Fig. 3.2 and $I_b = 1/4$ to 1 mA, $V_{ce(sat.)}$ is as low as 1 to 4 mV for most alloy transistors, germanium or silicon. Planar epitaxial values are about 5–20 mV. In the Fig. 3.1 circuit under optimum I_b conditions the best result is usually 10–20 mV for alloy and 50–100 mV for planar epitaxial.

At values of I_L above 100 μA $V_{ce(sat.)}$ increases and at 1 mA is usually 10 mV, corresponding to the best case above of 1 mV.

For much higher currents e.g. 10 A I_L with 1 A I_b, $V_{ce(sat.)}$ for Fig. 3.1 is usually from 0·3 to 1 V for alloy types. The connection of Fig. 3.2 is then generally unsatisfactory if V_n is reversed, since 5 A I_b may be required.

CUT-OFF CHARACTERISTICS

In order to reduce load current to zero in Fig. 3.1, the base–emitter junction must be reverse-biased or short-circuited, or at least the base must be more negative than the turn-on potential for the transistor. In Fig. 3.2 it is insufficient to drop V_b to earth potential—it must be more negative than $-V_n$. The only safe criterion is that for a *p–n–p* transistor *the base must be more positive than the emitter and collector* for turn-off. Similarly, for *n–p–n the base must be more negative than emitter and collector.*

When 'cut-off', currents still flow from the transistor; I_{cbo} flows out of the collector and I_{ebo} out of the emitter. Generally, but not necessarily, I_{ebo} is less than I_{cbo}, but if no I_{ebo} value is quoted by the manufacturer, it can only be assumed to be less than or equal to I_{cbo}.

The effect of these currents is shown in Figs. 3.3 and 3.4, which correspond to Figs. 3.1 and 3.2, respectively. In the first case I_{cbo} flows into R_L, giving a voltage drop $I_{cbo}R_L$ across the load. In the second, I_{ebo} causes a voltage drop $I_{ebo}R_L$ across the load.

Temperature affects I_{cbo} as described in Chapter 2, and I_{ebo} follows the same law, so that any calculation must be made at the highest

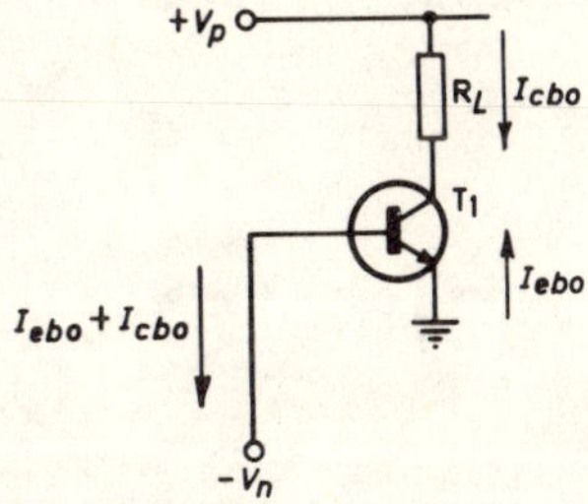

FIG. 3.3 Cut-off for circuit of Fig. 3.1

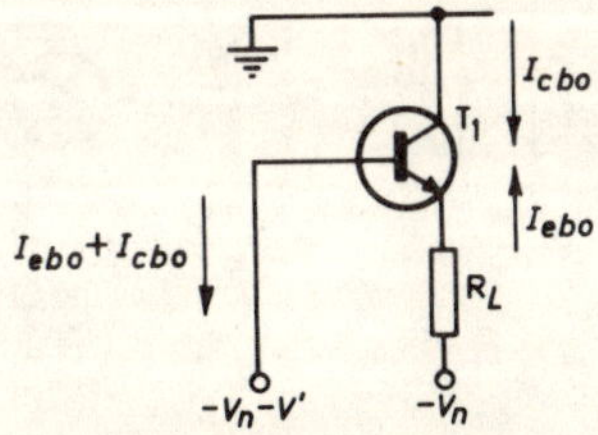

FIG. 3.4 Cut-off for circuit of Fig. 3.2

temperature of interest. In the majority of practical applications this effect rules out the use of germanium types for low-level circuits whenever R_L exceeds about 1 kΩ.

SWITCHING FROM 'ON' TO 'OFF'

When a transistor connected as shown in Fig. 3.1 or Fig. 3.2 is alternately in saturation and cut-off, the base conditions have to change from a circuit giving a known I_b to one giving a known V_b. A simple series resistor driven by a voltage source is usually satisfactory, but care must be taken that the potential V_1 reached on the positive swing is much greater in magnitude than V_{be} so that I_b is known to be $\approx V_1/R_b$ (Figs. 3.5, 3.6).

A capacitor-coupled drive circuit often causes difficulty and can only be used safely under special conditions. The snag is (*see* Fig. 3.7) that on the first positive swing, T_1 conducts, charging C as shown; on the next negative swing T_1 cuts off leaving C charged. On the second positive swing C charges further and still does not discharge on the next negati,e swing. Eventually C becomes charged to $(V_1 - V_{be})$ and T_1 base never turns on again; in other words, C and the transistor

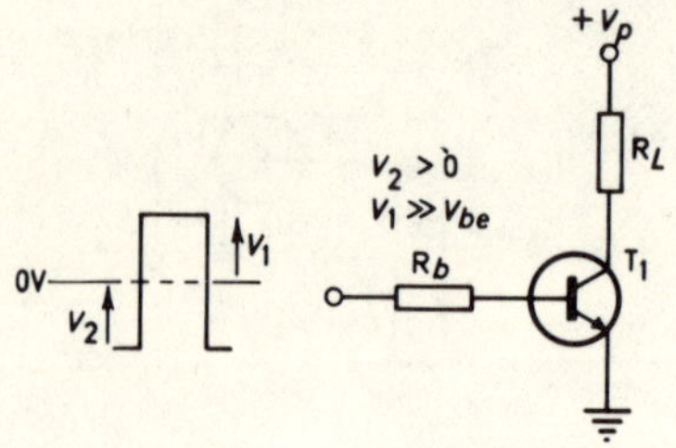

FIG. 3.5 Drive for saturation (circuit of Fig. 3.1)

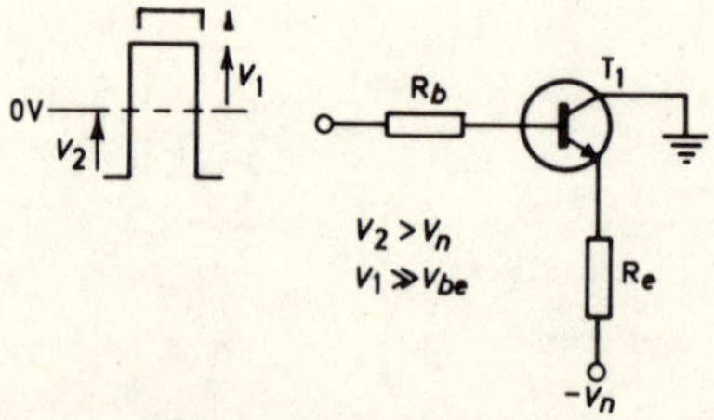

FIG. 3.6 Drive for saturation (circuit of Fig. 3.2)

have 'd.c. restored' the input waveform which appears on the base going negative only (see Chapter 1). One solution is to allow C to discharge through a resistor connected to a positive potential, while T_1 is off, so that T_1 will conduct on the next positive swing to replenish the lost charge (Fig. 3.8). This can be satisfactory if the input base waveform is regular in mark space, since I_b which flows to replenish C can then be calculated; on the whole however the situation is best avoided. If the earthed emitter circuit is being used, a simpler and better remedy exists. A diode between base and emitter enables C to discharge and recharge every cycle. This can clearly be done only in the earthed emitter circuit, since in the earthed collector version V_b must fall to at least $-V_n$.

Transformer coupling may be used as shown in Fig. 3.9 to avoid the problems of d.c. coupling and d.c. restoration. Some limitations are the feasibility of the transformer design in very-low-frequency

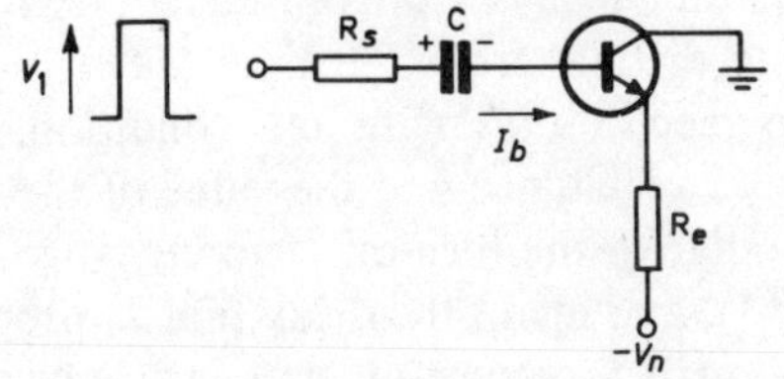

FIG. 3.7 Unsatisfactory drive

circuits (high inductance being required) and the loss through the transformer of the zero-frequency component of the drive waveform. The latter consideration can rule out the use of this method in cir-

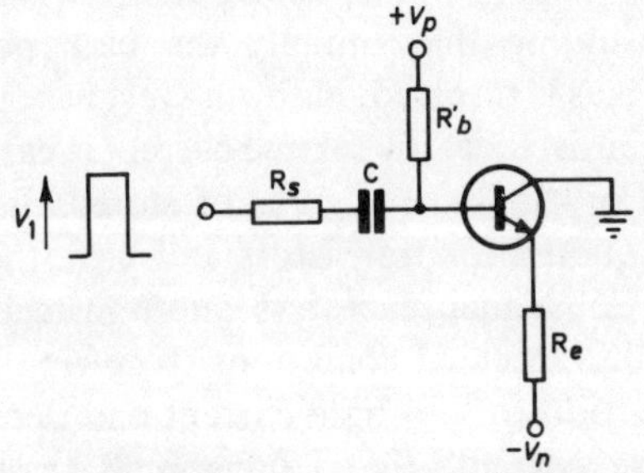

FIG. 3.8 Improved version of Fig. 3.7

cuits where it is intended to turn the transistor 'on' for the greater part of the cycle: the base waveform after passing through the transformer sits at such a level that the positive and negative voltage–time

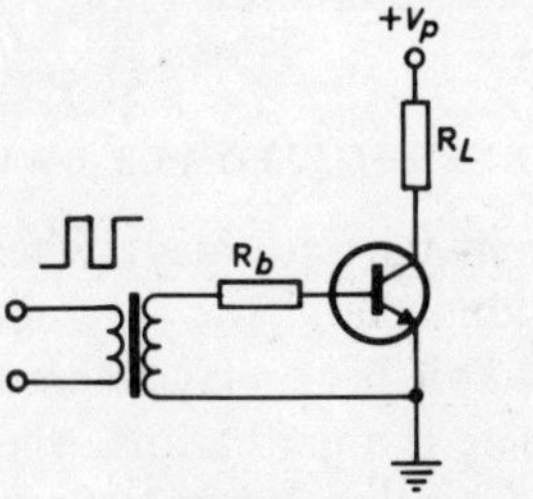

FIG. 3.9 Transformer-coupled drive

areas are equal, resulting in only low base current in the 'on' condition. As will be seen, the main use for transformer coupling is in 'floating' chopper circuits.

Transient Effects

When a transistor is saturated and its base–emitter voltage is then reversed, there is a time delay before the current decreases to leakage level. This is quite distinct from the time taken for the collector–emitter *voltage* to reach its ultimate 'off' condition, which is determined by the stray capacitance and the value of the load.

The effect is similar to the turn-off characteristic of a single junction (see Chapter 1), and again the delay time is proportional to the degree of saturation before turn-off and inversely proportional to the current which flows 'backwards' in the base circuit when turn-off is initiated. The degree of saturation refers not only to the emitter–collector current flowing when saturated but also to how much base current is used which is in excess of the minimum to achieve saturation. The reverse base current which flows when cutting off is calculated by assuming that initially the base potential remains unchanged when bias is reversed. Knowing this base potential and the source e.m.f. and resistance, the reverse current is easy to calculate.

As in the case of diodes, *Q* figures of stored charge in the base circuit are often quoted for transistors and have the same meaning (see Chapter 1). Some manufacturers quote instead delay or hole-storage times under specified conditions. Knowing that these times increase in proportion to 'on' base current and decrease in proportion to 'off' base current, the designer can readily convert the information to suit his circuit conditions.

Transistor inter-electrode capacitances also affect switching performance, and because of these and the other effects just described, transistors with f_0 of 1 MHz are useful as heavily saturating switches only up to a repetition rate of a few kilohertz.

TRANSISTOR POWER SWITCH

Transistors are used in the saturated switching mode at least as much as in any other connection. The most obvious use is the replacement of a mechanical switch or relay where moving parts have disadvantages. Where there is a fire hazard, where the equipment must operate under severe vibration, where a circuit must be switched by remote control, where high operating speed is required, the transistor can out-perform all but the most expensive mechanical arrangements.

The simplest application is perhaps the direct use of the circuit of

Fig. 3.1 where R_L is a load to be connected or disconnected from voltage V, T_1 taking the place of a relay. A relay has the advantage of alternately giving a good short- and a good open-circuit compared with a series $V_{ce(sat.)}$ and a leakage I_{cbo}, but has shorter life owing to contact wear and cannot generally be operated so fast.

Design is straightforward. Find the load current (V/R_L) and select a suitable transistor type (i.e. alloy or planar epitaxial, germanium or silicon, *p–n–p* or *n–p–n*, according to price, tolerable 'off' leakage, and supply polarity) having useful large-signal current gain β_L at $I_c = V/R_L$. Supply $I_b > V/\beta_L R_L$ to turn on, allowing an extra factor of 2 if convenient; ensure V_{be} at least reaches zero (preferably reverses) to turn off.

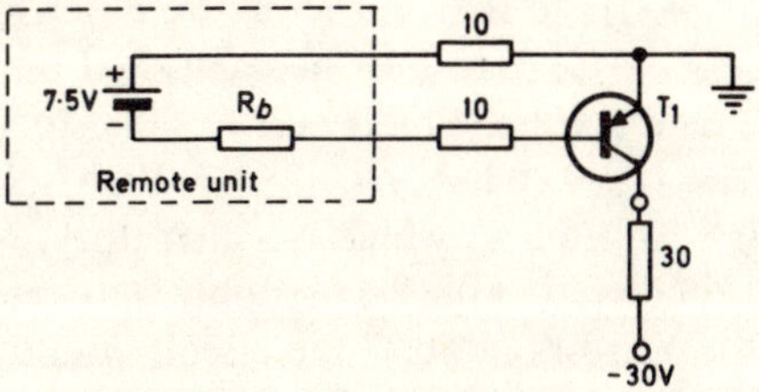

FIG. 3.10 Power switch

For example, one terminal of a 30 Ω load is permanently connected to −30 V d.c.; the other terminal is to be connected to and disconnected from zero potential by changing the voltage at the remote end of a twin cable attached to the switch. The current along the cable must not exceed 100 mA and its total resistance is 20 Ω.

This requires the circuit arrangement of Fig. 3.10, which is similar to 3.1 except that R_b is returned to a separate supply.

The transistor can be a *p-n-p* type rated at 1A, at least, and 30 V. An ideal device is the silicon planar ZT 211; alternatively an alloy germanium type may be used with the 'advantage' for illustrative purposes of a significant level of leakage I_{cbo}. A suitable type is the NKT403, which has a guaranteed β of at least 30 at 1 A, so that for the saturated condition I_b must exceed 1/30 A and could nominally be 1/15 A, i.e. 66 mA. The use of such a generous extra I_b saves the designer the tedious task of adding tolerances, because so long as V_s which supplies I_b is $\gg V_{be}$, saturation is certain even with wide component variations.

Now, V_{be} at saturation is quoted as 0·75 V, so a supply of 7·5 V is adequate for the base circuit. (Almost as good would be -5 V, and even better, -10 V.) Base current is therefore $(7{\cdot}5 - 0{\cdot}75)/(R_b + 20)$ and this must equal approximately 66 mA. Therefore

$$R_b = \frac{6750}{66} - 20 \approx 82\ \Omega$$

Connection of the 7·5 V supply in the direction shown in Fig. 3.10 therefore connects the load if $R_b \leqslant 82\ \Omega$. Because $V_{ce(sat.)} = 0{\cdot}75$ at 1 A, the load actually receives 29·25 V.

If the remote 7·5 V supply is reversed, then the transistor cuts off. To check this, first assume it to be true, then confirm that V_b is positive when leakage is taken into account. The base current is $(I_{cbo} + I_{ebo})$ and this causes the base to fall relative to emitter by $(R_b + 20)\ (I_{cbo} + I_{ebo})$, so that the actual base potential V_b will be $7{\cdot}5 - (R_b + 20)(I_{cbo} + I_{ebo})$. It must be confirmed that, at the maximum operating temperature, this is still positive or zero. At 80°C, for example, $I_{cbo(max.)} = 10$ mA, $I_{ebo(max.)} = 10$ mA, giving $V_b = 7{\cdot}5 - 102(20) \times 10^{-3} = +5{\cdot}5$ V, which is satisfactory. Note, however, that if in the cut-off direction the 7·5 V supply were reduced to +2 V, then with 'bad' transistors at 80°C the circuit would only just turn off the load current.

Transistor power in the 'off' state is $(V_n - I_{cbo}R_L)I_{cbo} \approx 300$ mW at 80°C and in the 'on' state is $I_L V_{ce(sat.)} + I_b V_{be}$, i.e. $0{\cdot}75(66 + 1000) \approx 800$ mW. Mean power depends on mark/space, but if the 'on' state lasts for more than a few milliseconds the transistor rating would be taken as 800 mW.

It is important to remember that, although these static transistor dissipation levels are low compared with the load power of 30 W, a slow transition between the two states can cause high transistor dissipation. Clearly the worst case is when about 15 V appear across the load, giving a current of 0·5 A, and a transistor dissipation of 7·5 W—as in the load. Care must therefore be taken either that the transition is over in one or two milliseconds or that a heat sink is used to enable 7·5 W to be dissipated. It is not sufficient that the transition is short compared with the on and off times: *it must be short compared with the transistor thermal time constant.*

STANDARD CIRCUITS USING TRANSISTOR SATURATION

Following the above principles of transistor switching, the design of several standard switching circuits becomes easy, provided the function of the circuit is clearly understood. As in the case of most switching circuits, the waveforms must be assessed by using the capacitor charging laws set out in Chapter 1. It is then possible to calculate values to ensure correct operation.

In the following examples *p–n–p* versions of the standard circuits have been used for illustration; *n–p–n* devices are also suitable and it is useful for the reader to become familiar with both forms of all the circuits he may encounter.

Free-running Multivibrator

The standard multivibrator of Fig. 3.11 is intended to operate with T_1 and T_2 alternately cut-off or saturated. Assume then that T_1 is saturated, so that V_{c1} is zero; if T_2 is to be cut-off, T_2 base must be positive and T_2 collector at, or approaching, $-V_s$.

The action from now on is that R_4 causes T_2 base to approach $-V_s$ at a rate determined by C_2R_4. When it reaches $-V_{eb2}$, T_2 begins to conduct and will eventually saturate provided $R_4 < \beta_2 R_3$. As T_2

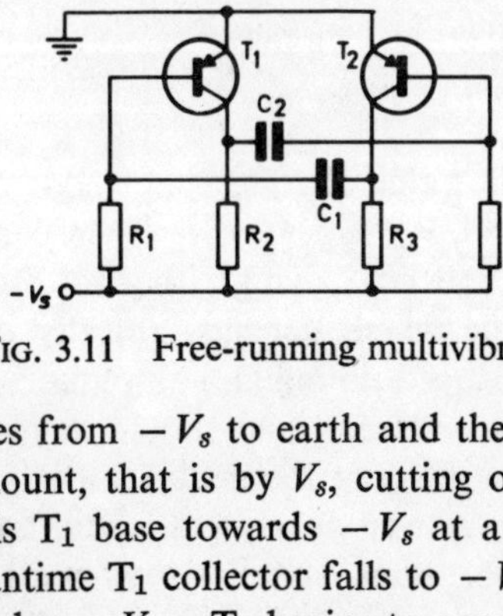

FIG. 3.11 Free-running multivibrator

saturates, V_{c2} rises from $-V_s$ to earth and therefore causes V_{b1} to rise an equal amount, that is by V_s, cutting off T_1. This state remains as R_1 pulls T_1 base towards $-V_s$ at a rate determined by R_1C_1. In the meantime T_1 collector falls to $-V_s$ at a rate given by C_2R_2. As V_{b1} reaches $-V_{eb1}$, T_1 begins to conduct and soon saturates, provided $R_1 < \beta_1 R_2$.

The whole action is then repeated, giving the waveforms shown in Fig. 3.12. The time taken for T_1 base to reach $-V_{eb1}$ when starting at $(-V_{eb1} + V_s)$ above earth, and aiming at $-V_s$ with a time constant C_1R_1, is given by

$$t_1 = C_1R_1 \log \frac{2V_s - V_{eb1}}{V_s - V_{eb1}}$$

This approximates to $t_1 = C_1R_1 \log 2$ provided $V_s \gg V_{eb1}$. This period is the time for which T_1 is cut off and T_2 is saturated;

the other 'half'-cycle has a period of $t_2 = C_2R_4\log 2$ provided $V_s \gg V_{eb2}$.

Note that for correct action $R_1 < \beta_1R_2$ and $R_4 < \beta_2R_3$, which is the normal static condition for saturation; as suggested previously, a further factor of 2 avoids all tolerance problems ($R_1 < \beta_1R_2/2$, $R_4 < \beta_2R_3/2$). Note also the need for $V_s \gg V_{eb1}$ and $V_s \gg V_{eb2}$; although the circuit would operate, the timing would depend on $V_{eb1\ and\ 2}$ if this condition were not obeyed.

The above equations enable simple design to be carried out. If external loads are negligible and if the circuit values are symmetrical ($R_1 = R_4$, $R_2 = R_3$, $C_1 = C_2$), all will be well. There are one or two tricky points which can arise when these conditions do not apply.

External load

If a load R_L is added from T_1 collector to earth, it is clear that when T_1 cuts off it falls only by a fraction of V_s given by $V_sR_L/(R_2 + R_L)$. When T_1 turns on again its change of collector voltage is $V_sR_L/(R_2 + R_L)$, and this is also the positive swing on T_2 base. The timing equation is now

$$t = C_2R_4 \log \frac{R_2 + 2R_L}{R_2 + R_L}$$

and so depends on the value of R_L.

If, on the other hand, R_L appears directly in parallel with R_2 then either T_1 no longer bottoms, giving either no oscillation or a frequency and amplitude which depend critically on β_1, or T_1 still bottoms, giving unchanged timing and amplitude.

Where possible, any external load should therefore be connected in parallel with R_2 or R_3 (not to 'earth') and R_1 or R_4 should be designed to maintain saturation with the additional collector current. If this cannot be done and the load must go to earth, an emitter–follower may be placed between the collector and the load (see Chapter 4) so that the influence of the load current is reduced by the β of the transistor.

Asymmetry

To appreciate the difficulties of asymmetrical operation, it is necessary to examine the collector waveforms more closely. In Fig. 3.12 it can be seen that when T_1 cuts off, T_1 collector falls with a time constant R_2C_2 (since the right-hand side of C_2 is fixed at just below earth by T_2 base), reaching its ultimate level of $-V_s$ after a time of about $4R_2C_2$ (within a few per cent). In the normal description of its action it is assumed that this state has been reached before T_1 turns on again.

Suppose that C_1 is only 1/100th of C_2 and that $R_1 = R_4$, $R_2 = R_3$; then a ratio of 100 : 1 would be expected in the on and off times of T_1. With the ratio given T_1 would be cut off for a time of about $C_1R_1 \log 2$ and T_1 collector would be falling during this interval, reaching $-V_s$ in a time $4C_2R_2$. Now, $C_2 = 100C_1$ and $R_1 \approx \beta R_2/2$, e.g. $15R_2$. This gives a cut-off time of 0·7 C_1R_1 and a collector time of 27 C_1R_1 to reach $-V_s$.

It is clear that T_1 collector does not have time to fall far before T_1 turns on and T_1 collector rises to earth again. This means that T_2

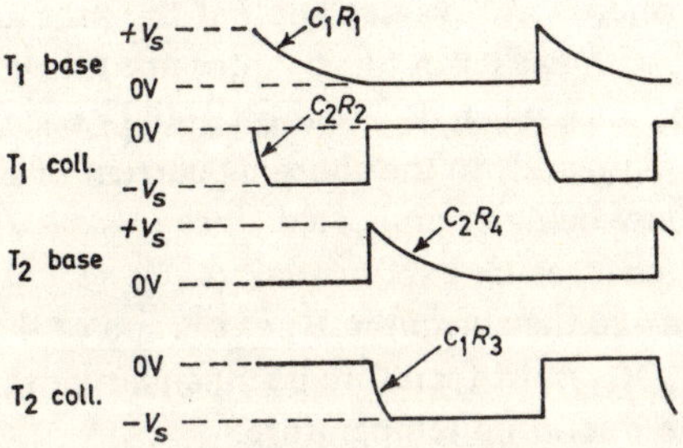

FIG. 3.12 Waveforms for free-running multivibrator (Fig. 3.11)

base rises only slightly and soon recharges, so that the 'T_2-off' interval is much shorter than anticipated.

The circuit therefore has a limit to the mark/space ratio which may be obtained by increasing C_2/C_1, given roughly by 0·7 C_1R_1 = 4 C_2R_2 = 8 C_2R_1/β, i.e. $C_2/C_1 = \beta/11{\cdot}5$. This may be exceeded by a factor of about 2 before serious drop of output from T_1 accompanied by inaccurate timing occurs, corresponding to a practical limit of 10 : 1 if $\beta \approx 50$.

Another method to change mark/space is to vary R_1 and R_2 in the same proportions while leaving C_1 equal to C_2. This avoids the above situation, since the fall time of T_1 collector is reduced in the same proportion as the cut-off time for T_1. Another difficulty now appears, however, because when T_2 saturates, its load R_3 has R_1 in parallel and it is T_2 collector which supplies the current V_s/R_1 which raises T_1 base to $+V_s$. (This is naturally true also when the circuit is symmetrical, but then R_1 is $(\beta/2)R_3$ and may be ignored.) The consequence of this may be seen by assuming $R_1 \approx R_3$ in an attempt to obtain a mark/space of $\beta/2$ to 1. Because the collector load of T_2 is halved at initial saturation, R_4 must be half its normal value, i.e. $R_3\beta/4$. This halves the cut-off period for T_2 and the resulting mark/space is β to 4, not β to 2.

The best method for high mark/space ratio is therefore to use a combination of these two methods, making $C_2/C_1 \approx \beta/11{\cdot}5$, $R_4/R_1 \approx 4/1$, $R_3/R_2 \approx 4/1$ giving a mark/space of about 20:1 if $\beta = 50$.

Temperature effects

Since the timing equation used earlier involves V_{eb}, change of temperature clearly affects the timing unless $V_s \gg V_{eb}$.

Other effects not mentioned in the equation are $V_{ce(sat.)}$ and leakage current. The former is always negligible if the precautions for good bottoming dealt with earlier in this chapter are taken. I_{cbo} is a more serious problem, since when T_1 is cut off and T_1 base is falling as C_1 charges, $(I_{cbo} + I_{ebo})$ adds to the charging current of R_1. Cut-off time is reduced, the law being complicated since $(I_{cbo} + I_{ebo})$ charges C_1 linearly and R_1 charges C_1 exponentially.

For correct design the current in R_1 at the end of the cut-off period of T_1 (namely V_s/R_1) must therefore be much larger than $(I_{cbo} + I_{ebo})$ at the maximum operating temperature.

A secondary effect of I_{cbo} is that T_1 collector swings from $V_{ce(sat.)}$ to $(-V_s + I_{cbo}R_2)$, not to $-V_s$, as assumed. This again affects timing and also reduces the output swing so that a second condition is that $I_{cbo}R_2 \ll V_s$. Since $R_1 \approx \beta R_2/2$ it is sufficient to satisfy the previous condition, which is $\beta/2$ times more stringent.

Temperature effects therefore dictate V_s/R_1 and $V_s/R_4 \gg (I_{cbo} + I_{ebo})_{max.}$

Voltage ratings for T_1 *and* T_2

From the circuit action it is evident that at saturation, dissipation in $T_{1,2}$ will be $(V_{ce(sat.)} \times V_s/R_{2,3} + V_{eb} \times V_s/R_{1,4})$, which will usually be negligible.

The base waveform goes positive by V_s, so that V_{eb} reverse must be rated at V_s. V_{ce} rating is also V_s, and the dissipation at cut off is $I_{cbo}(V_s - I_{cbo}R_2)$, which is very small.

The above shows how thorough knowledge of the operation of the transistor as a switch enables the standard multivibrator to be designed and the effects of parameter variation easily predicted. Such circuits can be designed to operate correctly with almost any transistor so long as the minimum β and the voltage ratings are known.

For example, a standard multivibrator is to drive a load of 1 kΩ, connected between T_2 collector and a 10 V negative supply. The frequency is 1 kHz and the mark/space is 1:1.

The simplest approach is to make $R_2 = R_3 = 1$ kΩ, where R_3 is the load. Select a transistor type in common use and find its minimum β at 10 mA, e.g. 20. Then, $R_1 = R_4 = 20$ kΩ/2 = 10 kΩ. Now, each half-period is given by $t_1 = t_2 = 0{\cdot}7\ C_1R_1 = 0{\cdot}7\ C_2R_4$ and this is to be 0·5 msec, giving $C_1 = C_2 = 0{\cdot}07$ μF.

Charging current near the end of a half-cycle is $V_s/R_1 = 1$ mA, so at the maximum temperature $(I_{cbo} + I_{ebo})$ must be less than $N/100$ mA, where N is the maximum percentage error which can be tolerated from this cause. For some applications (e.g. 1 per cent drift at a temperature of 60°C) this will dictate the use of silicon transistors; this must be rated at 10 V reverse V_{eb}, which is met only by alloy types.

Timing errors are naturally caused by resistor tolerances, only R_1 and R_4 being significant. Changes in the −10 V line would have no effect whatever on timing if V_{eb} were negligible, but would cause a proportionate increase in collector swing.

Use of Planar Transistors

One of the most frequently made design errors in this circuit is failure to observe the reverse V_{eb} rating. The result may be the catastrophic failure of both transistors if the coupling capacitors are large enough to store sufficient energy for transistor destruction. Values in excess of 1 μF are almost certain to achieve this. When this occurs the reason is soon appreciated and a more careful design is then calculated.

Much less obvious is the case where, although base-emitter breakdown occurs, it is non-destructive. The effect is that the capacitors discharge rapidly on reaching the reverse breakdown voltage and then discharge from that point in a normal manner. A superficial examination of circuit waveforms does not show up the peculiar mode of operation, but the timing cycle is seen to be shorter than predicted. This is naturally due to the smaller voltage through which the capacitor has to recharge (with a 10 V supply and 5 V breakdown, for example, the multivibrator period is halved).

The inexperienced designer assumes that the short timing is merely the result of yet another miscalculation of the type he keeps making and changes the capacitor in the prototype until the correct timing is obtained. Apart from being bad practice this adjustment to the calculated value can still give erratic timing due to variations in reverse breakdown voltage; moreover the transistors may fail after a few hours of operation even if the capacitors do not carry sufficient charge to give instant destruction.

Several remedies are possible and are discussed at length in the companion volume, *Circuit Consultant's Casebook* (Business Books, 1970). Where the accuracy of output swing is unimportant a simple cure is to insert diodes in each emitter lead; another is, of course, to use a safe supply voltage.

One-shot Multivibrator

This is one of the most useful circuits ever devised and seems to find application in every pulse system. It is also known as a 'flip-flop' (though many use this term to describe the 'bistable' or 'two-state' device), as a 'monostable multivibrator' and as a 'delay multivibrator'.

The circuit is basically a standard multivibrator in which one coupling is direct; it can take two forms, according to whether this coupling is by common emitter connection or from collector to base. Since the main intention of these examples is to familiarize the student with designing saturating circuits, the second form will be considered (Fig. 3.13).

Circuit function

In Fig. 3.13 the values are designed so that in the quiescent condition T_2 is saturated and, by virtue of the coupling to T_1 base, T_1 is cut off.

When an input pulse is applied which is sufficient to turn T_1 on, the resulting rise in potential of T_1 collector is coupled to T_2 base, taking it to $+V_n$, and T_2 cuts off. R_2 is now connected by R_4 to the negative line and the values of these resistors are designed to be low enough to cause T_1 to saturate.

The input pulse can now disappear and T_1 will remain in saturation until T_2 conducts again, which occurs when C has recharged through R_5 to $-V_{eb2}$ (a similar action to the free-running multivibrator). When this takes place, T_2 finally saturates, causing T_1 to turn off. After a further time of about $4CR_3$, T_1 collector reaches $-V_n$ and the original quiescent conditions are regained.

The collector waveform of T_2 is a negative pulse (*see* Fig. 3.14) of width determined by V_n, C, R_5 and V_{eb2}, and amplitude determined by V_p, V_n, R_1, R_2, and R_4. Neither of these depends on the magnitude or width of the input trigger pulse unless this is either so short that T_1 has no time to saturate through R_4, or so long that it continues to hold T_1 on after C has recharged. The minimum input signal to

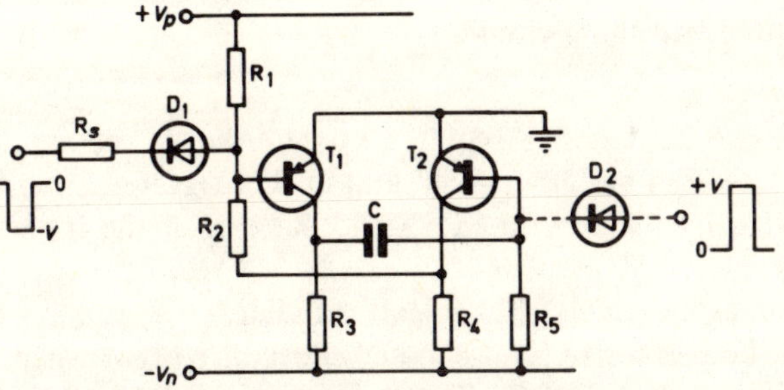

FIG. 3.13 One-shot multivibrator

trigger is simply that which can drag T_1 base negative in spite of R_1, R_2, and V_p.

Uses of the one-shot circuit

The collector waveform of T_2 has the property that its rising edge as the circuit reverts to its quiescent state occurs a certain time later

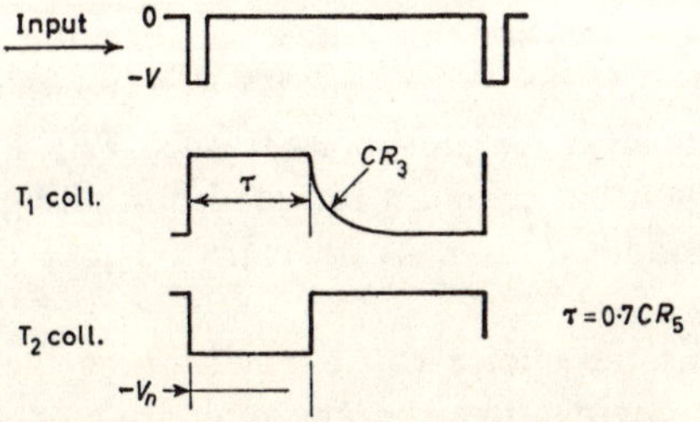

FIG. 3.14 Waveforms for one-shot multivibrator (Fig. 3.13)

than the input trigger negative-going edge. This time depends on circuit values, so that a variable delay unit can easily be made.

If the input consists of a series of pulses with minimum interval greater than the time for C to recharge, then the output consists of a set of pulses of uniform width and height. The average value of these pulses obtained by a smoothing circuit is directly proportional to the mean input frequency. The one-shot circuit is therefore an alternative to the diode or transistor pump used as a frequency discriminator.

When a pulse signal is intended to indicate its moment of arrival by operating a gate which requires an input of longer duration than the signal itself, the signal can usually be made to trigger a one-shot and the output can then operate the gate. The one-shot is then acting as a pulse-stretching circuit.

Design

The two basic design requirements are that in the quiescent state T_2 is saturated and T_1 cut off, and in the triggered condition T_1 is saturated by the current in R_2 and R_4 even if the input has been removed.

For reasons which will be dealt with later, it is usually convenient to run both transistors at similar values of current when saturated (to within a factor of 2). The minimum current is dictated by the useful β of the transistors and the demand of any external collector load. The maximum current depends on transistor and supply power ratings and the power available for triggering.

R_3 and R_4 are now decided and R_5 is given by $\beta\ R_4/2$ to ensure saturation of T_2. The value of R_1 is now determined by the I_{cbo} of T_1 and the $V_{ce(sat.)}$ of T_2, which both try to turn on T_1 in the quiescent state; provided $V_p > 2$ V (i.e. $\gg V_{ce2(sat.)}$), then R_1 current will simply be $\gg I_{cbo1}$, e.g. 0·5 mA for germanium or 1 μA for silicon at 50°C.

The triggered condition may now be examined, the required condition being that the current in $(R_2 + R_4)$ with T_2 cut off must saturate T_1. If R_1 were not present then $(R_2 + R_4)$ would be $\leqslant \beta\ R_3/2$ by the usual criterion for good bottoming; because of R_1 this must be modified to pass an extra 50 μA (or whatever the current in R_1 is).

There remains the choice of C, which obeys the same charging laws as in the previous free-running multivibrator, giving an over-and-back time τ of 0·7 CR_5.

As an example assume that R_4 is to be the load and is 2·2 kΩ, supply lines are +5 and −10 V, and silicon transistors having $\beta_{Lmin.}$ of 25 at 5 mA are to be used. The over-and-back time τ is to be 1 msec.

Then,

$$R_5 \leqslant \beta\ R_4/2 \leqslant 27{\cdot}5\ \text{k}\Omega, \text{ e.g. } 22\ \text{k}\Omega$$
$$R_1 \leqslant (V_p/50)\ \text{M}\Omega, \text{ e.g. } 82\ \text{k}\Omega$$
$$R_3 = R_4 = 2{\cdot}2\ \text{k}\Omega$$

The current in T_1 when saturated is $10/2{\cdot}2 \approx 4{\cdot}5$ mA, so that the base current for bottoming should be $\geqslant (2/\beta)4{\cdot}5 = 0{\cdot}36$ mA.

$(R_2 + R_4)$ must therefore supply 0·36 mA + V_p/R_1, i.e. 0·36 + (5/82) = 0·42 mA, giving $R_2 + R_4 \leqslant 10/0{\cdot}42 \leqslant 24$ kΩ, or R_2 = 18 kΩ.

C is now given by $\tau = 0{\cdot}7\ CR_5$ and for a time of 1 msec this gives

$$C = \frac{10^{-3}}{0{\cdot}7 \times 22 \times 10^{-3}} = 0{\cdot}065\ \mu\text{F}$$

Causes of timing error, transistor voltage ratings, temperature drift

In general, T_2 behaves in the same way as in the free-running multivibrator. Timing errors are caused as before by its V_{eb} variation and by its $(I_{cbo} + I_{ebo})$ which adds to the charging current of C. It must be rated as before since the base is driven to $+V_n$.

T_1 has less influence on timing; its V_{eb} reverse rating depends only on V_p, R_1, and R_2, and is usually less than 5 V.

Output levels

Output may be taken from T_1 or T_2 collector. On T_1 collector the falling edge as the circuit reverts to its original state is slow, the fall time being about $4R_3C$, which is about half the value of τ. It is therefore impossible to obtain from this waveform a precise pulse corresponding to the beginning of the slow fall, unless a very accurate voltage-sensitive trigger circuit is added.

The output from T_2 has well-defined edges but the output falls only to

$$-V_n \frac{R_2}{R_2 + R_4}$$

not to $-V_n$. The rate of the final rise is proportional to the rate at which CR_5 is falling as it begins to turn T_2 on. This is given by $dV/dt = i/C = (V_n/CR_5)$, and the rate of rise of I_{c2} is then $(g_m V_n/CR_5)$*, giving a voltage rate of

$$V_n \frac{g_m R_2 // R_4}{CR_5}$$

at the collector. In the above example this would be about

$$\frac{10 \times 1/50 \times 2{\cdot}2 \times 18}{0{\cdot}065 \times 22 \times 20{\cdot}2} \times 10^6 = 274\ \text{V/msec}$$

giving a rise time of

* $g_m = \left.\dfrac{\partial i_c}{\partial V_{eb}}\right|_V$

$$\frac{V_n}{274} \cdot \frac{R_2}{R_2 + R_4} = 37\ \mu\text{sec}$$

The rise time is therefore 3·7 per cent of the interval τ, and this will hold for any value of C until $3{\cdot}7\tau/100$ is comparable with transistor turn-on time.

Limitations in use

This standard one-shot circuit is simple to design but has certain limitations. If τ is to be variable, then C may be changed, but where times greater than a few tens of microseconds are required the large value of C makes its continuous variation impractical and it must be switched. R_5 may be varied but its maximum value is restricted to about $\beta_L R_4$ and its minimum to about R_3, since it presents an additional load on T_2 collector (see asymmetrical operation of the free-running multivibrator). This is why approximately equal values were chosen for R_3 and R_4.

A less obvious restriction applies when the input is a train of pulses with a minimum interval comparable with τ. The circuit will operate correctly on the first pulse, but if the second occurs *before* T_1 *collector has reached* $-V_n$, then the next value of τ will be smaller. If this continues and the input pulse spacing is constant, each output pulse after the first will tend to an equilibrium width which is less than τ. When the output from T_2 is smoothed and the resulting d.c. used as an indication of input frequency, the law of V_{out} against f_{in} will become non-linear when f_{in} is high enough to cause the effect in question. The reason that the effect is often overlooked is that the time constant R_3C is much less than R_5C and the designer feels that this is a good reason to forget it. Unfortunately the collector circuit aiming and final potentials are the same $(-V_n)$ whereas the base circuit aiming potential is $-V_n$ and its final level $-V_{eb}$. The collector circuit therefore recovers only after about $4R_3C$ compared with the base circuit time of 0·7 R_5C.

Any attempt to reduce the recovery time by reduction of R_3 requires lower R_2, thus reducing the output swing from T_2. The basic problem is to recharge C as rapidly as possible, and one simple method is the addition of an emitter-follower T_3 between T_1 collector and C (Fig. 3.15), its emitter load R_6 being less than R_3 by a factor of about $\beta_3/5$. This speeds up circuit recovery by a factor of 5 but adds only 20 per cent extra load to T_1 collector circuit.

An alternative method which is not quite so effective is to catch

T_1 collector by use of a diode at a potential more positive than $-V_n$, so that the time taken is only about $2CR_3$. This reduces T_1 collector swing and so reduces τ also, but the ratio between τ and circuit recovery is improved.

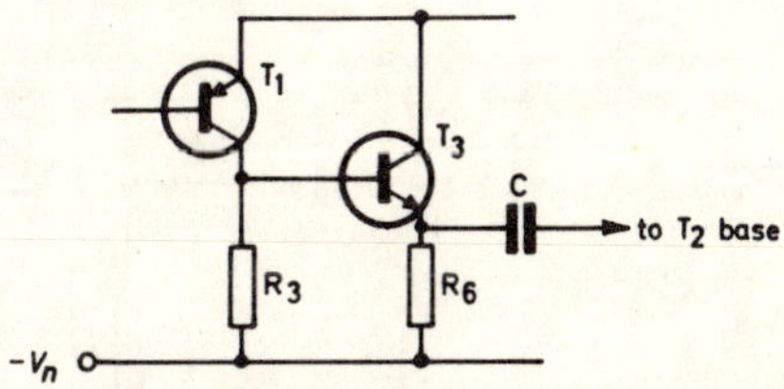

FIG. 3.15 Addition of T_3 (*see text*)

Note that if the input is present for a time greater than τ, recovery will not begin until the input is removed, although the output from T_2 collector will be normal.

Trigger requirement

The input signal must take T_1 base negative enough to raise T_1 collector sufficiently to cut off T_2. R_2 then takes over and the input is no longer required. The input current required from the source is just greater than V_p/R_1 (about 0·6 mA in our example) and its required potential swing is $(V_{eb1} + V_{FD1}$ (0·6 mA)), 2 V being a very conservative estimate.

The diode D_1 ensures that if the input returns to zero potential in a time less than τ (which is usually the case), then T_1 is not turned off, as this would immediately cause T_2 to turn on again. The output would then be a pulse identical to the input in width.

An alternative triggering input is at T_2 base, where a positive input coupled by a diode D_2 (Fig. 3.16) cuts off T_2. The diode then cuts off, allowing T_2 base to rise to $+V_n$.

Another trigger input, also shown in Fig. 3.16, is to T_1 collector, where a positive pulse drives T_2 off by means of C. Although this is a less sensitive input than the T_1 base trigger, it has several advantages and should be regarded as the preferred method whenever ultimate sensitivity is unimportant. Its main advantage is that spurious trigger pulses appearing after normal triggering and before completion of the one-shot action, have no effect on timing. An incidental point in its favour is that in sequences of timing-interval circuits, direct-coupling is often possible at the correct d.c. level for collector triggering, thus obviating the need for a coupling capacitor.

Triggering methods and their properties are discussed more fully in *Circuit Consultant's Casebook* (Business Books, 1970).

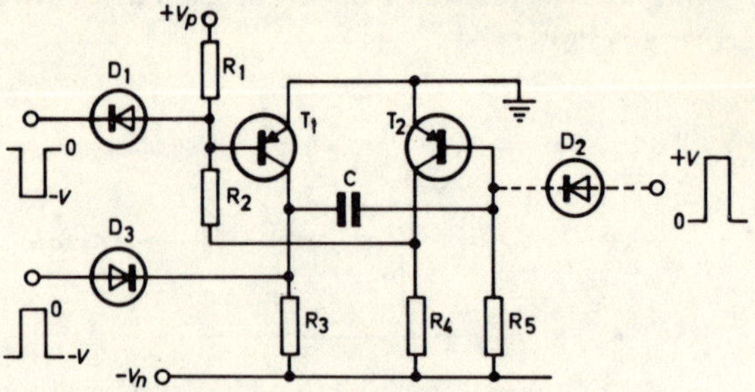

FIG. 3.16 Alternative trigger

The Standard Bistable

This circuit differs from the previous multivibrator in that both collector–base feedback paths are directly coupled (Fig. 3.17).

Circuit function

The circuit is designed so that if T_1 is cut off the current in (R_3 + R_5) minus the current in R_4 is sufficient to saturate T_2. When T_2 is saturated, the current in R_1 exceeds I_{cbo1} plus the current in R_2, so that T_1 is maintained in the cut-off condition, resulting in a stable state.

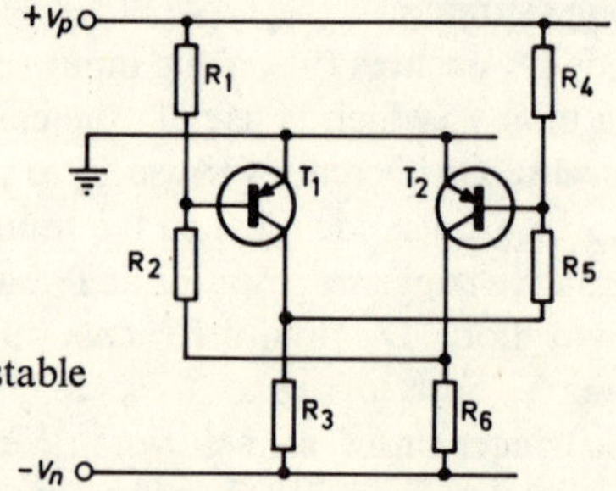

FIG. 3.17 Standard bistable

In the opposite condition, values are similarly chosen so that another stable state exists with T_1 saturated and T_2 cut off. This usually, but not inevitably, leads to a symmetrical design where $R_1 = R_4$, $R_2 = R_5$, and $R_3 = R_6$.

Use of the bistable

The most common use for this circuit is in binary counters in computer systems. A method of input trigger routing (described later) is added which makes the circuit change state alternately when an input pulse train is applied. The output from either collector is then differentiated, giving a pulse train at half the input frequency; by using this to trigger a second bistable the process of frequency division continues. If n such stages are used, the final binary changes state after a total of 2^{n-1} input pulses have been applied, so that the system may be used as a binary counter. There are many variations on these lines and the interested reader is referred to the many volumes devoted to computer systems and circuits.

The bistable is also useful in converting a momentary signal into a permanent state for use as an alarm or interlock signal.

In most modern equipment, discrete circuit bistables are seen only rarely owing to the advantages given by standard microcircuits. However, knowledge of the design procedure is important in fully understanding the counting process and in enabling the student to appreciate the complex operation of the circuits within a J-K or D-type bistable.

Design

Design follows the same lines as the previous circuits in this chapter. Assuming the symmetrical form, determine the required loads R_3, R_6. Design R_1, R_4 as before, so that $V_p/R_{1,4} = V_p/R_4 \gg I_{cbo1,2}$; then finally design R_2, R_5 to pass more than $V_p/R_4 + 2V_n/\beta R_3$ to ensure saturation.

For example, assume $V_p = 10$, $V_n = 15$, $R_3 = R_6 = 4{\cdot}7\ \text{k}\Omega$, and that silicon transistors with β_L of 30 min at 4 mA are to be used. Then

$$R_1 = R_4 \leqslant V_p/I_{cbo(max.)} \leqslant 200\ \text{k}\Omega,\ \text{e.g. } 18\ \text{k}\Omega$$

Current for saturation

$$2V_n/\beta\, 4{\cdot}7 + V_p/18 = 0{\cdot}20 + 0{\cdot}55 \approx 0{\cdot}75\ \text{mA}$$

$$R_2 + R_6 \leqslant V_n/0{\cdot}75$$

Therefore

$$R_2, R_5 \leqslant 15\ \text{k}\Omega,\ \text{e.g. } 12\ \text{k}\Omega$$

Temperature effects

When T_1 is cut-off I_{cbo1} tends to lower T_1 base potential and cause T_1 conduction, and the circuit fails unless the current V_p/R_1 exceeds I_{co} at the maximum temperature.

The output voltage swing is from $V_{ce(sat.)}$ (almost zero) to

$$-\frac{V_n R_2}{R_2 + R_6} + I_{cbo}\frac{R_6 R_2}{R_2 + R_6}$$

and so varies with I_{cbo}.

In most applications the first temperature effect is the more critical. If silicon transistors are used it is often possible to obtain satisfactory high-temperature operation even if V_p is zero, by ensuring that $I_{cbo1}R_1//R_2$ is less than about 50 mV so that the transistors cannot turn on owing to I_{cbo}. A similar criterion is often used for germanium types, but then R_1 would be very small (because I_{cbo} is large and the transistor turn-on voltage small) and would rob T_1 of much of its turning-on base current in the opposite state. This usually limits the reliable operating temperature to well below 50°C and the use of V_p, even if only 1 V, improves this situation considerably.

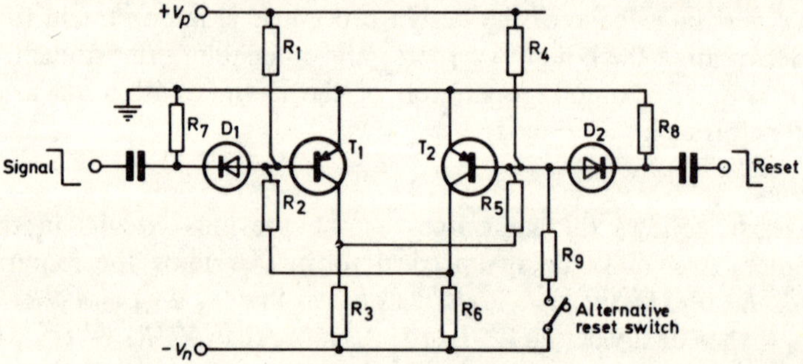

FIG. 3.18 Triggering a bistable

Triggering circuits

There are many methods for triggering the bistable, and only two will be considered here. If the circuit is to be used for alarm indication where an input signal pulse changes the state of the binary which is later reset by a different signal, the input may simply be coupled by a diode to either base, and the other signal to the other base (Fig. 3.18).

The circuit is 'set' so that the transistor connected to the signal is cut off and when the signal appears the circuit changes state, provided the input signal can supply more current than V_p/R_1 and enough voltage to bring the base below earth. This voltage is the forward diode drop plus $(V_pR_2)/(R_1 + R_2)$ which is the amount by which the base is positive.

The re-setting signal can be coupled in the same manner or can be manually operated as shown, where R_9 must pass more current than V_p/R_4, i.e. $R_9 \leqslant (V_n/V_p)R_4$.

A much more difficult situation arises when the bistable is to be used as a counter. Each successive input pulse must change the state of the circuit so that a 'steering' or 'routing' circuit must be added to direct the input pulses to each base alternately.

The usual arrangement is shown in Fig. 3.19, which has superficially a simple action. Supposing T_1 is on, then T_1 base is just below earth so that D_1 is near conduction; T_2 collector–base is at $+ V_p R_5/(R_4 + R_5)$ so that D_2 is cut off by a few volts. If a positive input pulse of a few volts is applied, this is immediately coupled to T_1 base and T_1 therefore cuts off. No pulse was applied to T_2 base, so T_2 is turned on by T_1 turning off. When a second input pulse is applied, the situation is completely reversed and again the state of the circuit changes. The bistable therefore operates as a counter or frequency divider as intended.

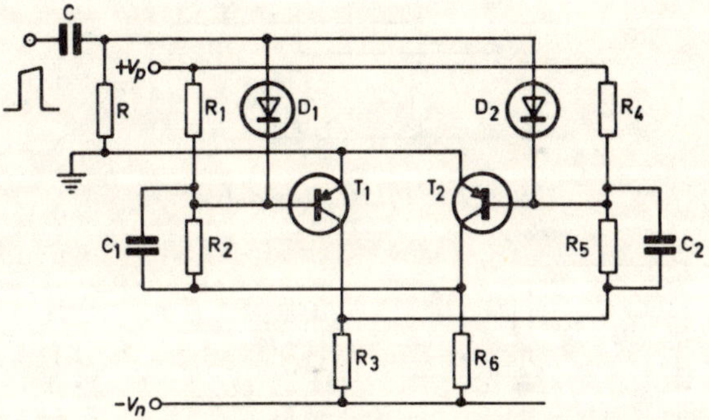

FIG. 3.19 Binary counter

The above explanation is by no means sufficient, because T_2 base cannot fall any lower than D_2 allows, so that if the input remains positive both transistors will cut off. When the input finally falls, there is no reason why the transistor which was previously cut-off should now turn on, since all circuit conditions are symmetrical. If any unintentional asymmetry exists, the tendency will be for one particular state to be preferred and the circuit may never change state. If, on the other hand, the signal rapidly cuts off T_1 and is removed before T_2 base has descended appreciably, then there is no reason why the changeover which has been initiated should continue; the more likely event is that T_1 will resume conduction.

It is only the addition of capacitors C_1 and C_2 which enables the desired action to take place. If T_1 is saturated and T_2 cut-off, the upper connection of C_2 is at $V_p/[R_5/(R_4 + R_5)]$ and its lower connection at $V_{ce(sat.)}$.

The upper connection of C_1 is at $-V_{eb}$ and its lower connection at about $-V_n/[R_2/(R_2 + R_6)]$. The charges on C_1 and C_2 are therefore as shown in Fig. 3.20, and it is clear that C_1 carries a greater voltage than C_2. (In the practical example, $V_p = 10$, $V_n = 15$; R_1, $R_4 = 18$ kΩ; R_2, $R_5 = 12$ kΩ; R_3, $R_6 = 4{\cdot}7$ kΩ, giving $V_{c1} = 10{\cdot}8$ V, $V_{c2} = 4$ V.)

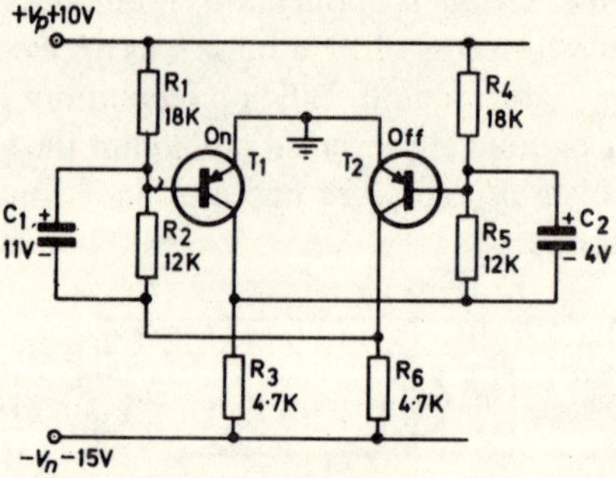

FIG. 3.20 Practical bistable

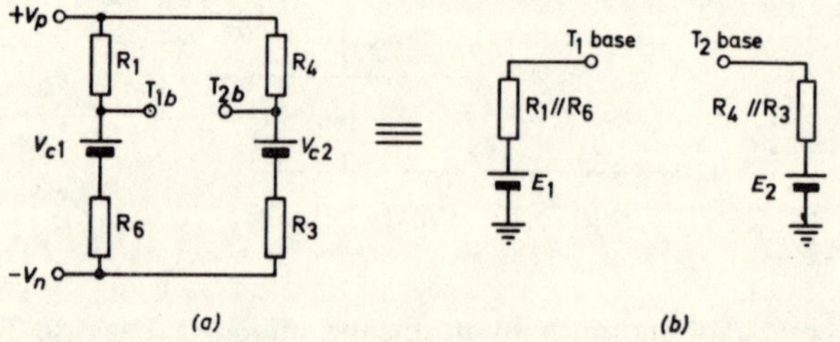

FIG. 3.21 (*a*) Base voltages in Fig. 3.20, (*b*) equivalent circuit of (*a*)

When the input pulse arrives, T_1 is cut off, and if the input remains positive, T_2 base descends only until D_2 turns on; both transistors remain with bases positive and are consequently both cut off. If the input now disappears rapidly *before capacitors* C_1 *and* C_2 *have appreciably altered their charge*, the circuit now appears as shown in Fig. 3.21 (*a*), where the capacitors are represented by batteries and the transistors are omitted, as they are still cut off at this time.

By Thévenin's theorem the circuit can be represented by Fig. 3.21 (*b*) from the point of view of the two bases. Both base circuits have the same source resistance, but the e.m.f. for T_1 base circuit is

$|E_1| = I_1R_1 - V_p$ and for T_2, $|E_2| = I_2R_4 - V_p$. Since $V_{c1} > V_{c2}$, $I_1 < I_2$ and therefore $|E_2| > |E_1|$ (assuming a workable design where V_p is not so large that saturation will never occur). In the example, $V_{c1} = 11$ V and $V_{c2} = 4$ V, giving $I_1 = 14/22{\cdot}7 \approx 0{\cdot}6$ mA, and $|E_1| \approx 1$ V; $I_2 = 21/22{\cdot}7 \approx 0{\cdot}9$ mA and $|E_2| \approx 6$ V. The source resistance to both bases is $R_1//R_6 = R_4//R_3 = 4{\cdot}7//18 = 3{\cdot}7$ kΩ.

As the input signal falls, the initial base current for T_1 will be $1/3{\cdot}7$ mA and for T_2 $6/3{\cdot}7$ mA. Provided $\beta(T_1)$ is not $> \beta(T_2)$ by a factor of 6, it will therefore be T_2 which conducts first, and as it does so its collector rises and reduces T_1 base current.

It is now clear that only C_1 and C_2 cause correct changeover action, and although they have also the effect of improving collector–base coupling during transition, their main purpose is the storing of a charge in accordance with the state of the circuit. For this reason they are often called 'memory' capacitors, as they remember the previous state.

The correct values are obtained from a study of the above action. The charge on C_1 and C_2 must remain from the moment of input triggering to the end of the trigger pulse. On the other hand, the charge must change when the binary settles in a new state before the next input trigger pulse begins. The time constant associated with C_1 is $C_1R_2//R_6$ when T_1 has just turned on, but is $C_1R_2//(R_1 + R_6)$ when T_1 is off. Similarly, C_2 time constant is $C_2R_5//(R_4 + R_3)$.

In the circuit given, the input trigger pulse length is determined by the differentiating circuit CR, where R includes circuit-loading with T_1 and T_2 cut off.

The criteria for correct memory are therefore $CR < C_1R_2//(R_1 + R_6) < T$, where T is the input signal period.

Note that owing to C_1 and C_2 the collector waveforms have slow falling edges, the fall time constant being equal to the memory time constant $C_1R_2//(R_1 + R_6)$.

Other Switching Circuits

There are many other circuits in which the transistor is used as a switch, and also several versions of the above standard circuits. The reader should experience no difficulty in their design provided that he makes certain of the precise mode of operation before calculating circuit values.

The use of a transistor as a low-level chopper switch requires a different design approach and is dealt with in Chapter 9.

The n-channel J-FET as a Switch

In many respects the *n*-channel J-FET operates as a switch in the same way as an *npn* bipolar transistor, but the detailed drive requirements are different.

Turning off the n-*channel J-FET*

For the 'off' state the gate must be driven more negative than the source and drain by an amount V_{GS}(off). Thus the actual voltage needed depends on which of the drain or source is the more negative at the time of interest. In Fig. 3.22(*a*), if v_{in} were zero then a gate voltage $V_o = V_{GS}$(off) would ensure turn-off. However, with $v_{in} = \hat{v} \sin \omega t$, V_D is negative for a maximum value $\hat{v}$, and V_o must therefore be more negative than this by V_{GS}(off), giving a required value of $-[V_{GS}(\text{off}) + \hat{v}]$.

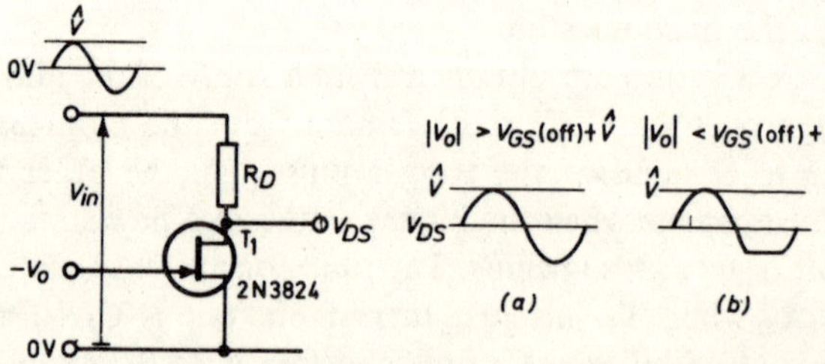

FIG. 3.22 Gate drive to switch off the J-FET

A smaller value of V_o would allow conduction for part of the input cycle if T_1 were on its specification limit for turn off [see Fig. 3.22(*b*)].

Turning on the n-*channel J-FET*

For the 'on' state it is necessary only to make the potential between the gate and either drain or source approximate to zero. It is neither required nor desirable to force current into the gate. In this way the FET has a low drain-source resistance for current in either direction and it does not generate offset voltages equivalent to the $V_{CE(sat)}$ of the bipolar transistor, so that for zero drain-source current it behaves as a passive resistor. If current is forced into the gate, then internal offset voltages will be produced; at the same time the 'on' resistance may be reduced, giving an apparent improvement at high input current levels. Resistance values under gate forward-drive conditions are, however, never specified so that the improvement is not guaranteed and the practice cannot be recommended.

The 'on' state drive circuit is easily designed because gate current is zero, allowing high values of external gate resistance to define the gate potential (see Fig. 3.23).

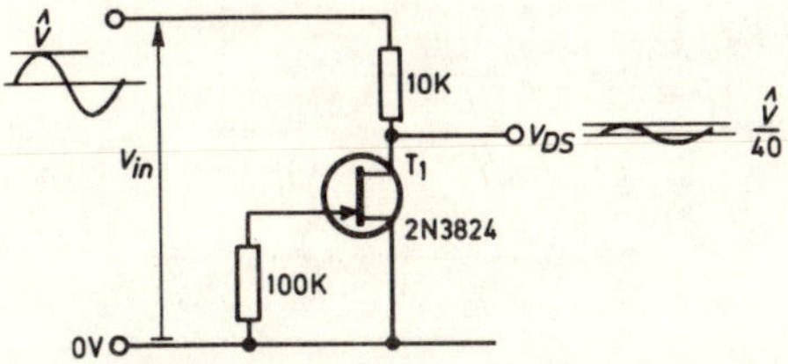

FIG. 3.23 Gate drive to switch on the J-FET

Switching between 'on' and 'off' states

Switching between states is easily performed by using bipolar drive transistors. Fig. 3.24 shows some examples of how various input drive signals can be made to switch states.

In Fig. 3.24(*a*) an input level of +7 V is to turn off the shunt FET and an input of −7 V is to turn it on. The circuit functions as follows: T_1 is driven on and off by the input. When on, T_1 is bottomed causing the FET also to be on since its gate-source potential is small. When T_1 is off, its collector falls to −15 V which is 10 V in excess of the most negative signal on the FET source or drain. This is 2 V clear of the required 8 V for the 2N3824, thus ensuring turn-off.

In Fig. 3.24(*b*) the switch drive is a typical TTL logic output. In logic 1 ($\geqslant$ + 2·4 V) the FET, T_3 is required to be cut off and in logic 0 ($\leqslant$ 0·4 V) is to be conducting. With the input at + 2·4 V, T_1 conducts and bottoms T_2 giving a gate potential of − 20 V on T_3. This exceeds the minimum drain potential by 15 V thus ensuring turn off (which demands V_{GS} more negative than − 10 V for the 2N4859A). With the input at 0·4 V, T_1 and T_2 are both cut off, leaving V_{GS} = 0 causing full conduction of T_3.

In Fig. 3.24(*c*) the switch drive is a typical operational amplifier output in which approximately + 14 V is to turn on the FET, T_3, and − 14 V is to turn it off. The simple diode coupling is adequate here; were the opposite sense of switching required then Fig. 3.24(*b*) could be used with the addition of a diode in series with T_1 emitter,

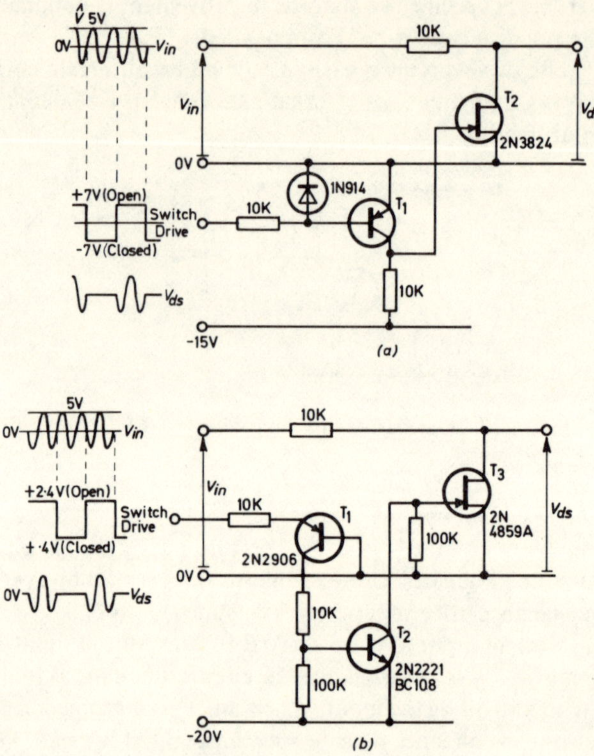

FIG. 3.24 Switching the J-FET

or in shunt with T_1 emitter-base connections to avoid excess reverse V_{BE} on T_1.

In Fig. 3.24(*d*) a series FET switch is used in a sampling circuit: capacitor C is to store whatever potential is present at the input when the switch drive rises to + 2·4 V (TTL logic 1). Here T_1 cuts off when the switch drive appears, cutting off T_2. Were the configuration of Fig. 3.24(*b*) or (*c*) to be used for driving T_3 gate, the exponential rise towards zero on T_3 gate would result in rather slow switching on in terms of the time scale shown. The method given allows T2 collector to rise towards + 15 V, reaching zero much sooner than in the other circuits. Diode D_1 then cuts off so that T_3 gate is not driven beyond zero. The off condition of T_3 occurs when the switch drive is at logic 0 (0 to + 0·4 V) causing T_1 to conduct and

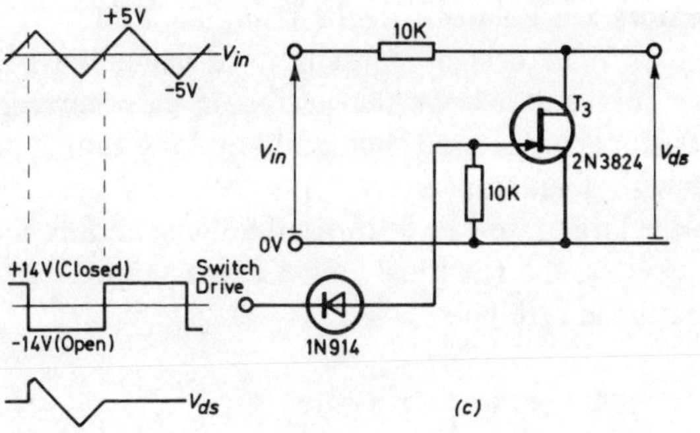

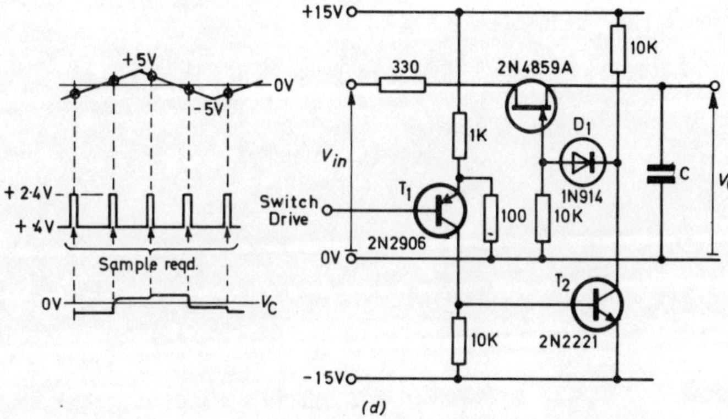

FIG. 3.24 *continued*

T_2 to bottom. T_3 gate is then at approximately − 17 V, an excess of 12 V over the most negative drain or source potential, thus ensuring cut off (10 V needed for the 2N4859A). The waveforms in Fig. 3.24(*d*) assume that the output capacitor C is buffered.

SUMMARY

The rules for switching on and off the J-FET are clear and easy to achieve. The only common error is to forget to examine the voltages present on both source *and* drain in the off state to ensure that gate drive is adequate.

Multivibrators and Related Circuits Using the FET

It is possible to design multivibrators using the FET with the advantage that, because of the negligible gate current, very high timing resistors can be used thus enabling long timing periods to be obtained with small capacitors. This seems an attractive idea but owing to the large gate-source turn-off voltage and its wide variation between devices, the resulting timing is inaccurate. For this reason such circuits are rarely used.

Chapter 4

Transistor equivalent circuits

To predict the performance of a transistor circuit, the circuit diagram is redrawn with the transistors replaced by representative combinations of voltage and current generators, resistors and reactive elements. Provided these representations or 'models' are correctly devised for the conditions of operation, Ohm's and Kirchhoff's laws enable performance to be calculated.

For manual analysis, quite simple models are generally used. The only difficulty is in finding from manufacturers' data the parameter values needed in the model. Manufacturers vary widely in the amount and type of data supplied. When figures are tabulated they may be in 'h' parameters, 'T' parameters or 'y' parameters; characteristic curves are frequently given instead of tabulations. Usually the designer can deduce the information he needs for initial analysis if curves are given. For 'worst-case' design, i.e. calculating the extremes of performance when all component tolerances conspire to give worst results, he must often make intelligent guesses at limit values in the absence of guaranteed figures.

Even in simple circuits the manual analysis of worst-case performance can be very tedious. It is not always easy to decide whether the increase in value of each of two resistors will have aiding or opposing effects, and this is essential knowledge for predicting worst-case limits. In complex circuits, especially at high frequencies, the number of circuit elements affecting performance (for instance the inductance and capacitance of connecting leads and the HF parameters of the transistors) also makes any analysis difficult.

In these cases computer analysis can be used with advantage. Many programs are available, each with features which are especially attractive in some respects and irksome in others. The simpler programs ignore non-linearities in the semiconductors but are easy to use, economic in computation time and sufficiently accurate for most uses. Other programs use complex models and purport to give more accurate results but are expensive in use and require obscure data for the models. Programs for transient analysis are particularly tricky and only regular usage enables sensible results to be achieved consistently, there being numerous pitfalls awaiting the beginner.

For his initiation into computer circuit analysis the designer can become familiar with ECAP (an IBM program) which has the advantage of being easily grasped even by the non-computer-minded engineer. If suitably disposed he may then proceed to more complex programs most of which demand some competence in computer programming. The danger is that the process of the analysis itself may become so absorbing that its actual purpose is neglected.

The form of model for either manual or computer analysis depends on the application which may be broadly classified as in Fig. 4.1. The complexity of the model depends on the accuracy required, the computing facilities available and on the values of the external components.

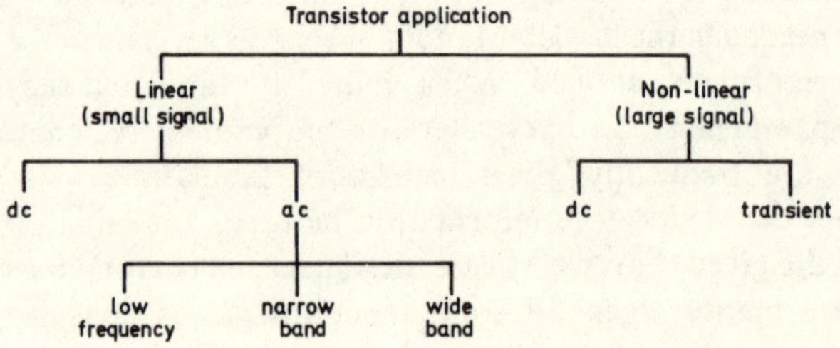

FIG. 4.1 Transistor model types

Linear Models

When the signal current in the emitter is less than 20% of the emitter bias current, a linear model is likely to be adequate. A d.c. analysis can then be carried out to establish bias conditions from which the parameters needed for the a.c. analysis can be determined.

The small signal model of Fig. 4.2 can be used to perform d.c. analysis. The current generator in the model delivers a current $I_B h_{FE}$, where h_{FE} is equal to the quantity β or $\alpha/(1 - \alpha)$ used in Chapter 2. The feedback resistance r_{bc} which is much larger than the other resistors in the model causes the effective input resistance into

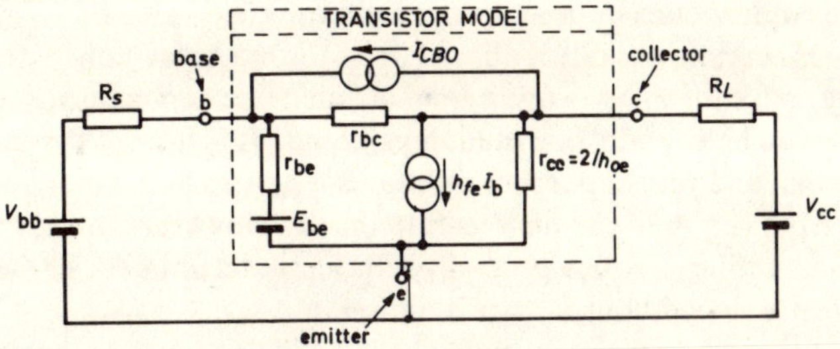

FIG. 4.2 Basic emitter follower/grounded emitter circuit

the base to be $r_{be}/2$ if $R_L \gg r_{ce}$. If however $R_L \ll r_{ce}$ (the normal case) the input resistance is unaffected by r_{bc}, and the only noticeable effect of r_{bc} is then to reduce the output resistance from r_{ce} to $r_{ce}/2$ if $R_s \gg r_{be}$. Thus r_{bc} may be omitted if r_{ce} is set at $2/h_{oe}$ for low R_s and $1/h_{oe}$ for high R_s. Since h_{oe} often has a spread of as much as 5:1 in the quoted value it is common practice to set r_{ce} to a value $1/h_{oe}$ for all R_s values. The simplified circuit is shown in Fig. 4.3; for PNP devices the emf and current direction are reversed.

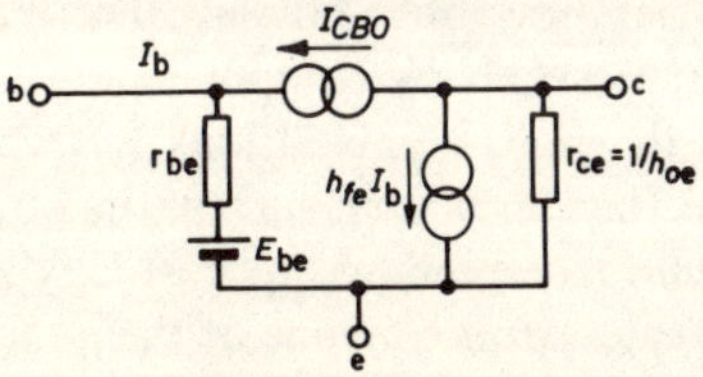

FIG. 4.3 Simplified version of Fig. 4.2

Obtaining Practical Values

1 h_{oe}: this is defined as the output admittance between collector and emitter with open-circuit base, at the bias current and voltage used in the cirucit. It is frequently quoted in tabular form, but $1/h_{oe}$ can also be obtained by measuring the slope of the V_{ce} against I_c characteristic curve for steps of base current drive. Typical values of $1/h_{oe}$ are 10 to 100 kΩ when the emitter current is in the milliampere region.

2 r_{be}: this is equal to $r_{bb'} + (1 + h_{FE})\, r_e$ where $r_{bb'}$ is the resistance between the base terminal and the region b' of the base where most

of the emitter current flows to the collector. The value of $r_{bb'}$ varies with bias conditions but is about 100 Ω for most low-power devices.

The value of r_e is proportional to absolute temperature and is $25/I_e$ Ω at 17°C (with I_e in milliamps). Since I_e is needed for calculation of r_e, and yet the purpose in knowing r_e is to find I_e, an iterative method is required for an accurate result. However, calculating I_e while assuming $r_e = 0$ is generally sufficiently accurate for deriving r_e and hence r_{be} without further iteration.

Applications of the d.c. Linear Model

For designing bias circuits where some pessimism is allowable so that generous limits of V_{be} can be used, the methods of Chapter 2 give sound answers in a simple way and they are the methods normally used. If conditions are such that worst-case limits using these methods are marginally unacceptable, a more realistic, less pessimistic, assessment may show that the proposed circuit is after all satisfactory. The d.c. model under discussion enables this analysis to be carried out.

When several directly coupled stages are involved, and especially if d.c. feedback loops are used, the worst-case combinations of component tolerances may not be obvious. These circuits can usually be analysed very successfully by computer programs such as ECAP which determines the right combinations for worst-case limits for any desired node voltage or branch current. Its output is a tabulation of all the limit values requested and its cost is typically less than the engineer's time in calculating any one of them.

In using such an analysis program, one must beware of obtaining an over-pessimistic result due to ignoring the tracking of related parameters. For instance, r_{be} in Fig. 4.3 is closely related to h_{FE} so that r_{be} cannot have its minimum value when h_{FE} is a maximum. The model of Fig. 4.4 is therefore preferable as the tracking effect is automatic. Fig. 4.5 gives an idea of the spreads in parameters for this equivalent circuit for a typical small-signal transistor.

With small silicon devices, I_{cbo} can often be ignored, but with germanium or power devices it must be included. This presents no problem manually but many computer programs require that any current generator must have an associated parallel resistor. Moreover there will be an upper allowable limit to the resistor value so that a figure of, say, 10^{10} Ω would be inadmissible. The presence of the resistor demanded by the program may well cause errors when

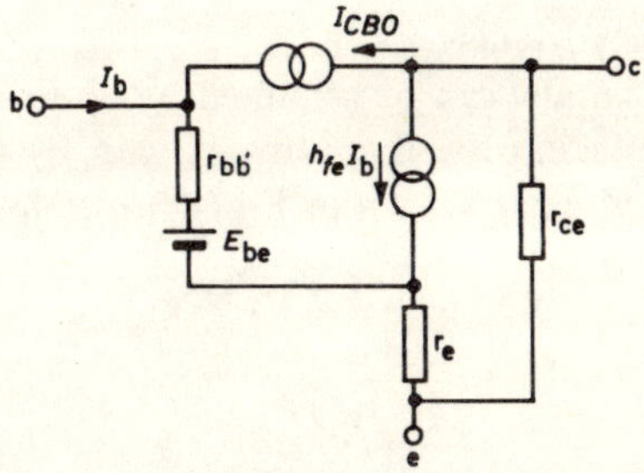

FIG. 4.4 d.c. linear model with tracking

	°C	−35	+ 25	+ 75
I_{CBO} (μA)		0.01	0.03	1.00
h_{FE}	min max	50 140	80 240	110 320
r_{ce}	min max (K)	50 360	40 200	35 175
E_{be}	min max (V)	.75 .85	.60 .70	.47 .57
r_e	min max (Ω)	15 24	20 30	23 35
r_{bb1}	min max (Ω)	50 150	50 150	50 150

FIG. 4.5 Typical parameter values

I_{cbo} is small, in which case the model of Fig. 4.6 should be used. Here a 10 MΩ resistor is used across generator I_{cbo} to satisfy the program. A further 10 MΩ resistor is added in parallel and the current in that resistor controls a current generator in parallel with it, producing twice the resistor current. The net effect is to cancel the error introduced by the first resistor. Such manoeuvres are often needed when using computer analysis programs.

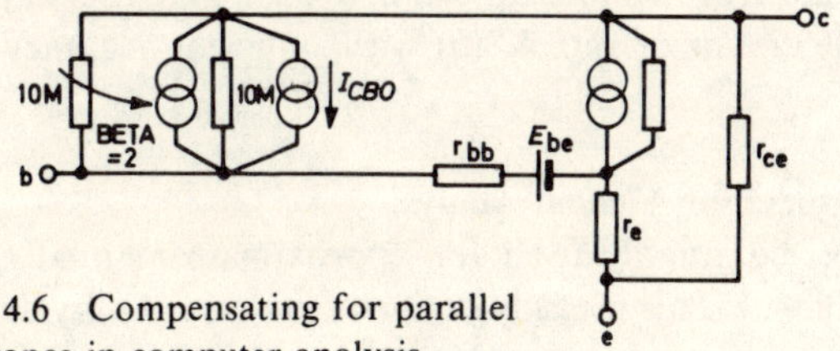

FIG. 4.6 Compensating for parallel resistance in computer analysis

Linear Low-frequency Models

A low-frequency model can be obtained by omitting E_{be} from the d.c. model and replacing the d.c. current gain by the incremental gain h_{fe} (often called β) as shown in Fig. 4.7(a). Now

$$i_b = v_{b'e}/(h_{fe} + 1)r_e$$

$$h_{fe}\,i_b = \frac{v_{b'e}\,h_{fe}}{(h_{fe} + 1)r_e} = \alpha_0 v_{b'e}/r_e$$

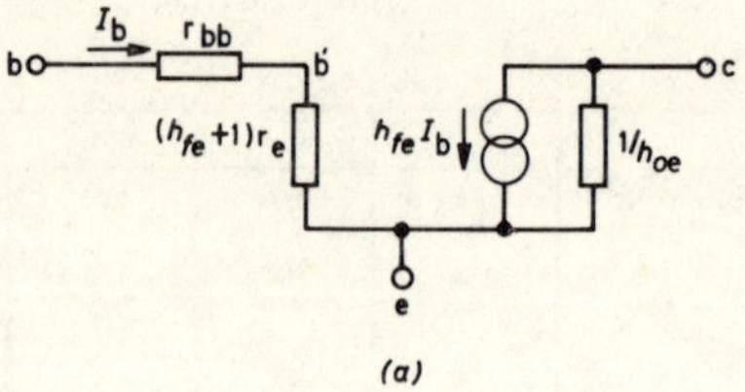

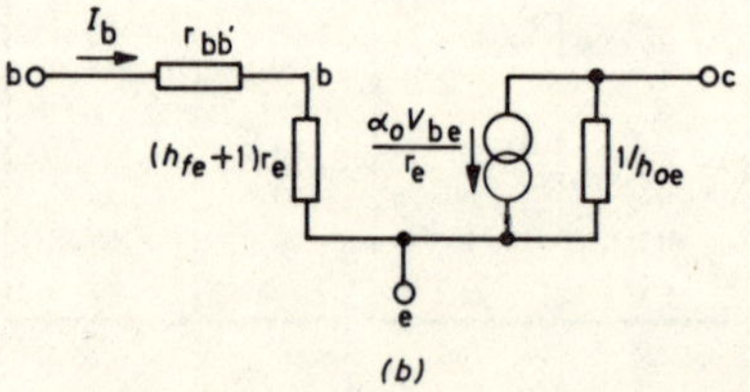

FIG. 4.7 Low frequency models

which leads to the model in Fig. 4.7(b).

This model can be used for all configurations (grounded base, emitter or collector), but manual calculations are easier if different models are used. The circuit shown in Fig. 4.8(a) is the common emitter circuit redrawn with grounded base and this can be converted to the circuit of Fig. 4.8(b) with sufficient accuracy for most purposes.

Simplified Model for Manual Analysis

The model commonly used for approximate manual analysis is shown in Fig. 4.9. This is easily derived from Fig. 4.7 by defining the current generator in terms of V_{be} rather than $v_{b'e}$, assuming that

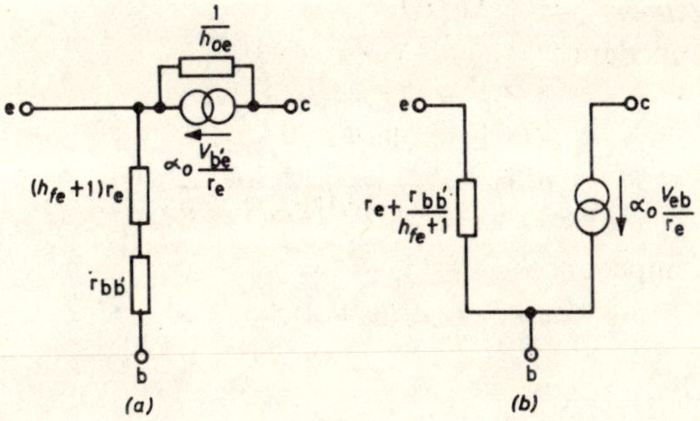

FIG. 4.8 Grounded base models

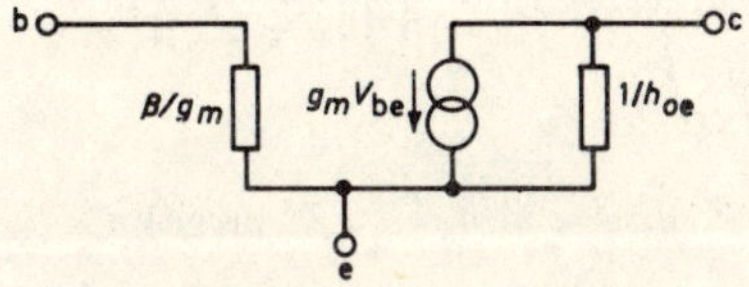

FIG. 4.9 Usual model for manual analysis

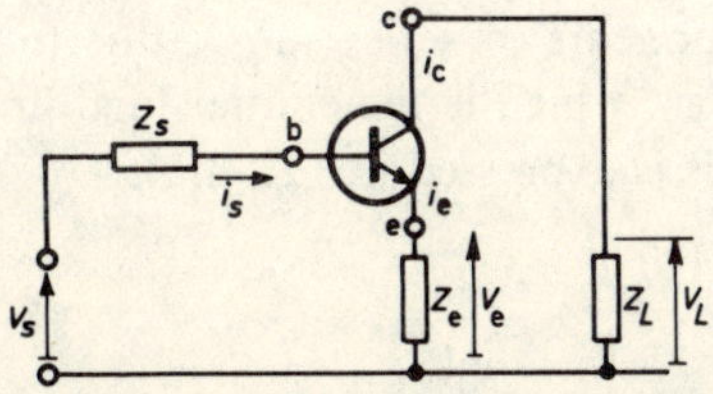

FIG. 4.10 Simple amplifier for analysis

$h_{fe} \gg 1$, and introducing g_m to equal $1/(r_e + r_{bb'}/(1 + h_{fe}))$: in line with common usage h_{fe} is written as β.

From the analysis of the circuit of Fig. 4.10 in which each electrode has an associated series external impedance, all the design equations usually required can be obtained and are listed below.

Emitter follower

Input impedance

$$Z_{in} = \beta[Z_e + 1/g_m] \quad (1)$$

Gain
$$v_e/v_s = \frac{Z_e}{Z_e + 1/g_m + Z_s/\beta} \quad (2)$$

Output impedance

$$Z_{out\ e} = Z_s/\beta + 1/g_m \quad (3)$$

Earthed emitter amplifier

Input impedance

$$Z_{in} = \beta[Z_e + 1/g_m] \quad (4)$$

Gain
$$v_L/v_s = \frac{Z_L}{Z_e + 1/g_m + Z_s/\beta} \quad (5)$$

Output impedance

$$Z_{out\ c} \approx \beta/h_{oe} \quad (Z_e \text{ large}) \quad (6)$$

$$\approx 1/h_{oe} \quad (Z_e \text{ zero}) \quad (7)$$

Note that the input impedance to both circuits is equal to the sum of two components in series: β/g_m, due to transistor internal resistances, and βZ_e, which is the emitter load impedance increased by a factor β. This is expressed in Fig. 4.11.

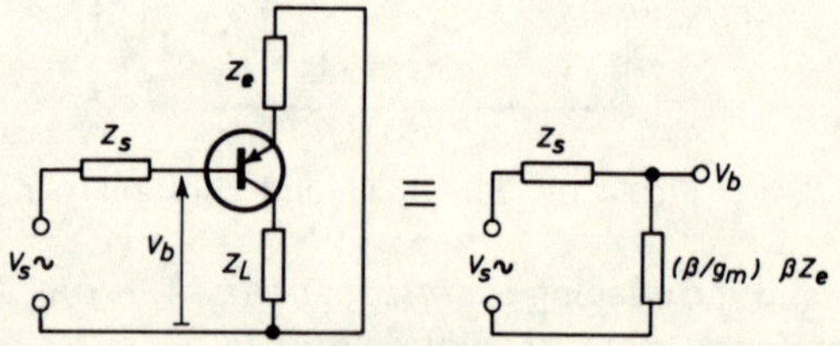

FIG. 4.11 Transistor input impedance as seen by source (enables v_b to be calculated)

Examination of equation (2) shows that the gain would be unity except for the terms $1/g_m$ and Z_s/β in the denominator. The emitter circuit therefore behaves like a generator of source e.m.f. v_s and internal impedance $1/g_m + Z_s/\beta$, as indicated in Fig. 4.12, giving

$$v_e = v_s \left[\frac{Z_e}{Z_e + (1/g_m) + (Z_s/\beta)}\right]$$

An alternative way of looking at this circuit is to calculate from Fig. 4.11 the voltage v_b actually reaching the base, namely

$$v_b = v_s \frac{\beta/g_m + \beta Z_e}{Z_s + \beta/g_m + \beta Z_e}$$

As far as the transistor is concerned, this is a zero resistance source connected directly to the base: the transistor functions according to

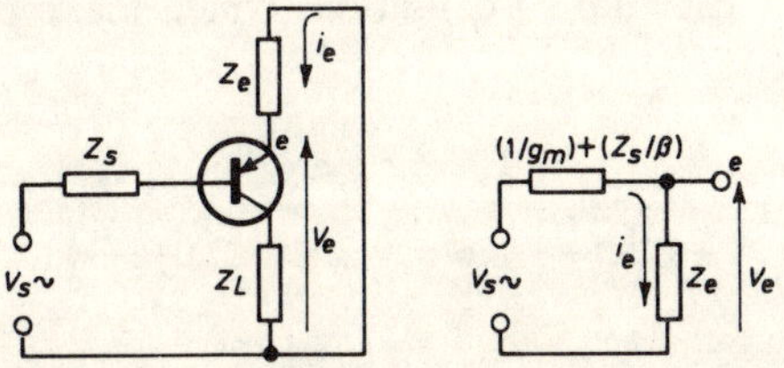

FIG. 4.12 Transistor emitter output as seen by Z_e (enables v_e to be calculated and, hence, i_e)

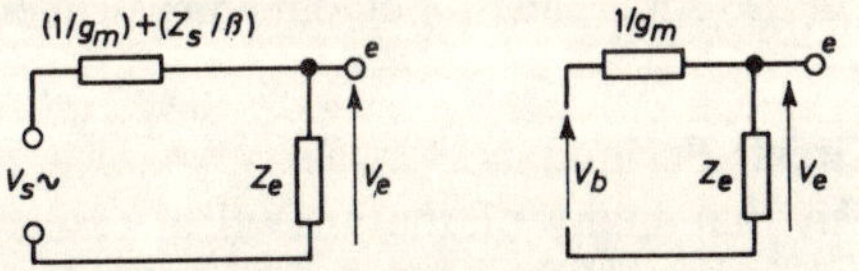

FIG. 4.13 Alternative forms of Fig. 4.5 give identical results

the voltages which are applied to it and it cannot know what lies beyond. Hence, the emitter circuit can be regarded as shown in Fig. 4.13. The emitter voltage using this circuit would be

$$v_s \frac{\beta/g_m + \beta Z_e}{Z_s + \beta/g_m + \beta Z_e} \times \frac{Z_e}{Z_e + 1/g_m} = v_s \frac{Z_e}{Z_e + 1/g_m + Z_s/\beta}$$

as before.

Having used the emitter equivalent circuit of Fig. 4.12 or 4.13, the emitter voltage can therefore be calculated, and so the emitter current is known by dividing by Z_e. Alternatively, the same answer is obtainable by dividing v_s by $[Z_e + 1/g_m + Z_s/\beta]$ or by dividing v_b by $[Z_e + 1/g_m]$; one of these two must be used if $Z_e = 0$ to avoid an indeterminate result.

By using one of the equivalent circuit forms given above, the input impedance, emitter voltage, and emitter current can therefore be calculated. By assuming that the collector signal current is approximately equal to emitter signal current (which is true for the conditions which lead to the simplified equations), the signal collector voltage is known by multiplying this current by Z_L.

This is illustrated by the dismembered form of equivalent circuit shown in Fig. 4.14, 4.14 (*a*) showing the input circuit and 4.14 (*b*) the output circuit. Using this representation, all linear circuits may be analysed very simply, although the full T-equivalent circuit has to be resorted to for exceptional conditions which make the approximations invalid.

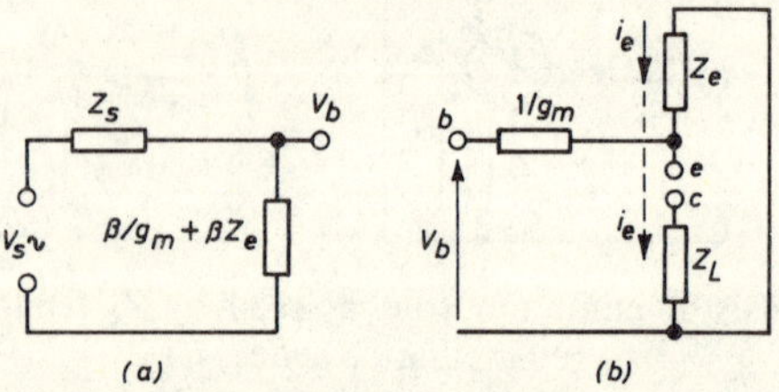

FIG. 4.14 (*a*) Input, and (*b*) output equivalent circuits

Variations in Circuit Performance

It will be seen that many of the analytical results are very dependent on transistor parameters, particularly β and g_m. These parameters are by no means constant between various specimens of one type of transistor or even for one particular transistor when the ambient temperature or operating current changes. The practice of selecting transistors from a batch to a tight specification giving, for example, a spread in β of 1·5 to 1 instead of the normal figure of perhaps 3 or 4 to 1, is unsatisfactory in at least two respects. It leads to higher transistor cost because of the work involved in selection and the possibly poor yield of acceptable transistors, and it leads to a great variety of transistors with only slightly different specifications.

The designer must therefore ensure correct circuit performance in spite of these variations and this aspect of design is at least as important as, and is indeed complementary to, his concept of the configuration to be used. One part of this process has been covered in Chapter 2, where the importance of the bias arrangements was stressed, in order to stabilize the operating currents and voltages.

This involves arrangements which reduce base voltage variations and also ensure by the use of large emitter resistors returned to a high voltage that any base variations have the least possible effect on emitter current (and therefore collector voltage). This second precaution can be regarded as keeping low the transistor stage gain from base to collector for very slow-moving signals, which is indeed achieved (see Fig. 4.14) by high R_e.

This provides a clue to a method for stabilizing the gain at higher frequencies. Instead of making Z_e zero at the signal frequency (by capacitor to earth) which would give the highest obtainable gain, a series resistor can be included as in Fig. 4.15 (*a*) and (*b*). The gain equation v_c/v_s is $R_L/[1g/_m + R_e']$ (if X_c is negligible at the signal

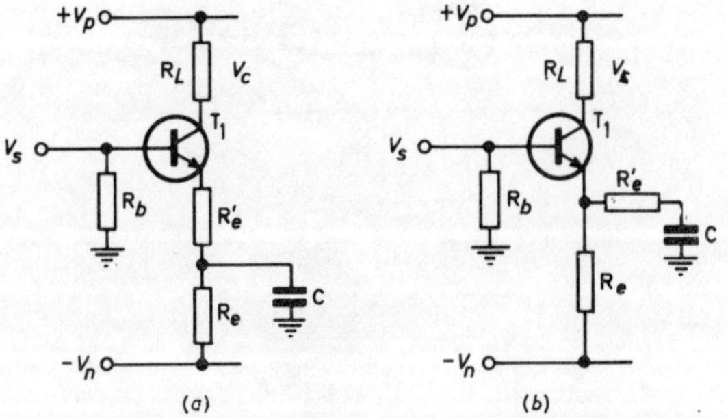

FIG. 4.15 Addition of R_e' to stabilize gain

frequency), and if $R_e' \gg (1/g_m)$, variations in $1/g_m$ have no effect on gain. Knowing the possible variations in $1/g_m$, R_e' can be designed to reduce their effect by any desired extent. Unfortunately, the gain itself is reduced by the same amount, so by varying R_e' the gain could vary from, e.g., 100 $\pm$ 30 per cent to 30 $\pm$ 10 per cent or 10 $\pm$ 3 per cent, etc.

This method of overcoming the effects of variables which are outside the designer's control by adding fixed components to swamp the variables is used in many designs.

In the present example it can be regarded as a form of negative feedback, a subject discussed in Chapter 9. The ideal is to make the circuit performance depend only on static components, the accuracy of which are well specified. In an actual design transistor parameters will inevitably affect performance but, in a correct design, only to an extent which can be allowed by the circuit performance specification.

The following chapters give procedures for the design of a variety of circuits using the above philosophy.

Emitter Coupled Pair

This circuit (Fig. 4.16) is a very useful amplifier having two input and two output terminals. Its bias arrangements present more problems than the simple earthed emitter amplifier and will be dealt with later in the chapter; only its small a.c. signal behaviour will be discussed here.

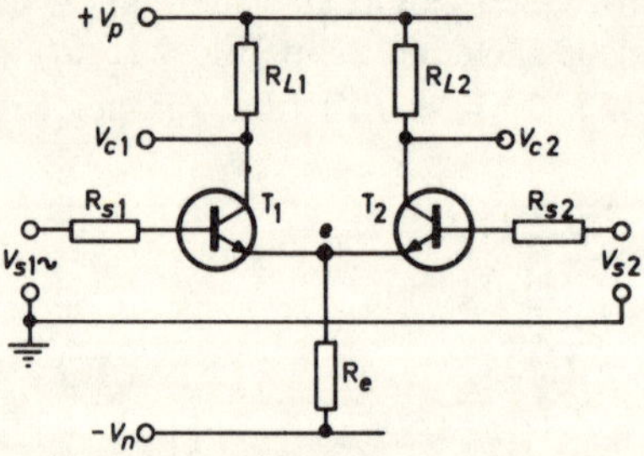

FIG. 4.16 Emitter-coupled pair

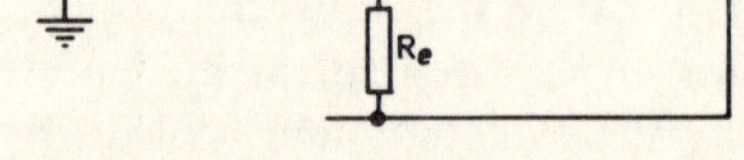

FIG. 4.17 Equivalent circuit for $v_{S2} = 0$

Since the signal currents are assumed to be small compared with transistor bias currents, operation is almost linear; therefore the outputs produced by the two inputs v_{S1} and v_{S2} can be assessed separately and the results added, by the principle of superposition.

If v_{S1} is present and v_{S2} short-circuited, the common emitter voltage v_E can be calculated from Fig. 4.17, showing v_{S1} connected to E through an apparent source resistance $(R_{S1}/\beta_1 + 1/g_{m1})$, E being loaded by R_E in parallel with the emitter impedance of T_2, namely $[R_{S2}/\beta_2 + 1/g_{m2}]$. For simplicity assume T_1 and T_2 are identical and that $R_{S1} = R_{S2}$, and further, that R_E is very large in comparison with $(R_S/\beta + 1/g_m)$. In this case the source resistance between v_{S1} and v_E and the loading on v_E are equal, giving $v_E = \frac{1}{2}v_{S1}$. Hence,

$$i_{E1} = (v_{S1} - v_E)/[R_S/\beta + 1/g_m] = \tfrac{1}{2}v_{S1}/[R_S/\beta + 1/g_m],$$

and

$$v_{C1}/v_{S1} = -\tfrac{1}{2}R_L/(R_S/\beta + 1/g_m)$$

(The reason for the negative sign is obvious from inspection of the direction of i_{E1} in Fig. 4.17.)

In this special case, where R_E is very large, T_1 and T_2, R_{S1} and R_{S2} are identical, and only v_{S1} is connected, *the voltage gain from v_{S1} to* T_1 *collector is* 1/2 that obtained from *a single transistor stage* having an earthed emitter.

Continuing the analysis, T_2 emitter current is also

$$\tfrac{1}{2}v_{S1}[R_S/\beta + 1/g_m]$$

so that

$$v_{C2}/v_{S1} = +\tfrac{1}{2}R_L/[R_S/\beta + 1/g_m]$$

the voltage gain from v_{S1} to T_2 collector is therefore equal but opposite in sign to the gain to T_1 collector.

The effect of a finite value of R_E is to change v_E from $v_{S1}/2$ to

$$v_{S1}\frac{[R_S/\beta + 1/g_m]//R_E}{2[R_S/\beta + 1/g_m]} = \tfrac{1}{2}v_{S1}\frac{R_E}{R_E + R_S/\beta + 1/g_m}$$

which is slightly smaller. The value of T_1 emitter current, i.e. $(v_{S1} - v_E)/[R_s/\beta + 1/g_m]$, is therefore increased and T_2 emitter current i.e. $v_E/[R_S/\beta + 1/g_m]$, is decreased. The gain to T_1 collector is therefore higher than to T_2 collector.

$$v_{C1}/v_{S1} \text{ being } -\frac{[\tfrac{1}{2}R_E + R_S/\beta + 1/g_m]R_L}{[R_E + R_S/\beta + 1/g_m][R_S/\beta + 1/g_m]}$$

and

$$v_{C2}/v_{S1} \text{ being } +\frac{\tfrac{1}{2}R_ER_L}{[R_E + R_S/\beta + 1/g_m][R_S/\beta + 1/g_m]}$$

Returning to the simple case ($R_E \to \infty$), the effect of source v_{S2} will be equal and opposite to that of v_{S1}, so that

$$v_{C1}/v_{S2} = +\tfrac{1}{2}R_L/[R_S/\beta + 1/g_m]$$

and

$$v_{C2}/v_{S2} = -\tfrac{1}{2}R_L/[R_S/\beta + 1/g_m]$$

With both inputs present the collector voltages are, by superposition,

$$v_{C1} = -v_{S1}\tfrac{1}{2}R_L/[R_S/\beta + 1/g_m] + v_{S2}\tfrac{1}{2}R_L/[R_S/\beta + 1/g_m]$$

$$= \frac{-(v_{S1} - v_{S2})\frac{1}{2}R_L}{R_S/\beta + 1/g_m}$$

and

$$v_{C2} = \frac{-(v_{S2} - v_{S1})\frac{1}{2}R_L}{R_S/\beta + 1/g_m}$$

The significance of these results is that if $v_{S1} = v_{S2}$, then $v_{C1} = v_{C2} = 0$; and if $v_{S1} = -v_{S2}$, then

$$v_{C1} = -v_{S1}\frac{R_L}{R_S/\beta + 1/g_m} \quad \text{and} \quad v_{C2} = +v_{S1}\frac{R_L}{R_S/\beta + 1/g_m}$$

This amplifier therefore produces equal and opposite outputs dependent only on the difference between the two input signals and not on the sum. Thus, if

$$\frac{R_L}{R_S/\beta + 1/g_m} = 30$$

and two sine-wave signals were applied, identical in frequency and phase, v_{S1} being 100 mV peak, v_{S2} being 95 mV peak, then v_{C1} would be $(100 - 95)30 = 150$ mV peak and in antiphase with v_{S1}, and v_{C2} would be 150 mV peak and in phase with v_{S1}.

This result is easy to understand by considering the current flow caused by v_{S1} and v_{S2}. If these are identical signals and R_E is infinite (see Fig. 4.17), v_E will also move identically with them. (If it did not, then the currents flowing out of the v_E junction into T_1 T_2 would change so that the current in R_E must change, which is impossible; therefore v_E must have moved in such a way as to cause no current change in T_1 or T_2, i.e. v_E must have followed v_{S1}, v_{S2} identically.) No signal current flows in T_1 or T_2, and no collector voltage change occurs.

When v_{S1} and v_{S2} differ, then a current flows along the $(R_{S1}/\beta_1 + 1/g_{m1})$ and $(R_{S2}/\beta_2 + 1/g_{m2})$ paths, causing a current change which is one way for T_1 and the other for T_2, giving equal and opposite polarity collector voltages.

A finite R_E clearly changes the first result, since if $v_{S1} = v_{S2}$, the current in T_1 and T_2 does not now remain constant but the total current changes by v_{S1}/R_E. If T_1 and T_2 are identical and are sharing the current equally the result is a change of current in both T_1 and T_2 of $\frac{1}{2}v_{S1}/R_E$. Each produces an output voltage of $-\frac{1}{2}v_{S1}R_L/R_E$ and each output is in antiphase with v_{S1} and v_{S2}.

The presence of a finite R_E therefore results in an appreciable output even when $v_{S1} = v_{S2}$.

When an amplifier is to be used for measuring the small difference between two large signals, the criterion for good performance is the ratio between 'push–pull' gain $v_{out}/(v_{S1} - v_{S2})$ and 'push–push' gain $v_{out}/(v_{S1} + v_{S2})$, often called 'rejection ratio'. In the emitter-coupled pair this is proportional to R_E, as shown above, and is typically between 20 and 200.

The use of a constant-current device (Chapter 6) instead of R_e improves this figure to between 2000 and 20 000.

Biasing the emitter-coupled pair

In designing a practical emitter-coupled pair amplifier, the bias conditions present a special problem. If the basic circuit is used as it stands, both bases being returned through resistors to a fixed potential (e.g. 'earth') and the common emitter connection taken to a suitable supply, then although the current in R_E is well defined, the ratio in which it divides between T_1 and T_2 is unknown. With normal transistor spreads for V_{eb} against I_e, one transistor may be cut off while all the current from R_E passes into the other. This is no exaggeration, as a glance at production spreads of I_c for constant V_{eb} soon confirms.

The corresponding valve circuit, the cathode-coupled pair, behaves in the same way if two pentodes are used. This is the main reason why triodes are preferred, since by using equal anode loads any tendency to cut off causes the appropriate anode voltage to rise, thus increasing the current. With pentodes, as with transistors, a change in anode or collector voltage has little effect on the current and the unbalance remains.

For the amplification of alternating signals only, which is a very common application, the zero-frequency gain should be reduced as shown in Figs. 4.18 and 4.19. In Fig. 4.18 each transistor is separately biased in a normal manner and C_e must be a low reactance compared with the emitter circuit impedance at the lowest angular operating frequency ω_L, i.e.

$$\frac{1}{\omega_L C_e} \ll R_{e1}//\left(\frac{1}{g_{m1}} + \frac{R_{S1}}{\beta_1}\right) + R_{e2}//\left(\frac{1}{g_{m2}} + \frac{R_{S2}}{\beta_2}\right)$$

In calculating rejection ratio, R_e must be taken as $R_{e1}//R_{e2}$. At frequencies much below ω_L the rejection ratio will be degraded, since the push–pull gain falls greatly, and the push–push gain falls only by a factor of 2 if $R_{e1} = R_{e2}$.

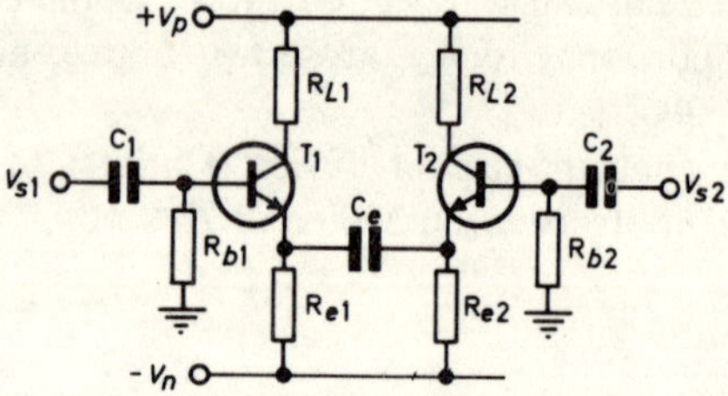

FIG. 4.18 Bias circuit for emitter-coupled pair

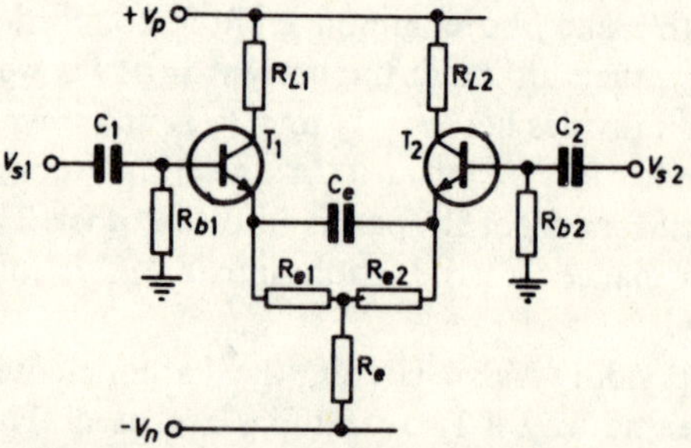

FIG. 4.19 Alternative bias circuit for emitter-coupled pair

In Fig. 4.19 R_{e1} and R_{e2} are arranged to drop a voltage greatly in excess of V_{eb1} and V_{eb2}. For a low-gain application C_e may be omitted, but since R_{e1} and R_{e2} reduce the push–pull gain much more than the push–push gain, rejection ratio deteriorates. If C_e is added, then a similar criterion applies, as stated for Fig. 4.18. The equation is slightly altered and becomes

$$\frac{1}{\omega_L C_e} \ll (R_{e1} + R_{e2})//\left(\frac{1}{g_{m1}} + \frac{R_{S1}}{\beta_1} + \frac{1}{g_{m2}} + \frac{R_{S2}}{\beta_2}\right)$$

The emitter-coupled pair may be used for differential amplification, its rejection ratio being as high as 20 000 (86 dB) per stage if the emitter feed is a constant-current source. Special precautions must be taken in fixing the operating point, otherwise severe unbalance occurs.

Models for Higher-frequency Operation

The models so far described take no account of the internal capacitances of the transistor or of the fall of h_{fe} at high frequencies, h_{fe} reaching unity at f_T. To calculate performance when these effects cannot be ignored requires a more complex model. If the frequency band of interest is narrow compared with the mean operating frequency a model based on 'y', i.e. admittance parameters is convenient, mainly because these parameters are easy to measure. For wide-band calculations the 'hybrid Π' model is normally used, being a modified version of the linear low-frequency model of Fig. 4.7. The necessary data for transistors intended for these applications is generally provided by the manufacturer. Thus an IF amplifier device will have y parameters quoted, and a video device's 'h' parameters will be given.

Narrow-band Models

At frequencies above f_T/h_{fe} the model must represent the phase lag between the output and input currents as well as collector-base capacitance which causes appreciable coupling between input and output. In narrow-band applications the transistor parameters can be measured at the frequency of interest with correct bias conditions. The model in Fig. 4.20 uses 'y' parameters that are useful for frequencies up to a few hundred megahertz. They are all measured with the input or output short-circuited to a.c. which minimizes the effect of external stray capacitance. For example, if the input is short-circuited and a voltage is applied to the output, then

$$y_{re} = i_b/v_o \quad \text{and} \quad y_{oe} = i_o/v_o$$

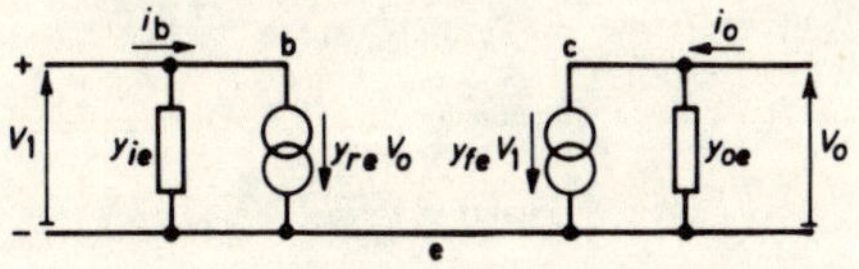

FIG. 4.20 Narrow-band model for a transistor

Similarly with the output short-circuited

$$y_{ie} = i_b/v_1 \quad \text{and} \quad y_{fe} = i_o/v_1$$

The y parameters for common collector or common base connections can be derived from the common emitter values or measured directly. As an example consider the circuit shown in Fig. 4.21(*a*). The y equivalent circuit is given in (*b*) which can be reduced to that

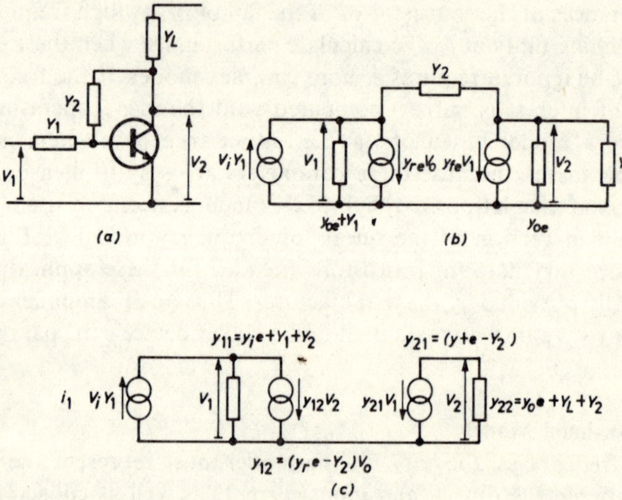

FIG. 4.21 Narrow-band model for an amplifier

shown in (*c*) by comparing short circuit admittances. The resulting equations are:

$$i_1 = v_1 y_{11} + v_2 y_{12}$$
$$i_2 = v_1 y_{21} + v_2 y_{22} = 0$$

If for example the voltage gain v_2/v_i is required, this is obtained from

$$v_1 = -y_{22} v_2 / y_{21}$$
$$v_i y_1 = i_1 = -(y_{22} y_{11}/y_{21}) v_2 + y_{12} v_2$$

Hence

$$v_2/v_i = y_1 y_{21}/(y_{12} y_{21} - y_{22} y_{11})$$

At frequencies above a few hundred megahertz it is difficult to produce an effective short-circuit, so that although the theoretical model is still valid, the required parameters cannot be obtained. In such cases 50-Ω terminations are used for the measurements and a model involving '*s*' parameters is used for the analysis.

Wide-band Models

Analysis over a wide frequency range is often required, for example, in assessing the stability of a feedback loop. In such cases it is necessary to use a model in which the parameters are independent of frequency. The hybrid Π model shown in Fig. 4.22 is suitable up to a frequency of about $f_T/5$. At higher frequencies errors in the input

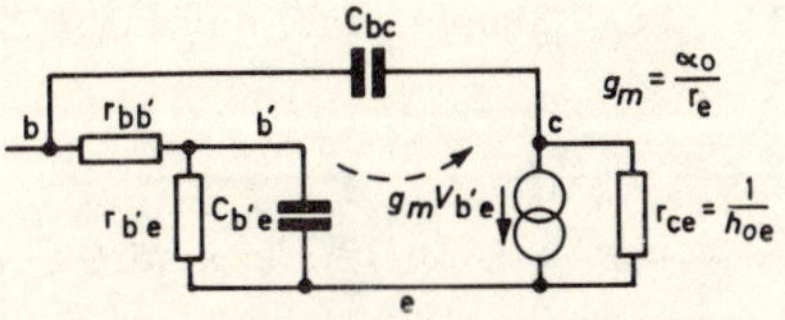

FIG. 4.22

impedance become significant. At low frequencies this model is the same as that of Fig. 4.7. Strictly, a fraction of C_{bc} should be shown between b' and c but the additional complexity is seldom warranted. The transfer conductance g_m is approximately $1/r_e$. Thus g_m = 40 mA/V at an emitter current of 1 mA and is directly proportional to emitter current, for all devices. At currents above about 25 mA in a small signal device it becomes necessary to include extrinsic emitter resistance since r_e is then only 1 Ω. $C_{b'e}$ is determined from $(C_{b'e} + C_{bc}) = 1/(2\pi f_t r_e)$. Greater precision than given by this formula is unnecessary because of the limited accuracy with which f_T is known. Since $C_{b'e}$ is usually noticeably greater than C_{bc}, it follows that $C_{b'e}$ is roughly proportional to $1/r_e$ and therefore proportional to I_e. The collector base capacitance C_{bc} is itself proportional to $V_{bc}{}^{-\frac{1}{3}}$.

An approximate analysis using a hybrid Π circuit is given below. Consider a single transistor stage operating at 1 mA with a source impedance of 500 Ω and a load of 1 kΩ. Let h_{fe} = 100, f_T = 100 MHz and C_{bc} = 3 pF. Neglect $r_{bb'}$ and r_{ce}. From f_T and r_e we obtain $C_{b'e}$ = 60·7 pF. A base voltage v_b produces an output voltage $-(R_L/r_e)v_b = -40v_b$ giving $41v_b$ V across C_{bc}. The same input current would flow in a capacitor $41C_{bc}$ connected between base and emitter. Neglecting the effect of this current on the load the circuit may be drawn as shown in Fig. 4.23. By inspection the gain at zero frequency is $(2{\cdot}5/3) \times 40 = 33{\cdot}3$. The gain is 3 dB down at $f = 2{\cdot}1$ MHz where 500 Ω in parallel with 2·5 kΩ equals the reactance of 183·7 pF.

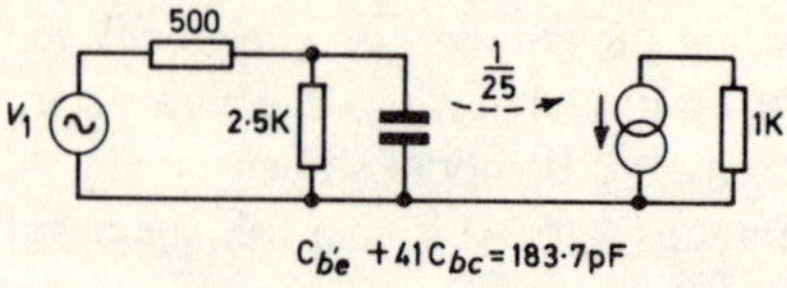

FIG. 4.23 Simplified form of Fig. 4.22

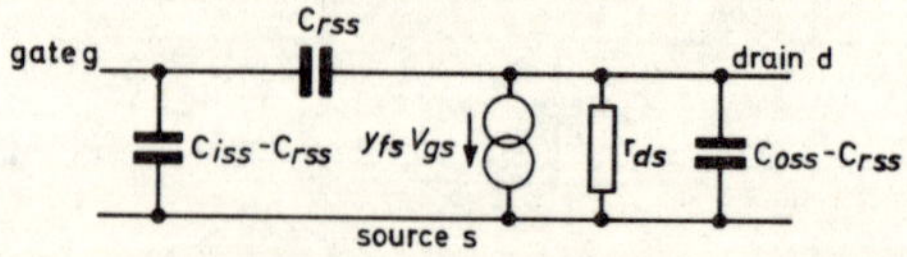

FIG. 4.24 J-FET model

To conclude this section an equivalent circuit of a J-FET is given in Fig. 4.24. The FET does not suffer the fall-off in internal gain characterized by f_T in the bipolar device, and its input resistance is virtually infinite. The only frequency dependent terms are therefore those due to inter-electrode capacitance. Calculations are simpler and more accurate and the only common pitfall is in omitting r_{ds} (equivalent to $1/h_{oe}$) which is usually too significant to ignore at low and medium frequencies.

Non-linear d.c. Model

A transistor consists basically of two diodes connected 'back-to-back' (as the illogical common phrase has it) with the base as the common electrode. Usually the emitter-base diode is forward-biased and the collector base diode reverse-biased. The base region is made so thin that most of the minority carriers entering the base from the emitter, due to the forward bias, are accelerated across the collector junction producing a high ratio of collector to base current. The Ebers and Moll d.c. model represents the transistor as two such diodes together with two dependent current generators representing normal and inverse current gains α_N and α_I. This model is shown in

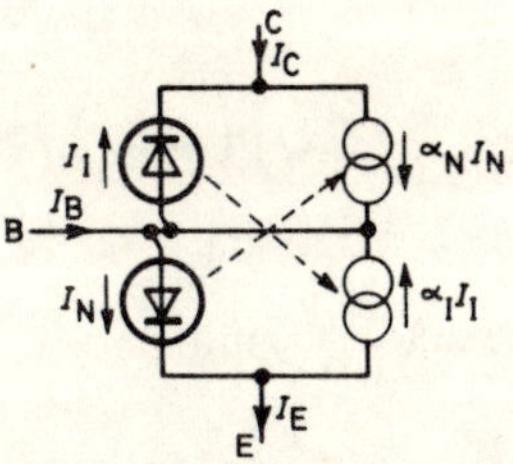

FIG. 4.25 Non-linear d.c. model

Fig. 4.25 for an *n-p-n* transistor. Here

$$I_N = I_{ECS}[\exp(qV_{BE}/M_E KT) - 1]$$
$$I_I = I_{CES}[\exp(qV_{BC}/M_C KT) - 1]$$

From Fig. 4.25,

$$I_C = \alpha_N I_N - I_I$$
$$I_E = I_N - \alpha_I I_I$$

As an example the model is used below to calculate the collector voltage of a transistor when the collector is open and the base current is 1 mA. The calculation is repeated with the collector and emitter interchanged to obtain performance in the inverse mode. The transistor is assumed to have the following characteristics: $\alpha_N = 0{\cdot}98$ $\alpha_I = 0{\cdot}5$.

$$\text{At } V_{CE} = 5\text{ V},\ I_{E1} = 1\text{ mA at } V_{BE1} = 605\text{ mV}$$
$$I_{E2} = 2\text{ mA at } V_{BE2} = 625\text{ mV}$$
$$\text{At } V_{EC} = 5\text{ V},\ I_{C1} = -1\text{ mA at } V_{BC1} = 586\text{ mV}$$
$$I_{C2} = -2\text{ mA at } V_{BC2} = 606\text{ mV}$$

In the forward active region the collector base diode is reverse-biased and $I_I \approx 0$, hence $I_N \approx I_E$. From the expression for I_N,

$$V_{BE} \approx M_E \frac{KT}{q} \log_e(I_N/I_{ECS})$$

hence

$$M_E = \frac{V_{BE2} - V_{BE1}}{\dfrac{KT}{q} \log_e \dfrac{I_{N2}}{I_{N1}}}$$

Since $KT/q = 25$ mV at $+17^1$C,

$$M_E = \frac{625\text{–}605}{25\text{ mV} \log_e 2} = 1{\cdot}15$$

and

$$I_{ECS} = I_{NI}\exp\left(-V_{BE1}/M_E\frac{KT}{q}\right)$$
$$= 10^{-3}\exp(-605/28{\cdot}75) = 7{\cdot}26 \times 10^{-13}\text{ A}.$$

Similarly in the inverse active region $I_N \approx 0,\ I_I \approx -I_C$

and $$V_{BC} \approx M_C\frac{KT}{q}\log_e(I_I/I_{CES})$$

hence $$M_C = \frac{606 - 586}{25\log_e 2} = 1.15$$

and $$I_{CES} = 10^{-3}\exp[-586/28{\cdot}75] = 1{\cdot}4 \times 10^{-12}\text{ A}$$

The currents flowing in the model diodes are shown in Fig. 4.26. In the normal mode $V_{CE} = V_{BE} - V_{BC}$, so that for the currents

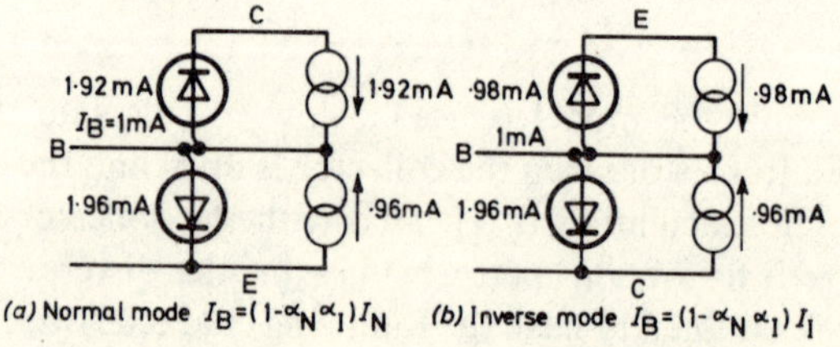

FIG. 4.26 Diode current, normal and inverse

shown $V_{CE} = 624{\cdot}3 - 604{\cdot}8 = 19{\cdot}5$ mV. In the inverse mode $V_{EC} = V_{BC} - V_{BE} = 605{\cdot}5 - 604{\cdot}4 = 1{\cdot}1$ mV.

Non-linear Transient Model

The Ebers and Moll transient model has two more current genera-

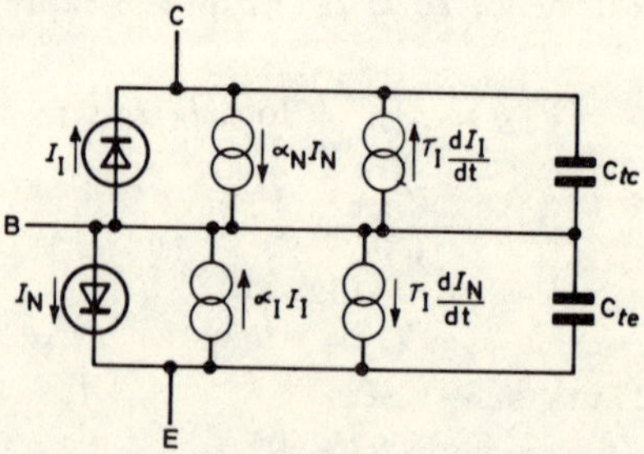

FIG. 4.27 Ebers and Moll transient model

tors than the d.c. model. Their currents depend on the rates of change of currents in the basic diodes. Junction depletion capacitances C_{tc} and C_{te} are included as shown in Fig. 4.27. For high-current modelling, extrinsic resistances must also be included.

Chapter 5

Operational amplifier characteristics

Circuit designers have available to them integrated circuits in addition to the discrete devices described in Chapters 1 and 2. These comprise complete circuits on a single chip housed in a multi-lead package which can be regarded as a single circuit component.

Logic systems are invariably constructed in this way and the study of logic device families and their applications is beyond the scope of the present volume.

Linear circuits for signal processing functions of all kinds are also in common use, and again it is impractical to attempt to cover the whole of this field in the detail necessary to be really useful to the designer.

Probably the most widely used of all linear circuits is the operational amplifier, and the most popular of these is the μA741 and its equivalents SN72741, LM741, etc.

Rather than attempt to deal generally with all linear circuits, detailed attention has, therefore, been concentrated on this chapter on the '741'.

The designer will find that familiarity with the properties and usage of this device leads to the ready understanding of most other linear integrated circuits.

Properties of the μA741 Operational Amplifier

In principle, a typical operational amplifier such as the μA741 is a development of the amplifier circuits described in Chapter 9, e.g. Figs. 9.4 *et seq.* The input stage is a differential pair in which the emitter load is an active circuit giving a high 'tail' resistance. Operating currents are very low so that the input base currents are in the micro-amp region, and several gain stages follow the differential pair, culminating in a low-resistance output stage.

The overall result is (for low frequencies) an amplifier with low-current differential input, very high gain, low output resistance and good isolation from variations of supply voltage. When these properties are required from a circuit, the complexity of designing and making a discrete version is so great as to be impracticable and the μA741 or similar integrated circuit is the obvious choice, especially when its low price and small size are taken into account.

Although these, and related devices, are extremely useful, there are still pitfalls to be avoided and these are discussed in Chapter 12.

Parameters of the μA741

The available connections for a μA741 are shown in Fig. 5.1. The terms '*inverting*' and '*non-inverting*' refer to the voltage polarity relationship between that input connection and the output.

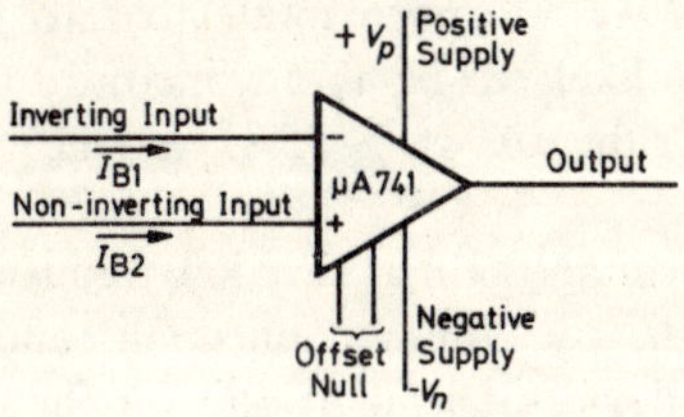

FIG. 5.1

Incidentally it is good practice always to draw operational amplifiers consistently. Changing the upper input connection to 'non-inverting' causes wiring and design errors even though the pin numbering for the particular package used may be correctly designated. Errors can also result from reversing the drawing to show inputs on the right and the output on the left since it is not immediately obvious whether the upper or low input connection is now non-inverting.

The important parameters for operational amplifiers are:—

1 Input offset voltage, v_{os}.
2 Input bias current, I_B.
3 Input offset current, I_{os}.
4 Gain, A_o.
5 Frequency response, F_o.
6 Slew rate, $V_{o/sec}$.

1 The input offset voltage v_{os} is defined as the input voltage required between the input pins to cause the output voltage to reach a level midway between the positive and negative supplies.

In a perfect amplifier this would be zero but small differences between the two V_{be} values of the input emitter coupled pair give values of v_{os} of 0 to 6 mV in either direction for most operational amplifiers. This represents an error of at least this amount in any measurement of a direct input voltage.

2 The input bias current I_B is the current taken by either of the input connections. It varies with temperature since its value depends on the β of the input transistors, and is 0 to 1·5 μA for the μA741 (−55°C to +125°C) in its military version.

3 The input offset current I_{os} is the difference between the two input bias currents; this would approximate to zero in a well balanced input circuit but is likely to be about one-third of the bias current, i.e. 0 to 500 μA for the μA741 (−55°C to +125°C).

4 Gain A_o is the voltage gain at very low frequencies from an input applied between the two input connections and the output. It is temperature- and device-dependent and for the μA741 varies from 25,000 to more than 200,000 (−55°C to +125°C) with no top limit specified.

5 Frequency response is properly defined by the break frequencies associated with successive stages of amplification, but in the case of the μA741 a large dominant lag is obtained by a built-in capacitor. This gives a low-frequency 3-dB point at about 5 Hz, so that at 1 kHz the gain is typically 1000, compared with the typical very-low-frequency gain of 30,000. The remaining lags begin to affect the response only above 1 MHz and can be ignored for most applications of the μA741.

6 The slew rate is the rate of swing at the output in volts per second due to an infinitely fast input sufficiently large to cause the output to reach its limit value. This figure is not calculable from F_o since it depends on available charging currents within the amplifier. The small-signal value of F_o thus corresponds to a much higher slew rate than can be obtained at high level. Slew rate limitations often dictate

which amplifier is required to meet a given specification even though a superficial examination of F_o could appear to yield several satisfactory alternatives.

Using the μA741

The first requirement, as in any amplifier, is to arrange the bias conditions. Fig. 5.2 is an attempt to use the μA741 as a linear amplifier, but it fails on two counts. First, the v_{os} of 0 to ± 6 mV means that with both inputs returned to the same potential (as in Fig. 5.2 if the effect of I_B is ignored) the output can sit at a level of $\pm 6 \text{ mV} \times \text{A}$, i.e. $\geqslant \pm 15$ V. In this state the output is at one or other supply voltage level (or as near as the output circuit will achieve) and input signals would have to reach ± 6 mV to cause the output to move.

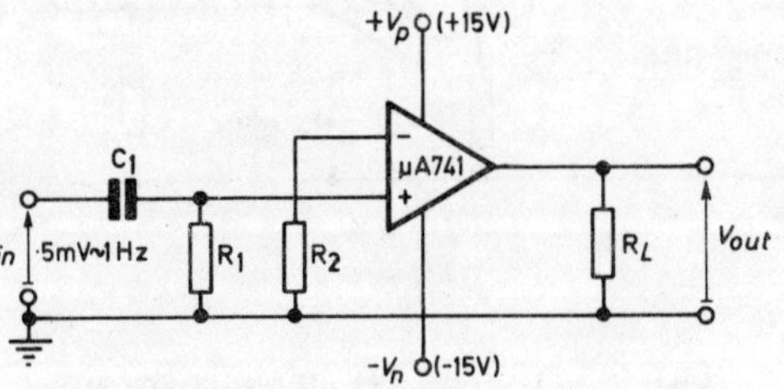

FIG. 5.2 Unsound 741 amplifier circuit

Secondly, even if an ideal specimen of a μA741 has been used so that V_{out} sits near zero, the input signal of 0·5 mV(pk) may attempt to swing the output by $0{\cdot}5 \times 25 = 12{\cdot}5$ V, or by $0{\cdot}5 \times 200 = 125$ V according to the value of A. Since the maximum possible swing is only ± 15 V, the first case will give a passable sine wave output but the second will resemble a square wave since the ± 15 V levels will be reached soon after the input departs from zero.

It is therefore required to bias the device so that the output standing level is comfortably inside the excursion limits. It is also necessary to ensure correct operation whether the gain A be as low as 25,000 or greater than 2,000,000. Correct bias is achieved by negative d.c. feedback and gain stability by negative feedback at the signal frequency (see Chapter 8).

Fig. 5.3 shows acceptable arrangements where the standing output level of V_{out} is held close to zero by feeding it back via R_3. Any departure of V_{out} from 0 V causes a signal to appear at the input in

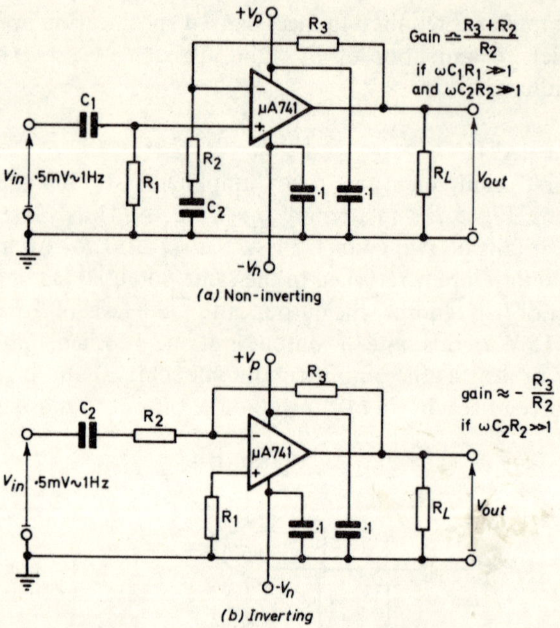

FIG. 5.3 Correct 741 amplifier circuits

the direction to correct the departure. If the input offset voltage v_{os} reaches ± 6 mV, then (neglecting I_B) V_{out} only moves by ± 6 mV. The signal gain V_{out}/V_{in} at a value close to $(R_3+R_2)/R_2$, is also stabilized for Fig. 5.3(*a*) and at $-R_3/R_2$ [Fig. 5.3(*b*)], since this is a negative feedback circuit of the type described in Fig. 8.1.

If operation down to zero frequency is required, C_1 and C_2 may be replaced by a short circuit. In that case the gain remains unchanged but the offset voltage is now amplified to $v_{os}(R_3 + R_2)/R_2$.

Other causes for V_{out} departing from zero are the input bias currents I_{B1} and I_{B2}. To assess their effect in this type of circuit, work out first the potential at the non-inverting input since this is unaffected by the feedback. Here the level is $-R_1I_{B1}$. (For the μA741 the input devices are *n-p-n* with low leakage: therefore the predominant current flow is *into* the input terminal.)

Now assume the inverting input leg will reach the same potential, $\pm v_{os}$. This is equivalent to assuming an open-loop gain sufficiently high that it can be regarded as infinite. This is justified, since the lower limit of gain is 25,000 and with a maximum conceivable output offset of ± 15 V this represents 15/25,000 V, i.e. 0·6 mV at the input, a negligible extra offset.

In the forms given in Fig. 5.3, the inverting input is therefore offset by $-R_1 I_{B1} \pm v_{os}$, and the current in R_2 is known to be zero with C_2 present. Now I_{B2} flows into the inverting input so it must flow through R_3, making V_{out} more positive by $I_{B2}R_3$ than the inverting input's potential of $-R_1 I_{B1} \pm v_{os}$. Therefore

$$V_{out} = I_{B2}R_3 - I_{B1}R_1 \pm v_{os}.$$

Now, $I_{B1} - I_{B2} = \pm I_{os}$, which is always specified for operational amplifiers to be noticeably less than I_{B1} or I_{B2}.

It is therefore advantageous for minimum offset to make $R_3 = R_1$, giving $V_{out} = \pm R_1 I_{os} \pm v_{os}$. In the case where C_2 is replaced by a short-circuit, similar reasoning shows that for minimum offset $R_1 = R_2//R_3$, and for the multiple case in Fig. 5.4, the optimum

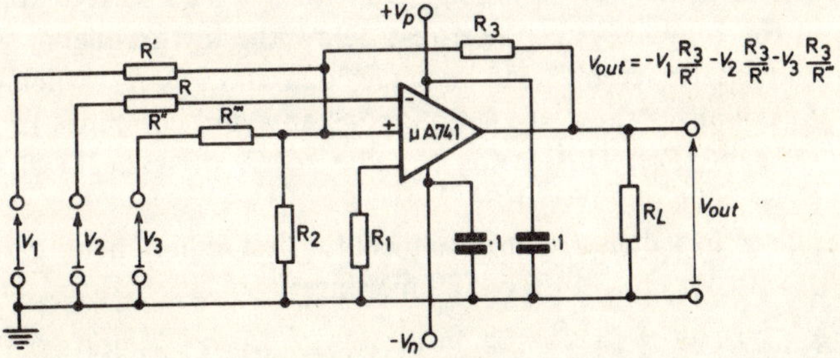

FIG. 5.4

condition is given by $R_1 = R_2//R_3//R_1^1//R_1^{11}//R_1^{111}$ and so on for any number of input resistors, assuming all sources $V_{1,2,3}$, etc., are of zero internal resistance. Note that in this multiple case the effect of v_{os} is multiplied by the gain experienced by voltage v_{os} in series with an input pin, namely

$$\frac{R_3 + R^1//R^{11}//R^{111}//R_2}{R^1//R^{11}//R^{111}//R_2}$$

In some cases V_{out} may have to be kept to a specific level for direct coupling to a following circuit. In others where a.c. coupling follows, it may be sufficient to ensure that adequate signal swing can be obtained at the output before the supply limits begin to affect the performance. In the first case an output offset as low as 20 mV may be required; in the second, 5 V may be quite satisfactory.

As an example, two bias circuits for an approximate gain of × 100 will be derived for the two requirements set out above, i.e. 20 mV or 5 V offset, with a low frequency cut off at 10 Hz.

Assuming the configuration of Fig. 5.3 (*a*) the output offset (see above) will be $\pm V_{OS} \pm I_{OS}R_1$, if $R_1 = R_3$. Now $v_{os} = 6$ mV and $I_{OS} = 0{\cdot}5\,\mu$A (over the whole temperature range -55°C to $+125$°C), so that for a maximum offset of 20 mV, $0{\cdot}5 \times 10^{-6}R_1 = (20 - 6)10^{-3}$. The maximum value of R_1 ($= R_3$) is therefore 28 kΩ, so that a suitable standard value would be 22 kΩ, giving a maximum offset of $(6 + 0{\cdot}5 \times 22)$ mV = 17 mV.

To achieve a gain of 100, $(R_3 + R_2)/R_2 = 100$, giving $R_2 = 22 \times 10^3/99 \simeq 220\ \Omega$.

The cut-off frequency due to C_2 approximates to $1/(2\pi C_2R_2)$, and that due to C_1 is $1/(2\pi C_1R_1)$. One of these may be made to dominate the other, or both may be made significant giving a steeper roll-off slope in the frequency response just below the low-frequency 3 dB point. In the latter case C_1 can be made equal to C_2R_3/R_2, which with $R_3 = R_1$ equalizes the two l.f. time constants and combines to give $f_o = \sqrt{2}/(2\pi C_1R_1)$. This yields a value of C_1 of $\sqrt{2}/(2\pi f_oR_1) = 1\ \mu$F. $C_2 = 100C_1 = 100\ \mu$F.

Therefore the values obtained to meet the first requirement (gain ≈ 100, $f_o = 10$ Hz, V_{out} offset <20 mV) are:

$$R_1 = R_3 = 22\ \text{k}\Omega;\ R_2 = 22\ \Omega,\ C_1 = 1\ \mu\text{F},\ C_1 = 100\ \mu\text{F}.$$

In the second requirement, V_{out} offset is allowed to be 5 V, which is often possible if a.c. coupling follows the amplifier, and it is worth checking whether this circumstance enables C_2 to be omitted. In this case one may be tempted to deduce that the new offset will be $(R_3 + R_2)/R_2$ larger, since this is the gain at zero frequency. It is, however, not quite so simple as this: the statement is true for the effect of v_{os}, but in considering I_{os}, R_1 should equal $R_2//R_3$ (not R_3) for that amount of offset, and if R_1 remains at 22 kΩ, this requires a change in the values of R_2 and R_3 to 22 kΩ and 2·2 MΩ, respectively. The offset due to I_{os} is now $I_{os}\ R_3 = 0{\cdot}5 \times 2{\cdot}2$ MΩ = 1·1 V. Thus the total offset is 6 mV × 100 + 1·1 V = 1·7 V which is well within the 5 V limit specified above.

C_2 can, therefore, be omitted in this case with consequent saving in components and improved l.f. response.

SUMMARY

The use of a μA741 as a low frequency amplifier requires the use of feedback techniques both for arranging the d.c. bias and for defining the gain. The built-in high gain and low input currents enable values to be calculated simply for most applications.

When input currents are too high or the bandwidth too low for the proposed application, there are many other available amplifiers which may fill the need. In particular, the input currents of the μA740 are in the nanoampere region owing to the use of J-FETS at the input.

This chapter has dealt with the simplest of applications of the μA741—the low frequency amplifier.

Chapter 12 illustrates the versatility of this device and its close relatives and also highlights some of the precautions that must be taken.

Chapter 6

Linear-sweep and constant-current circuits

A linear voltage sweep can be produced across a capacitor which is changing its charge. The charging equation for a capacitor is $i = C\mathrm{d}V/\mathrm{d}t$, and for the special case where $\mathrm{d}V/\mathrm{d}t$ is constant, i.e. the voltage sweep waveform is linear, this reduces to $i = CV/t$.

For a constant rate of voltage per second, i.e. a linear sweep, it is therefore necessary to charge the capacitor with a constant current. There are several circuit configurations such as the Miller integrator, constant-current source, bootstrap sweep generator, which produce a linear voltage sweep, and although they are at first sight quite different, they all work on this principle of constant-current charging of a capacitor.

In the first part of this chapter the bootstrap version is described in its simplest form and a typical design procedure is given with a numerical example. The second part describes the constant current circuit, the term usually applied to a circuit which supplies direct current, the magnitude of which is independent of the load into which it flows. This property can in practice hold only over a certain range of load conditions since there will always be a limit to the magnitude of the load voltage before the constant-current source limits.

Sometimes such a source is required to deliver current which is also constant with respect to changes of supply voltage and temperature, and although the same name is given to such a circuit and the configuration is similar, detailed design is more complicated.

THE BOOTSTRAP SWEEP CIRCUIT

Fig. 6.1 illustrates how use of the bootstrap principle can produce a constant charging current for a capacitor C.

Assume initially S_1 rests in the position shown, that C' is very large and that the box marked 'X1' is a unity-gain amplifier having high input impedance and low output impedance.

Initially, B is therefore at zero potential, A is at $VR_2/(R_1 + R_2)$, and the current in R_1 and R_2 is $V/(R_1 + R_2)$.

On opening S_1, the current $V/(R_1 + R_2)$ which had been flowing through S_1 now charges C at an initial rate given by $\mathrm{d}V/\mathrm{d}t = i/C$, that is $V/C(R_1 + R_2)$ V/sec. None of this current passes to the amplifier, since its input impedance is assumed to be very high, and if the amplifier output were not connected through C′ to A, capacitor C would charge exponentially towards $+V$, because as V_c rises the charging current decreases.

With the connection to A made as shown, however, point A rises exactly in step with B, provided C′ is very large. This means that the potential difference V_{AB} is the same as in the original state with S_1 closed, namely $VR_2/(R_1 + R_2)$, so that the current in R_2 remains $V/(R_1 + R_2)$ throughout the sweep and the charging rate for C is constant, being $V/C(R_1 + R_2)$ V/sec.

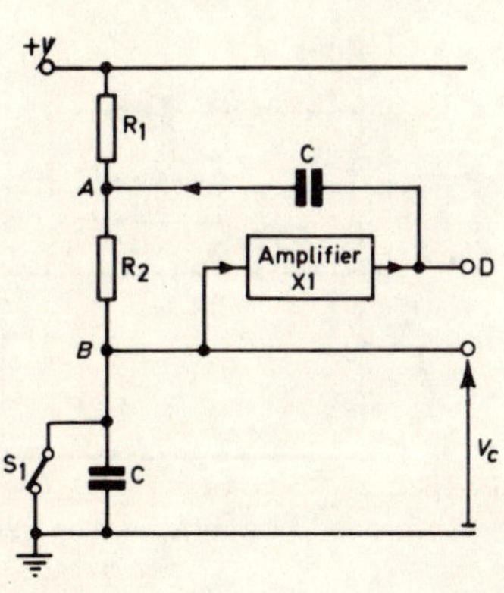

FIG. 6.1 Principle of bootstrap sweep generator

The output voltage waveform V_c is therefore a linear ramp until either S_1 is closed, giving a fast return to zero output, or the amplifier ceases to operate, which will occur in practice when V_c exceeds some fraction of the supply rail voltage of the amplifier.

Before considering practical design, it is important to note one or two points which can become confusing later if not completely understood at this stage.

First, it is quite possible for point A to rise to a voltage higher than $+V$, even though no higher supply rail is used for the amplifier. The source of the energy which produces this effect is the capacitor C′ which has been pre-charged to a voltage $VR_2/(R_1 + R_2)$.

Secondly, it is clear, especially after appreciating the first point, that R_1 is redundant as soon as the sweep process begins. Its only function is to help determine the initial charging current which the bootstrap loop then maintains. When the sweep begins, R_1 constitutes a load on the amplifier through C′, making the design of the amplifier more difficult and increasing the required value of C′. The circuit is therefore improved if R_1 is replaced by a diode (Fig. 6.2) which cuts off when A rises.

Thirdly, the linearity can be spoilt in three main ways. (1) The amplifier may modify the charging current received by C if its input impedance is too low. (2) C′ may be inadequate, the effect being that the proportion of voltage lost across C′ (by discharge through $R_1//R_2$) during the sweep, compared with the initial voltage on R_2 ($VR_2/(R_1 + R_2)$) will produce that same proportion of nonlinearity.

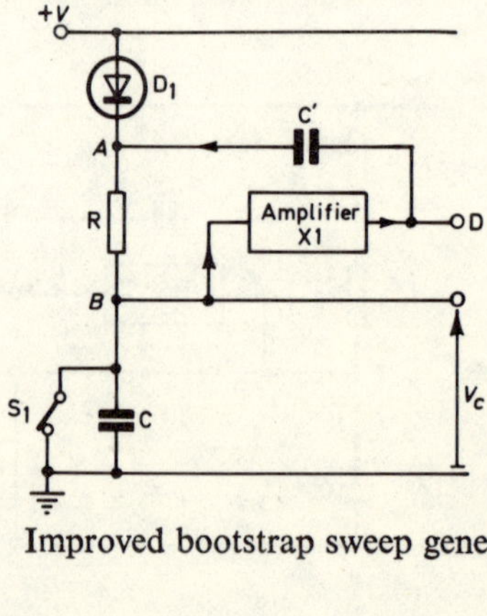

FIG. 6.2 Improved bootstrap sweep generator

(3) The amplifier gain may not be held closely to unity throughout

the sweep, the amount of departure from unity producing the same proportion of non-linearity.

Finally, the output can with advantage be taken from point D since the waveform is similar to that across C, and the load will not then affect significantly the charging current for C.

Bearing the above points in mind will make the design of the practical circuit of Fig. 6.3 easy to understand.

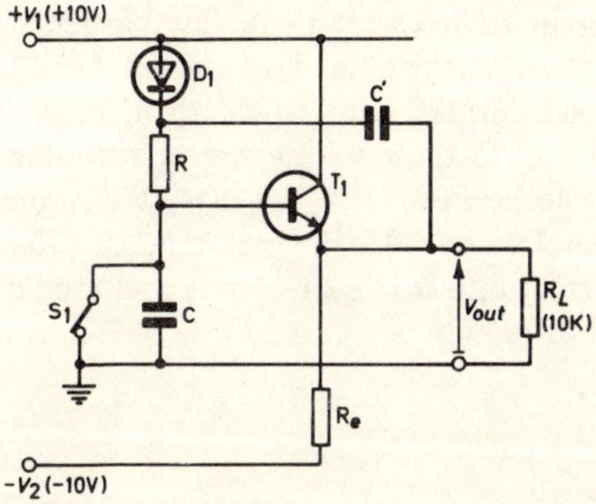

FIG. 6.3 Practical bootstrap circuit

Assume for example that a sweep output is required, after opening S_1, having a rate of 5 V/msec, into a 10 kΩ load from + and − 10 V rails. Typical values will be calculated and the expected linearity assessed using a single emitter-follower T_1 as shown.

Since the quiescent output level is just below zero volts and the maximum possible output will be +10 V (T_1 will then be bottomed) the maximum load current is 1 mA, reached at the end of the sweep.

Effect and Choice of R_E

Resistor R_E can be omitted completely; this means that while C charges for the first few hundred millivolts T_1 remains cut off and V_{out} remains at zero. When V_C has reached the turn-on voltage for T_1, conduction begins and V_{out} follows the sweeping waveform on C. There are two snags in this action: there is a delay between operation of S and commencement of V_{out}; and the transitional period as T_1 is beginning to conduct will imply a high output impedance from T_1, so that the bootstrap loop will have noticeably less than unity gain, giving non-linearity near the start. In many applications these may be unimportant defects and R_E and the -10 V line can be omitted.

Assuming these effects are to be avoided, R_E should provide a standing current of about the same order of magnitude as the maximum load current, so that when the output rises, the change of current in T_1 (due to the load current) is not large. It is unfortunate that the currents in R_E and R_L both increase as the output rises causing parameter variations in T_1 which change its gain. If the negative supply were many times the value given, or if a constant-current device were used, this particular effect could be overcome by making the standing current for T_1 many times the maximum load current. This would then, however, cause further difficulties owing to the resulting rise in base current leading to a vicious circle to be solved only by the use of more transistors.

R_E will therefore be given the (non-critical) value of 10 kΩ, giving an initial current into $R_E//R_L$ of 1 mA and a final current of 3 mA when V_{out} is at $+10$ V.

Choice of T_1

A tentative specification for the transistor T_1 must now be assumed before the design can proceed. The designer may have a particular type in mind, or may be prepared to search for one which is ideal for the purpose, depending mainly on how critical is the specification for the performance of the circuit.

For the purposes of this example, a non-linearity of 1 per cent has been taken as the tolerable limit, and the experienced designer will immediately realize that a high-gain transistor will be required. For the moment, a minimum β of 50 will be specified over the range of operating conditions which apply throughout the sweep.

T_1 may be silicon or germanium, the important differences in this circuit being the higher V_{be} for silicon, so that the output swings from about $-0{\cdot}7$ V to $+9{\cdot}3$ V, rather than $-0{\cdot}2$ to $+9{\cdot}8$ V, and the leakage of germanium which robs C of some of the current in R. Since I_{cbo} varies greatly with temperature, the use of germanium for T_1 will cause drift in the rate of sweep with temperature, the magnitude of which depends on the value of I_{cbo} compared with the charging current through R, which is not yet decided.

The voltage rating for T_1 is 10 V for BV_{ceo}; the power rating cannot be decided until R is chosen.

Choice of *R*

This component is the only one in the circuit which causes difficulty, and this is because almost any value would seem to be satisfactory. How is the designer to know whether a charging current of 1 mA or 50 mA would be preferable?

In such cases the possible values should be taken to opposite extremes in order to arrive at more reasonable limits. Here, a charging current of 1 μA (giving $R = 10$ MΩ) can be taken as one extreme and 1 A ($R = 10$ Ω) as the other. The consequences of these clearly wrong values can be evaluated and from the result it will be obvious which values are really possible.

If $R = 10$ MΩ, the charging current is not in fact 1 μA, but is 1 μA minus the base current of T_1. From the (reasonable) assumption that β is always at least 50, this base current will be at most 1/50 mA at the beginning of the sweep and 3/50 mA at the end, assuming a sweep to +10 does take place. It is clear that no sweep occurs, because the charging current is actually negative and C will charge negatively with a current starting at 1 μA minus 1/50 mA. *R* cannot therefore have this high value and in fact must be low enough to ensure that the current it passes greatly exceeds 1/50 mA.

On the other hand, if $R = 10$ Ω, charging current is about 1 A, as intended, but it must not be forgotten that the transistor emitter current, which varies from 1 to 3 mA due to R_E and R_L, includes the current in R as soon as D_1 cuts off, i.e. after a few hundred millivolts of sweep. Base current is therefore 1/50 (1 mA + 1 A) at the beginning and 1/50 (3 mA + 1 A) at the end, excepting for the first few hundred millivolts, when it is 1/50 (1 mA). This small value of *R* therefore produces a large enough current to swamp the base currents caused by R_E and R_L, but there is a sudden step in current when D_1 cuts off when the charging current changes from 1 A to $\left(1 - \frac{1}{50}\right)$ A. To avoid this effect R_E could be reduced to give a comparable current to that in *R* but then the original difficulty returns because R_E current varies during the sweep. Another snag in using a low value for *R* is, of course, the power rating required of the transistor (about 5 W) and the drain on the power supply.

The compromise is easy to make: *R* must be small enough to avoid base current variations of 1/50–3/50 mA during the sweep causing more than the allowed non-linearity (note that base current caused by *R* itself, namely $V/50R$ is constant throughout the sweep and causes no non-linearity); R must be large enough to avoid unnecessarily high dissipation in the transistor.

In the example under consideration, a value of 10 kΩ for R gives a nominal charging current of 1 mA. Base current is at most 1/50 mA initially, changes to 2/50 mA when D_1 cuts off, and ends at 4/50 mA at +10 V output. Charging current therefore varies from $\left(1 - \frac{1}{50}\right)$ mA to $\left(1 - \frac{4}{50}\right)$ mA, a variation of 6 per cent. If the first phase is ignored, variation is 4 per cent.

A value for R of 2·2 kΩ gives nominal charging current of 10/2·2 = 4·5 mA. Base current begins at 1/50 mA, changes to (1/50 + 4·5/50) = 5·5/50 mA when D_1 cuts off, and ends at (3/50 + 4·5/50) = 7·5/50 mA at full output. Total variation in charging current is therefore 6·5/50 parts in 4·5, i.e. 3 per cent, or ignoring the first phase, 2/50 in 4·5, or 0·9 per cent.

This circuit in its simple form always suffers from this initial non-linearity until D_1 cuts off, as is shown by the above calculations. Since this phase occurs only for a few hundred millivolts, however, it scarcely affects the overall linearity figure and the second figure for each case above is the relevant one. The value of 2·2 kΩ appears to be satisfactory, and although it would be tempting to make R smaller still, to meet the linearity requirement easily, discretion must be used: the letter of the specification may well be met but the 'kinky' appearance of the initial rise may be important to the user (though not specified) and dissipation will be raised without real justification.

Choice of D_1 and C

The charging rate for C depends slightly on the forward drop of D_1, and so the type of diode should be specified before calculating C. The dependence of the charging rate on D_1 is so small that for practical purposes it is only necessary to know whether D_1 is to be silicon or germanium, giving a typical voltage drop of 0·5 V or 0·1 V, respectively. The reverse leakage of D_1 has little effect on performance and merely adds slightly to the current which tends to discharge C′ during the sweep (which mainly consists of the current in R).

Assuming a germanium type is to be used (almost any signal diode is satisfactory, such as OA10, HG1005), a nominal allowance of 0·1 V drop can be made in calculating the charging current for C, giving $i = 9{\cdot}9/2{\cdot}2 = 4{\cdot}5$ mA. To obtain the required sweep rate of 5 Vm/sec, C must therefore be

$$\frac{4{\cdot}5 \times 10^{-3} \times 10^{-3}}{5} \text{ F}$$

or 0·9 μF.

Choice of C'

C' must now be determined, and since it has been assumed that linearity is important, it will be allowed to degrade linearity by only $\frac{1}{8}$th per cent. The effect of C' in practice is rather different from that of changing base current in T_1; both cause the same kind of non-linearity because the charging current for C is made to fall during the sweep; but whereas a 'worst-case' design for T_1, i.e. assuming its gain to be minimum, usually is pessimistic, the calculation for C' will be correct and the designer has no right to expect practical performance for a given value of C' to be better than predicted. The current for C′ is, as above, 4·5 mA. The voltage across C′ during the sweep must fall by no more than 1/8 per cent of the voltage across R, i.e. (1/800) 9·85 V. The total sweep time is about 2 msec (10 V at 5 V/msec), so that, using $i = C' \, \mathrm{d}V/\mathrm{d}t$,

$$4{\cdot}5 \times 10^{-3} = C' \frac{9{\cdot}85}{800} \times \frac{10^3}{2}$$

giving $C' = 730$ μF. In view of the wide tolerances of electrolytic capacitors C' must be given a nominal value of 1000 μF.

Tolerance and Temperature Effects

The design is now complete except for calculation of tolerances and temperature effects. Because of the simplicity of the charging equation for C, errors in the +10 V line, and in R and C all produce the same percentage errors in sweep rate. Temperature effects on these same parameters give sweep rate errors in the same way, and the only effects not so far included are the change in the forward drop of D_1, the β of T_1 and the V_{eb} of T_1. The first changes the current in R calculated from $(10 - V_F)/R$, knowing that V_F changes at −2·5 mV/degC at worst. The second causes linearity to improve and sweep rate to increase with temperature because base current is reduced. The third gives a d.c. output change of +2·5 mV/degC (positive because the base voltage is unchanged and the emitter–base voltage decreases as temperature rises).

Summarizing, the suggested design has a tolerance on rate of sweep of 7 parts in 250 plus the effects, in proportion, of +10 V variation and *R* and *C* tolerance. Linearity is better than 3/8 per cent. Provided transistor β over the range 5 to 7 mA exceeds 50 at the lowest operating temperature, the only detrimental effect of temperature change is a d.c. output voltage change of at worst +2·5 mV/degC.

The sweep rate tolerance (2·8 per cent even with perfect positive line, *C* and *R*) is too great for many applications but this can be improved in this simple circuit only by a higher-β transistor or a decrease in charging current, leading to worse linearity.

A much-improved circuit results if T_1 is replaced by the complementary emitter follower circuit described in Chapter 10. Circuit design is the same as above but β becomes $\beta_1\beta_2$, thus greatly reducing the effect of amplifier current on the charging current for C.

Practical Problems

In making practical use of the circuit just designed, it would normally be necessary to use an electronic switch for S_1, such as a transistor driven between saturation and cut-off. Depending upon the rate at which this switch is operated, the output is then a continuous sawtooth waveform or a succession of sawteeth each separated by a waiting period. Both waveforms are often required for time-base generation for oscilloscopes, and the operation of electronic switch S_1 to its closed condition, which terminates the sweep, is derived partly from trigger circuits operated by the signal being examined and partly from the waveform V_{out} itself.

Design of such a loop becomes complex when many different time-base frequencies have to be generated, especially when the required speeds approach the limit for available transistors.

A full appreciation of the problems would require detailed discussion of all the elements comprising the loop, but some insight can be obtained by noting the following points.

The design procedure above was concerned only with the generation of the ascending linear sweep and T_1 was taken to be *n–p–n* in order to illustrate the way in which the cathode of D_1 is driven more positive than any supply rail. If T_1 is *p–n–p*, then when V_{out} reaches +10 V, it is evident that a more positive line has to be available merely to supply load current, thus masking the above phenomenon.

When the closure of S_1 is considered, at the end of the sweep, all appears to be well provided that it is realized that some small series resistance in S_1 and C is inevitable. C discharges rapidly to earth (infinitely rapidly and with infinite current if no resistance is assumed) and V_{out} follows.

In practice R_L will always have in parallel with it some stray capacitance C_S and when the voltage across C descends to zero in a very short time, V_{out} can follow at the same rate only if current is available from some source which is sufficient to charge C_S at the high rate (dV/dt) required. In Fig. 6.3 the only sources which can charge C_S towards zero are the paths through R_E and R_L, since any transistor emitter current, other than the leakage, will flow in the opposite direction and can only charge C_S positively.

It is clear that if the sweep rates involved are such that C becomes as small as a few hundred picofarads, then typical load strays which are rarely less than 5 pF will cause the flyback time (i.e. V_{out} returning to zero) to be an appreciable proportion of the sweep. Reducing R_E enables faster discharge of C_S to take place but degrades the linearity of the sweep as shown in the initial design.

The root of the difficulty is that T_1 is unable to pass current in the right direction to discharge C rapidly and a *p–n–p* type would be more appropriate (*see* Fig. 6.4). Here the situation is reversed: R_E must supply the charging current for C and the load current. Maximum total current is at +10 V when R takes 4·5 mA (as always) and R_L takes 1 mA. The voltage across R_E at this time is least, being (V_2 − 10) or 10 V if V_2 = +20 V. R_E must therefore be at most 10/5·5 = 1·8 kΩ. This leaves no margin for tolerance and also leaves

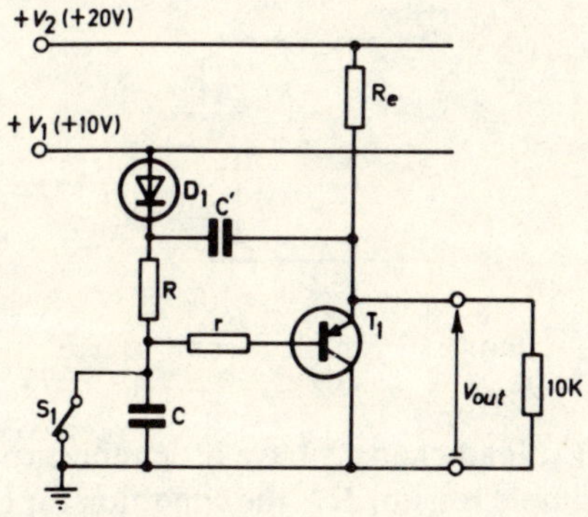

FIG. 6.4 Bootstrap circuit, *p–n–p* version

T_1 just cut off when V_{out} reaches $+10$ V. A safe value is 1·2 kΩ.

When S_1 closes and C discharges, C_S is now charged by as much current as T_1 emitter can supply. Since T_1 base has dropped by 10 V, the current which could flow if T_1 emitter did not follow would be virtually unlimited; C_S therefore discharges to zero virtually as quickly as C.

General Application of the Above Results

The calculations required in the design of the above circuits are typical of many non-linear circuits in that the main consideration is the provision of adequate currents to obtain the required voltage swings. Several spurious effects combine in an attempt to frustrate the designer by adding unwanted voltage drops, stray capacitance, and temperature drifts.

Knowing the semiconductor properties outlined in the first and second chapters, only Ohm's law and the charging equation $i = C\mathrm{d}V/\mathrm{d}t$ are normally required to complete these designs satisfactorily, provided the function of the circuit is clearly understood.

THE CONSTANT CURRENT CIRCUIT

The simplest circuit to produce constant current is a voltage source in series with a resistor. The current in the load is then given by $(V_1 - V_L)/R_1$ (*see* Fig. 6.5), so that, provided V_L is always small

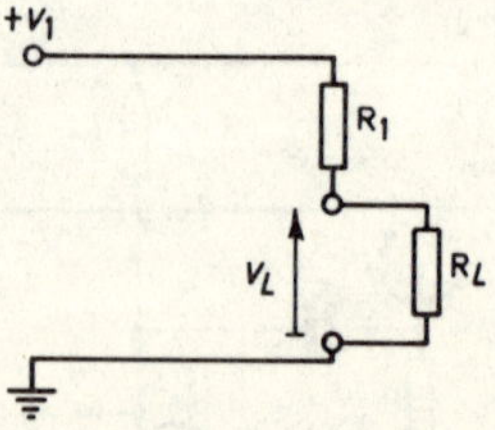

FIG. 6.5 Simple current source

compared with V_1, load changes have little influence on the current. If the load is a simple resistor R_L, the condition for constant current is $R_L \ll R_1$; if the load is a capacitor, the current remains constant until the capacitor voltage becomes comparable with V_1. The departure from constant current is in fact $100V_L/V_1$ per cent.

In many applications V_L is required to be several volts in magnitude and the current is to be held to within a few per cent. This can only be achieved in the simple circuit of Fig. 6.5 if V_1 is a few hundred volts, which is often inconvenient.

Figure 6.6 shows how the use of a transistor solves this problem even with low supply voltages. In this circuit T_1 base is held at a potential $V_1R_3/(R_2 + R_3)$ and T_1 emitter will be a few hundred millivolts more positive. The current in R_1 is therefore approximately $(V_1/R_1)[1 - R_3/(R_2 + R_3)]$, i.e. $V_1R_2/[R_1(R_2 + R_3)]$. T_1 collector current and, hence, the load current are therefore $V_1R_2/[R_1(R_2 + R_3)]$, provided T_1 is not saturated, i.e. provided V_L does not exceed $V_1R_3/(R_2 + R_3)$.

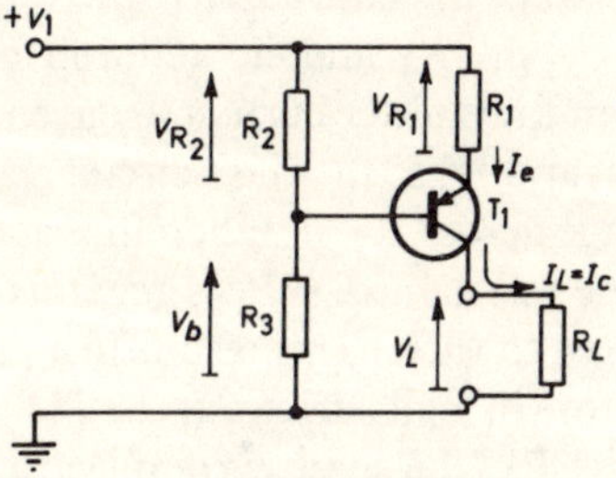

FIG. 6.6 Constant-current source

For different loads obeying this condition the collector voltage of T_1 varies, but this causes only a very small change in collector current. In fact, T_1 behaves like a source having resistance β/h_{oe} if R_1 is large, tending to $1/h_{oe}$ when R_1 is zero. Since β/h_{oe} is typically 1 MΩ and T_1 can pass several milliamps, the simple non-transistor circuit would require a supply of a few kilovolts to equal this performance.

Design of Constant-current Device

Circuit values are dictated by the available supply voltage V_1 and the maximum load voltage for which current is to remain constant. When possible, as in cases where $V_{L(max.)}$ is much smaller than V_1, but not quite small enough to allow a simple resistor supply to be used, the voltage on T_1 base relative to 'earth' should be as small as possible, i.e. just larger than $V_{L(max.)}$.

This leads to the highest possible value for R_1, giving the highest effective source resistance from T_1 into the load; and also makes the actual current less dependent on transistor V_{eb} variations, since these are a smaller proportion of the voltage across R_1 which determines the current.

Normal considerations for biasing a transistor must also be observed, in particular the base current of T_1 multiplied by $R_2//R_3$ must produce negligible voltage drop compared with the voltage across R_2 (or R_1). If this is not observed in design, the load current will be less than intended and will vary appreciably for different transistors and with temperature.

Temperature drift, again as in any normal bias circuit, is caused by V_{eb} changes and β and I_{co} changes. V_{eb} drift causes the voltage across R_1 to change by $+2{\cdot}5$ mV/degC even if the base voltage remains constant. β and I_{co} change the base current and move the base by this change multiplied by $R_2//R_3$ and in addition cause the collector current to differ from the emitter current as given by $I_c = \alpha I_e + I_{co}$.

Supply voltage variations in this circuit cause proportional changes in the voltage across R_2 and therefore across R_1. (This may be easier to see if the rail named $+V_1$ is regarded as stationary and the 'earth' line is taken to vary.) This results in a proportional change in emitter and therefore in load current. If the V_{eb} of T_1 is comparable with the drop across R_1, the load current will change by a larger percentage than line changes.

Typical Design

Assume V_1 is 20 V and a constant current of 1 mA $\pm$ 5 per cent is required for a load which may have any value from zero to 5 kΩ. Supply voltage V_1 is to be assumed constant.

A simple resistive supply is clearly inadequate since, if the resistor is chosen to give 1 mA into zero ohms, i.e. 20 kΩ, it will deliver only 4/5 mA into a 5 kΩ load. The necessary criterion for the success of this circuit, that the maximum load voltage (5 V) shall be only 5 per cent of the supply voltage, is not met (the percentage being 25).

The circuit of Fig. 6.6 will therefore be used, and since $V_{L(max.)}$ is $+5$ V, the voltage across R_3 will be made slightly greater, e.g. $+8$ V.

The load current is 1 mA, and if a transistor is used having a minimum β of 25, maximum base current will be 1/25 mA.

Values for R_2 and R_3 can now be calculated: the ratio $R_3/(R_2 + R_3) = 8/20$, and if the base current (1/25 mA) is allowed to cause a change of, e.g., $\frac{1}{2}$ per cent of the voltage across R$_2$ (0·06 V) the parallel resistance R_2 and R_3 is given by $R_2R_3/(R_2 + R_3) \leqslant 0{\cdot}06 \times 25$ kΩ. This suggests

$$R_2 \times 8/20 \leqslant 0{\cdot}06 \times 25\ \text{k}\Omega$$

i.e.
$$R_2 \leqslant 3{\cdot}75\ \text{k}\Omega$$

and
$$R_3/R_2 = 8/12 = 2/3$$

Now comes a form of juggling peculiar to the circuit designer, made necessary by the 'standard value' system of resistor manufacture. By use of a slide rule it is easy to find standard values for R_2 and R_3 which are near to the ratio 2/3, but if this nearness is in error by more than a few per cent the voltage across R$_2$ will no longer be 12 V and this would mean that R_1 (given by $(V_{R1} - V_{eb})/(1\ \text{mA})$, or nearly $V_{R1}/1$ mA) could not be given the convenient standard value of 12 kΩ.

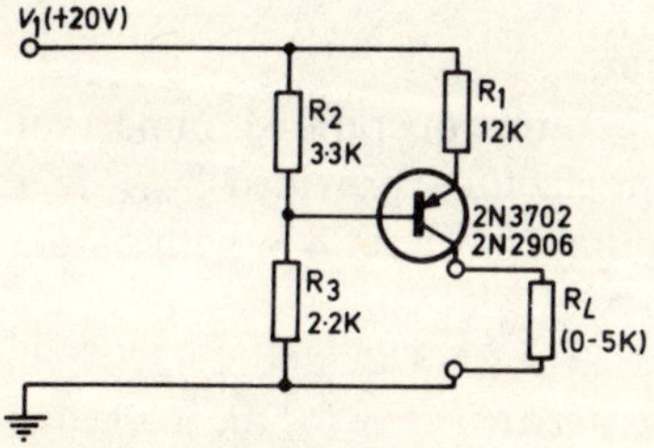

FIG. 6.7 Typical design of constant-current source

A careful study of the table of 5 per cent standard values (and resistors of this tolerance or better would have to be used) together with a slide rule search of 2/3 yields $R_2 = 3{\cdot}3$ kΩ, $R_3 = 2{\cdot}2$ kΩ. This gives the desired ratio, so that $V_b = +8$ V and $V_{R1} = 12 - V_{be} \approx 12$ V; hence, $R_1 = 12$ kΩ (see Fig. 6.7).

Having now designed the circuit, the designer must calculate its performance with respect to supply and temperature changes, even if this is not called for in the specification. This practice is always desirable, as it immediately reveals the shortcomings of a bad circuit, or bad choice of values, which can lead to vast performance changes for small supply or ambient temperature variations.

As indicated previously, a change of, e.g., 10 per cent in the +20 line causes a change in output current of 10 per cent, i.e. 0·1 mA in this design. A temperature rise of 10 degrees reduces V_{be} by 25 mV, thus increasing the current in R_1 by $25 \times 10^{-3}/R_1 = 25/12\ \mu A = 2\ \mu A$, thus increasing load current by 2 μA. β can rise by 2 per cent per degree C, giving a change of β from 25 to 30 for 10 degrees rise. This causes V_{R2} to rise by $(1/25 - 1/30)R_2//R_3$, where R_2 and R_3 are in kilohms, giving $1/150 \times 1{\cdot}32$ V = 8·8 mV. This in turn increases V_{R1} by the same amount, and I_e by $8{\cdot}8 \times 10^{-3}/R_1 = 0{\cdot}7\ \mu A$.

For the sake of example, it will be assumed that I_{cbo} rises by 0·5 μA. Then V_{R2} rises by $0{\cdot}5 \times 3{\cdot}3 \times 2{\cdot}2/5{\cdot}5 = 0{\cdot}66$ mV, increasing the current in R_1 by $0{\cdot}66/12 = 0{\cdot}05\ \mu A$. In addition, the load current is directly increased by 0·5 μA, giving a total increase of 0·55 μA due to I_{cbo}. Summarizing temperature drift, the output increases 2 μA due to V_{be}, 0·7 μA due to β for 10 degC rise, and typically 0·55 μA due to I_{cbo}.

Choice of Transistor

Unless high-temperature operation is important, most *p–n–p* small-signal types are satisfactory, provided $\beta_{min.}$ is at least 25 at 1 mA and that its V_{ce} rating is at least 8 V with base circuit resistance of 1·3 kΩ.

Suitable types are the 2N3702 and 2N2906.

If the supply line were negative (or the load connected to the supply instead of to earth), the circuit could be inverted and an *n–p–n* type used; suitable types are the 2N930 and BC108.

Stabilized Current Source

Although temperature-stability is adequate, the output of the circuit of Fig. 6.6 is no more stable than the supply, $+V_1$. A simple method to overcome this defect is shown in Fig. 6.8, where R_2 is replaced by Zener diode ZD_1. To understand the action of this circuit, assume ZD_1 is a perfect Zener diode, that is it behaves like a zero resistance battery. Then the voltage across R_1 is almost equal to the Zener voltage V_Z, provided this is much larger than V_{be}, and so the load current will be V_Z/R_1, which is independent of V_1.

This will not be strictly true with an actual Zener diode having series resistance R_Z, and in such a case the effect of changing V_1 is easily seen if the V_1 line is taken to be fixed while the 'earth' connection moves by a fraction of V_1. Any change will appear at T_1 base reduced by the factor $(R_3' + R_Z)/R_Z$ and this will then change I_E and I_L as before. The improvement is given by comparing $(R_3 + R_2)/R_2$ with $(R_3' + R_Z)/R_Z$, giving a stability against supply changes in the new circuit which is better by a ratio of

$$\frac{(R_3' + R_Z)R_2}{(R_3 + R_2)R_Z}$$

over the original.

A typical value of R_Z for a small Zener diode of 12 V nominal V_Z,

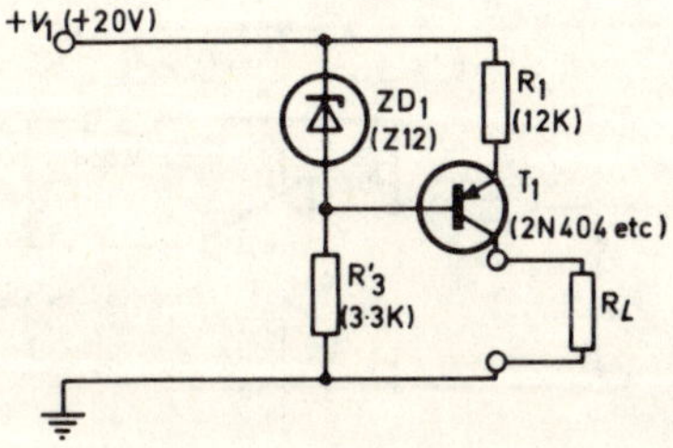

FIG. 6.8 Stable current source

run at 5 mA would be 20 Ω and R_3' would be 6/5 = 1·2 kΩ. The improvement ratio is therefore

$$\frac{(1200 + 20)}{5500}\frac{3300}{20},$$

i.e. 36·6 to 1. A 10 per cent increase in V_1 will now produce about 3 μA change in load current instead of 100 μA.

The circuit of Fig. 6.8 has another virtue: the effective value of R_2 is now a few tens of ohms instead of 3·3 kΩ. This reduces the effects of base current variation because less voltage change results.

These improvements are to some extent offset by the additional temperature drift caused by the Zener diode. Its temperature coefficient is about +0·07 per cent per degree C, so a 10 degC rise gives 0·7 per cent rise in V_Z and, hence, in load current. If necessary, this could be improved by adding one or two forward-biased diodes in series with ZD_1 and using a 10 or 11 V Zener diode. However, although the temperature coefficient is improved, the absolute value of load current would be less certain.

Operational Amplifier Constant-current Circuits

Where greater precision in either linearity or absolute accuracy is needed advantage can be taken of the high gain and low offset voltages and currents available from integrated circuit operational amplifiers.

Constant Current into a 'Floating' Load

If the load need not be joined to signal common, as is often the case with a pen recorder galvanometer or a panel milliammeter, the simple circuit of Fig. 6.9 can be used. In this circuit a perfect operational amplifier would deliver a current into the load given by

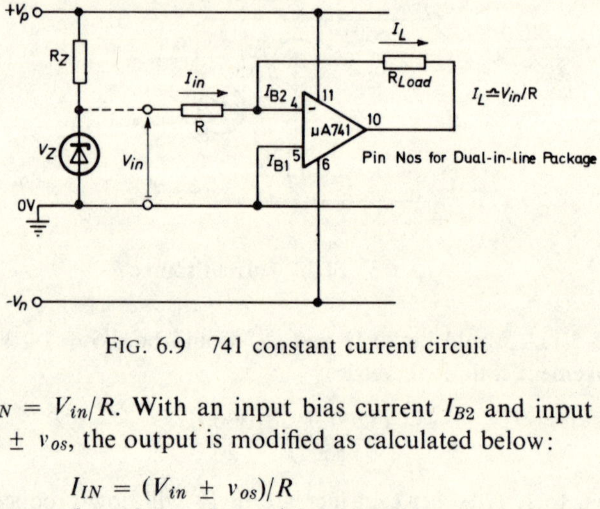

FIG. 6.9 741 constant current circuit

$I_L = I_{IN} = V_{in}/R$. With an input bias current I_{B2} and input offset voltage $\pm\ v_{os}$, the output is modified as calculated below:

$$I_{IN} = (V_{in} \pm v_{os})/R$$
$$I_L = I_{IN} - I_{B2} = (V_{in} \pm v_{os})/R - I_{B2}$$

The input bias current I_{B2} will therefore cause an error in I_L of $(I_{B2}/I_1)100\%$.

The error due to v_{os} is equal to the percentage that v_{os} is of V_{in} and may be of either polarity.

Using this circuit the specification previously considered, namely the definition of a current of 1 mA into a load of 0 to 5 kΩ, is easily met by using a μA741 and taking $V_p = V_n = 15$ V, $R = 15$ kΩ and V_{in} as a fixed voltage of 5·1 V derived as shown in Fig. 6.11. Apart from errors in V_{in} and R, variations over the temperature range and with long term drift are, as indicated earlier $\pm(v_{os}/V_{in})100\%$ due to offset voltage and $(I_{B2}/10^{-3})100\%$ due to input bias current.

For a 'worst case' μA741 these are $\pm 6/51 = \pm 0{\cdot}01\%$ and $[(1{\cdot}5 \times 10^{-6})/10^{-3}]100 = 0{\cdot}15\%$, respectively. Thus the only significant errors are those caused by the Zener diode and resistors.

The input resistance of the circuit of Fig. 6.9 is very nearly equal to R, so that the current taken from the source is equal to the load current. When V_{in} is a low-power signal rather than a fixed Zener diode source, a high input resistance circuit may be required, and the alternative circuit of Fig. 6.10 may then be used. The input resistance here is R_1 in parallel with the very high resistance of the μA741 complete with its feedback (many megohms); the load current is in the opposite sense to the version of Fig. 6.9.

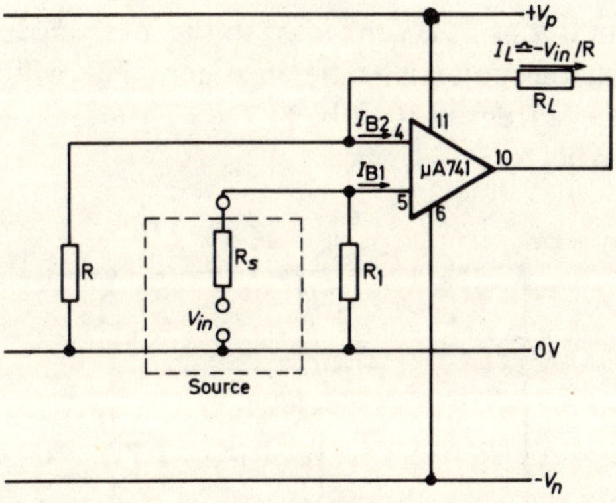

FIG. 6.10 High input resistance version of Fig. 6.9

The value assigned to R_1 depends on the application and upon the source resistance. If the source is directly coupled and is always connected whenever the μA741 is energized then R_1 may be omitted: the error resulting from input bias current is then a voltage $-I_{B1}R_s$ at the non-inverting input. The inverting input will reach the same voltage, $\pm v_{os}$, giving a resulting current into R and in the load of $(+I_{B1}R_s \pm v_{os})/R$. The input bias current to the inverting input, I_{B2} causes another error of $-I_{B2}$ into the load. For the case where $R_s = R$, the offset error due to the input bias currents cancels to zero if $I_{B1} = I_{B2}$, i.e. $I_{os} = 0$, regardless of the value of R_L.

This result is in contrast to the use of the same circuit to produce a *voltage* output from the amplifier in which case I_{B1} and I_{B2} cancel only if $R_s = R//R_L$.

In circumstances where the source may at times be disconnected and where the load current is then to approximate to zero, R_1 is essential. Its presence with the source connected then imposes an attenuation on the source voltage so that, with a perfect μA741, $I_L = -V_{in}R_1/[(R_1 + R_s)]R$ instead of simply $-V_{in}/R$. A high value of R_1, however, gives a greater offset error when the source is disconnected so a compromise must be sought.

Constant Current for Earthed Load

Often the floating-load connection of Fig. 6.9 is inconvenient and one of the load terminals is to be earthed. In such cases the circuits in Fig. 6.11 may be satisfactory. These are shown with V_{in} replaced by a Zener reference V_z giving a defined current to the load. In both circuits the emitter of T_1 is connected to the inverting terminal of the μA741 so that the potential difference across R will be $V_z \pm v_{os}$. This forces the current in R to be $(V_z \pm v_{os})/R$ and the majority of this flows into R_L.

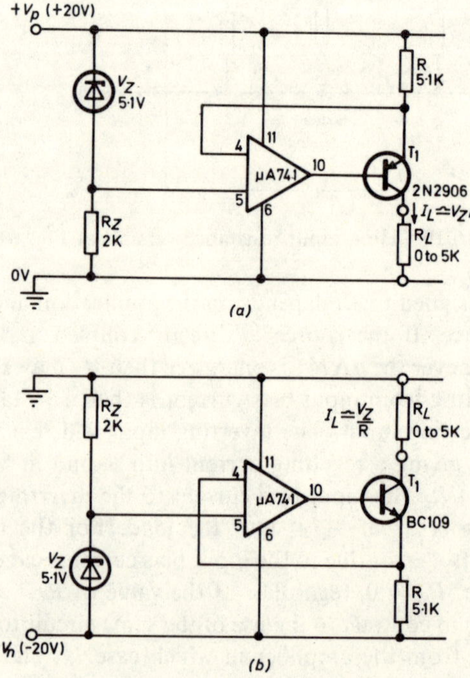

FIG. 6.11 741 constant current circuits for earthed loads

As in the previous circuit there are errors of $\pm(v_{os}/V_z)100\%$ and $(I_{B1}/I_L)100\%$ due to the μA741 parameters. There is an additional error due to the finite h_{FE} of T_1 which causes a loss in load current of $(100/h_{FE})\%$. This would be satisfactory for the present specification: if better performance is required, then T_1 should be replaced either by a Darlington-connected pair or by a J-FET. In the latter solution care must be taken that the gate can be driven adequately negative. For instance, if T_1 in Fig. 6.11 (*b*) were simply replaced by an N-channel device 2N3823, its source would be at 5·1 V above $-V_n$ and at a drain-source current of 1 mA the gate-source voltage may need to reach −6 V. As drawn the μA741 cannot provide this drive and its negative supply terminal would need to be taken to a more negative line such as −25 V.

If this is done, some precaution must be taken against the possibility of the −20 V supply being present with the −25 V absent. This could occur at switch-on or switch-off if these supplies are separate. Even if the −20 V line is derived from the −25 V, any decoupling capacitor attached to the −20 V circuit could allow it to remain after switching off the −25 V supply. The result would be a voltage on the non-inverting input more negative than the supply then present on the negative supply terminal, causing damage to the μA741 (see Chapter 12). A current-limiting resistor in series with the non-inverting input lead (about 4·7 kΩ) will give adequate protection.

Another configuration for constant-current drive is shown in Fig. 6.12, enabling an *n–p–n* transistor or an N-channel J-FET to be used to drive positively into an earthed load. There is a pitfall to be avoided here. The principle of operation is that the non-inverting input will be at a potential V_Z below V_p when the loop has settled (if this did not happen, a large output from the μA741 would result, turning T_1 further on or off as required). The current in R is therefore V_Z/R and this current must flow through the load R_L, the μA741 output adjusting itself to whatever potential gives this result.

The principle is correct but in the case where R_L is less than R the transistor stage has gain between base and collector approximating to $-R/R_L$. The μA741 is guaranteed stable only with unity (or less) feedback. As R_L is reduced towards its lower limit, oscillation is practically certain. Moreover the μA741 has to reach a low output level approaching 0 V which it cannot achieve with its negative supply also at 0 V.

The remedy, shown in Fig. 6.12 (*b*) ,is to add a resistor equal to R

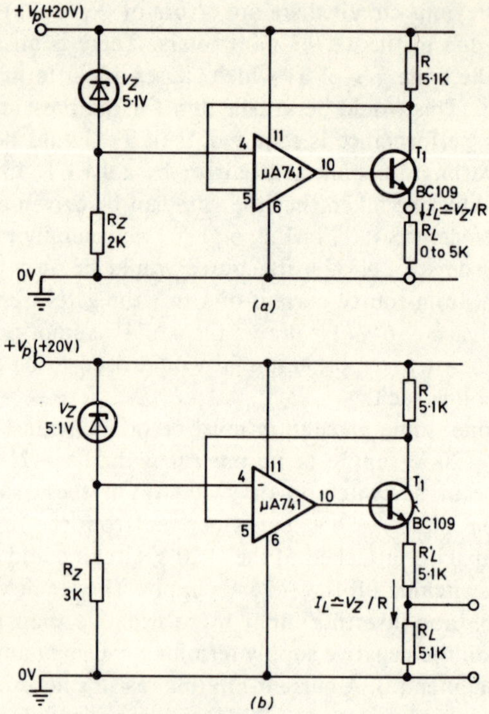

FIG. 6.12 Alternatives to Fig. 6.11

(5·1 kΩ) in the load circuit. This ensures that the gain in T_1 is always less than unity, with the snag that more voltage is needed at T_1 emitter in order to give the correct load current. With $R_L = 5{\cdot}1$ kΩ, 10 V is now required at T_1 emitter. This is still well within the capability of the circuit when $V_p = 20$ V as specified.

Circuits with Input and Output Referred to Earth

The circuits of Figs. 6.11 and 6.12, though useful for providing a fixed current to the load have only limited use in signal-processing systems where the input is variable and, like the load, is referred to earth. The circuit of Fig. 6.13 (*a*) may then be suitable. Although it is difficult to analyse this circuit by simple inspection, the application of Kirchhoff's laws easily proves that in theory the load current is given by $I_L = -V_{in}/R_1$ for any value of R_L. In practice there is an upper limit to R_L determined by the magnitude that $R_L\ I_L$ and V_O would

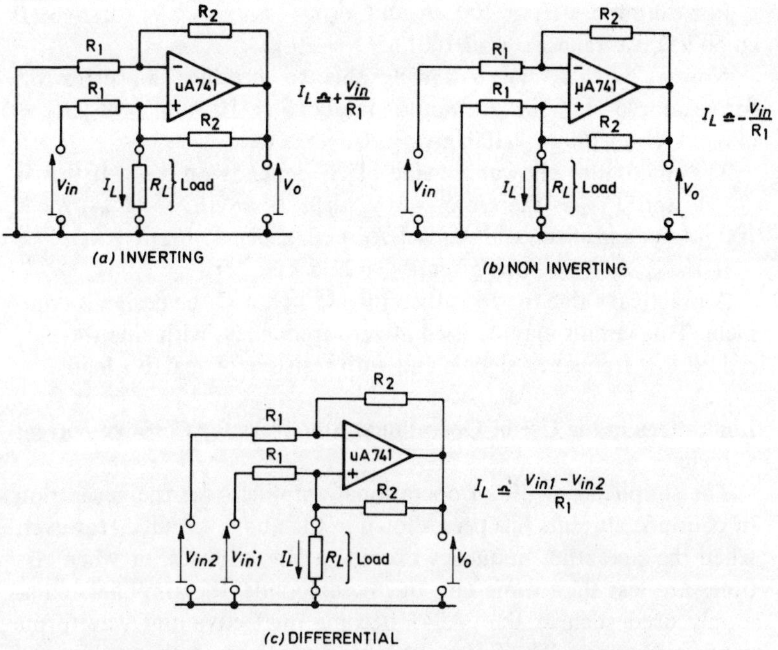

FIG. 6.13 Constant current circuits with input and output referred to earth

then reach. As in the previous circuits, if V_O reaches the amplifier limit, feedback is broken and the equation is untrue.

An alternative form giving no inversion, i.e. $I_L = +V_{in}/R_1$, is shown in Fig. 6.13 (*b*) and a differential-input version, $I_L = (V_{in1} - V_{in2})/R_1$ is shown in Fig. 6.13 (*c*).

As well as the ability to drive an earthed load from an earthed source, these circuits are also able to drive current into the load in either direction according to the differential input polarity. In designing circuit values, R_1 is fixed by the available input $(V_{in1} - V_{in2})$ and R_2 is constrained between two limits. The upper limit depends on the maximum load resistance R_L and the capability of the amplifier (about ± 14 V). The lower limit is fixed by the allowable error due to input offset v_{os} which gives an error equivalent to an unwanted differential input of $v_{os}(R_2 + R_1)/R_2$.

For example, if $I_1 = 100\ \mu A$ and $V_{in1} - V_{in2} = 2$ V, for $R_1 = 0$ to 50 kΩ, we take $R_1 = (2/100)$ MΩ = 20 kΩ.

Now $v_{os} = \pm 6$ mV. Allowing this to produce an error of, for example 1% in I_1 would imply $6 \times 10^{-3}(R_1 + R_2)/R_2 = (V_{in1} - V_{in2})/100 = 2/100$, giving $R_2 = 8{\cdot}5$ kΩ.

The maximum allowable value of R_2 depends on V_{in1}. If this is -1 V and V_{in2} is therefore -3 V, then $I_L = (V_{in1} - V_{in2})/R_1 = 100\ \mu A$ as specified, and $V_o = [(R_1 + R_2)I_L R_L - V_{in2}R_2]/R_1$.

If $V_{o\,max} = 14$ V, this gives $R_2 = 22{\cdot}5$ kΩ.

A practical value for R_2 is therefore 15 kΩ, and the design is complete. This circuit may be used at zero frequency, with sine wave or pulsed low frequency signals and with resistive or reactive loads.

Limitations in the Use of Operational Amplifiers for Constant-current Circuits

The simplicity of use of operational amplifiers for the generation of constant currents has been shown in the above circuits. However, when the operating frequency exceeds a few kiloherz, or when rise times are less than some tens of microseconds, the amplifiers commonly used such as the μA741 become ineffective and waveforms become distorted. The faster available devices generally need special frequency compensation networks to avoid spurious oscillation, are expensive and may still become unstable with certain types of load. In such applications discrete circuits are often to be preferred.

SUMMARY

Constant-current circuits are widely used in signal-processing systems, in time bases, stabilizers, ramp generators, deflector-coil drives and integrators. For d.c. and l.f. applications an operational amplifier circuit may often be used, but above a few kilohertz the bandwidth and slew-rate limitations may dictate the use of a discrete circuit.

The most common design error in this class of circuit is to allow the load voltage or other voltage in the circuit to reach a level which renders the current-defining mechanism inoperative.

Chapter 7

Practical design of simple amplifiers

This chapter combines the results of Chapters 2 and 4 to give practical design procedures for the most used amplifier circuits.

EMITTER FOLLOWER

This is the configuration of Fig. 7.1. To begin design, the first thing the designer needs to know is the peak output voltage $\hat{V}_{out}$. This determines the peak output current $\hat{I}_{out} = \hat{V}_{out}/R_L$, and R_e must be designed to ensure that this current can be supplied by the circuit. It also determines the minimum value for V_n, which must exceed $\hat{V}_{out}$ by at least 1 V to prevent saturation on negative signal peaks.

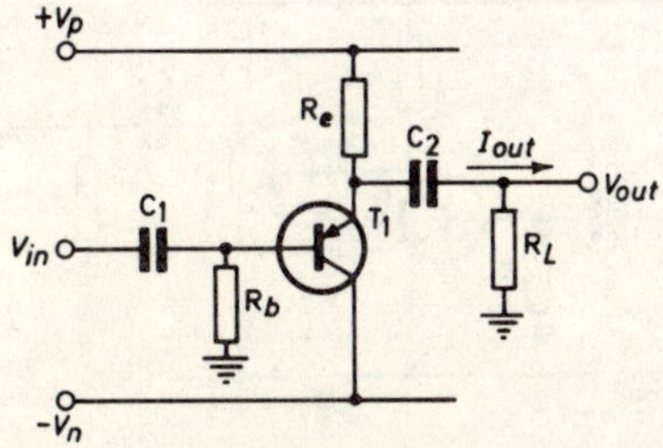

FIG. 7.1 Practical emitter follower

When the output reaches $\hat{V}_{out}$, current $\hat{I}_{out}$ is flowing through C_2 into the load and the transistor emitter current is given by $(V_p - \hat{V}_{in})/R_e - \hat{I}_{out}$, i.e., the difference between the current passing through R_e at that instant and $\hat{I}_{out}$. It is essential to the circuit action that this nett emitter current be greater than the useful minimum for the transistor. If $V_p \gg \hat{V}_{in}$ the condition can be restated: *the quiescent emitter current must exceed the peak load current.* The excess must be sufficient to allow for all tolerances, so that even in the worst case the transistor cannot approach cut-off.

When V_p is comparable with $\hat{V}_{in}$, the above statement is insufficient and can be modified to (*i*), *the current in* R_e *at the positive peak of* V_{out} *must exceed peak load current;* or (*ii*), *the quiescent emitter current must exceed peak load current where* R_e *is to be considered to be part of the load.*

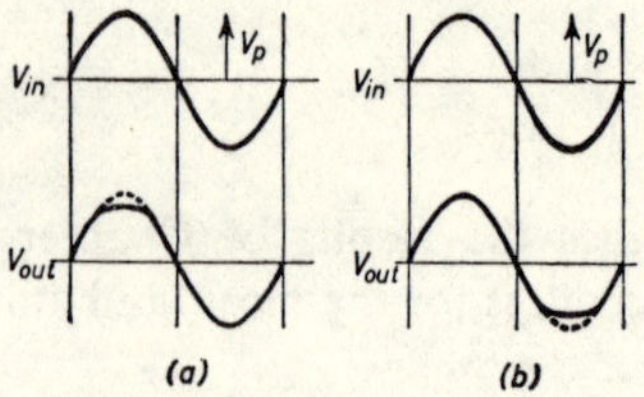

FIG. 7.2 Insufficient standing current: (*a*) *p–n–p*, (*b*) *n–p–n* (resistive load)

Failure to meet this condition results in distortion and low output, as shown in Fig. 7.2.

Having determined the operating current, the circuit values and resulting performance may be calculated (Chapters 2 and 4). It may

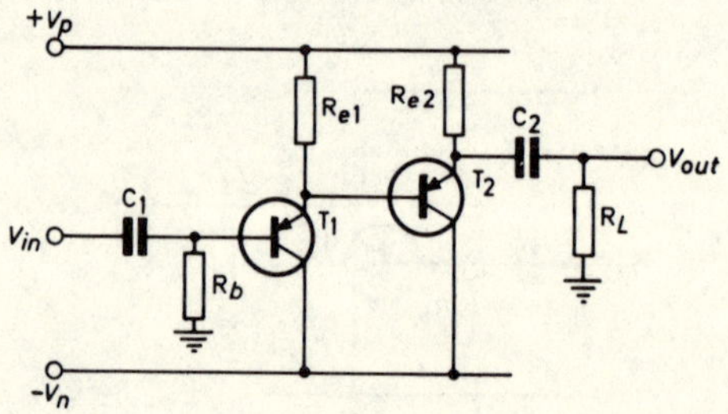

FIG. 7.3 Directly coupled emitter followers

be that the input impedance $R_b//\beta(R_L + 1/g_m)$ is too low for the application and a second emitter follower may be added (Fig. 7.3). This may usually be operated at lower current, since it does not have to supply I_{out}, and this results in a larger R_b and higher input impedance.

Provided the extra V_{eb} voltage drop is unimportant, it may be directly coupled as shown. It is common practice to omit R_{e1} so that

T_2 base current is also T_1 emitter current. This is satisfactory only if this current is sufficient to operate T_1 and is in the right direction!

If T_2 is germanium, especially a power type, it is quite normal at high temperatures for the base current to flow inwards for a *p–n–p* transistor. This is also quite possible with silicon transistors of very high β when operated at rather low emitter current. The omission of R_{e1} would in these cases cause T_1 to cut-off and the only safe procedure is to design R_{e1} to pass, when V_{in} is at its positive peak, at least I_{cbo2} plus a reasonable minimum current for T_1 emitter.

A similar criterion naturally holds for a pair of *n–p–n* transistors; a comparable situation arises when complementary types are used in the same cascaded system (Fig. 7.4). In this case the base

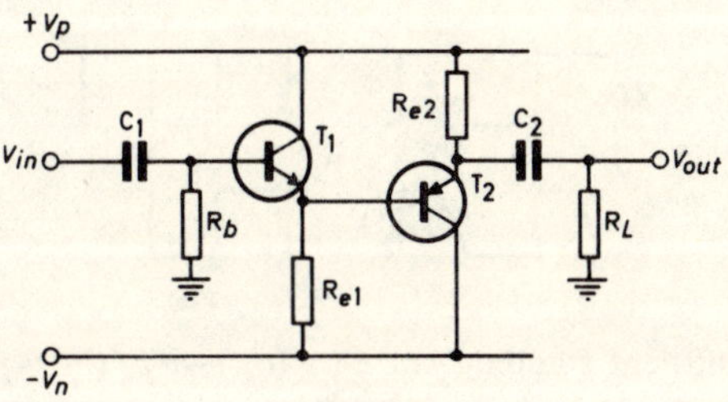

FIG. 7.4 Complementary emitter followers

current of T_2 when in its 'normal' direction, i.e. outwards, reduces T_1 emitter current. R_{e1} must therefore be designed so that $(V_n - \hat{V}_{in})/R_{e1} - I_{e2}/\beta_2$ is large enough for T_1 emitter current. The danger of T_1 cut-off increases in this circuit at low temperatures when β_2 becomes low.

Practical Example (Fig. 7.1 and 7.3)

Assume $V_p = +10$, $V_n = -10$, $R_L = 1\ \text{k}\Omega$, $\hat{V}_{in} = 3$ V, and input impedance is to be $\geqslant 10\ \text{k}\Omega$. Then, peak output current if gain is unity = 3 mA. When V_{in} is at its positive peak, the voltage across R_e is $(7 - V_{eb})$, and at this instant the current in R_e must exceed 3 mA, e.g. 4 mA. Therefore, $R_e \leqslant 7/4\ \text{k}\Omega \leqslant 1{\cdot}75\ \text{k}\Omega$, e.g. $1{\cdot}5\ \text{k}\Omega$. This gives a standing current of about 6·5 mA. A suitable transistor is chosen (e.g. 2N2906) with minimum β of, say, 50, so that standing base current can be as large as 130 μA. If maximum temperature is 50°C, I_{cbo} may be 4 μA, and at this temperature for a high-β

sample I_e/β may be negligible. Maximum variation in I_b is therefore from $+130$ to -4 μA, i.e. 134 μA.

About 1 V shift of base voltage from zero could be tolerated without reducing the current in R_e below its critical value, so that $R_{b(max.)} = (1/134)$ MΩ = 7·5 kΩ.

This clearly gives too low input impedance from R_b alone, so an additional transistor of the same type will be added, as in Fig. 7.3.

R_{e2} is decided as before (though the extra V_{eb} due to T_1 drops R_{e2} current a little) and R_{e1} must pass I_{co2} plus enough current to operate T_1, e.g. 1/2 mA total, giving $R_{e1} \leqslant 20$ kΩ, e.g. 18 kΩ. T_1 base current is therefore at maximum $1/\beta_1$ of $(500 + 170)$ μA, i.e. about 14 μA and at minimum is -4 μA, a variation of 18 μA. If T_1 base variation of 1 V is allowed, $R_b \leqslant (1/18)$ MΩ, e.g. 47 kΩ.

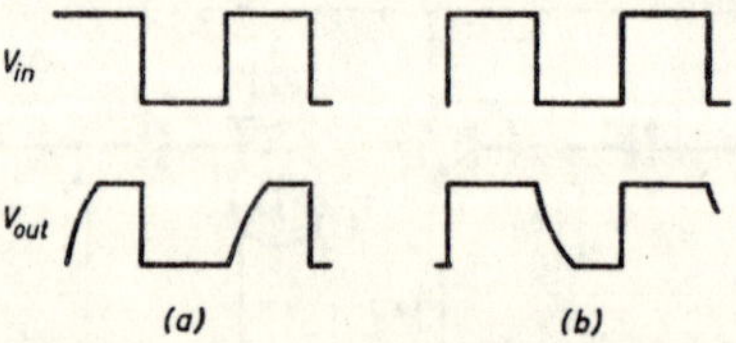

FIG. 7.5 Insufficient standing current: (*a*) *p–n–p*, (*b*) *n–p–n* (capacitive load)

Input impedance can be calculated from the formula given earlier, but when working out an initial design it is a good idea to check roughly that the design is not grossly in error. The reasoning in this case would be: the total load on T_2 emitter is $R_{e2}//R_L = 1$ kΩ// 1·5 kΩ ≈ 660; the load on T_1 emitter is therefore $\beta_2(660) = 33$ kΩ, in parallel with $R_{e1} = 18$ kΩ, giving about 11 kΩ. The input impedance to T_1 base is $R_b//(\beta_1 \times 11$ kΩ$) = 43$ kΩ which is well above the specified minimum of 10 kΩ.

The designer should now recalculate all currents, bearing in mind component and supply tolerances to confirm the soundness of the design. C_1 and C_2 are finally calculated such that $1/\omega C_1 \ll$ input impedance and $1/\omega C_2 \ll R_L$ at the lowest operating frequency.

Special care must be taken when the load is a shunt capacitance to earth, since the value of $\hat{I}_{out}$ then depends on dV_{out}/dt, not simply on V_{out}. When the waveform is a pulse, the rate of rise or fall (depending on whether the transistor is *p–n–p* or *n–p–n*) determines I_{out}, which is $C(dV/dt)$. If the standing current is inadequate, the transistor cuts off and C charges from R_e until $\hat{V}_{out}$ is reached, giving a slow transition. On the opposite edge of the pulse the transistor merely takes extra current and reproduces the input correctly, provided the resulting power dissipation during the transient is not destructive (Fig. 7.5).

EARTHED EMITTER AMPLIFIER

The bias arrangements here are similar and the criterion for minimum standing current still applies (Fig. 7.6).

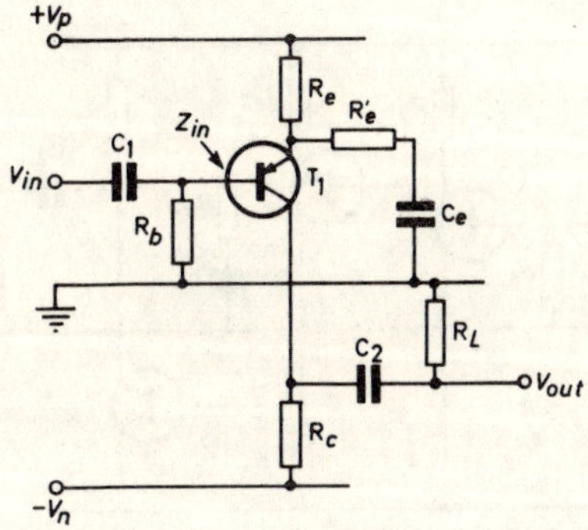

FIG. 7.6 Earthed emitter amplifier

Additional complications arise because T_1 must not be allowed to saturate when V_{out} swings to its positive peak, so that the standing collector–base voltage must exceed $\hat{V}_{out}$ by at least 1 V.

R_e' is now designed to give the required mid-frequency gain (where C_e is assumed to have zero reactance), by using the simplified formula

$$\frac{V_{out}}{V_{in}} = \frac{R_C//R_L}{R_e' + 1/g_m}$$

Note that when $R_e' \gg 1/g_m$

$$\frac{V_{out}}{V_{in}} \approx \frac{R_C//R_L}{R_e'} \qquad (R_e \gg R_e')$$

which is independent of the transistor, and the h.f. cut-off of the amplifier is approximately the f_o of the transistor. If $R_e' = 0$, gain has a maximum value of $g_m R_C//R_L$ but its h.f. cut-off frequency is only f_o/β.

C_e must be chosen so that its reactance $1/\omega_L C$ at the lowest signal frequency is low compared with $R_e' + R_e//(1/g_m)$; when $R_e' = 0$, $(1/\omega_L C)$ must be $\ll R_e//(1/g_m)$, not simply $\ll R_e$. At the frequency when $1/\omega_L C = R_e' + R_e//(1/g_m)$, the gain will be reduced by a factor of $\sqrt{2}$, i.e. 3 dB.

Finally, C_1 and C_2 are chosen so that $(1/\omega_L C_1) \ll Z_{in}$ and $(1/\omega_L C_2) \ll R_C + R_L$.

As in the emitter-follower circuit, it may be that Z_{in} is too low, in which case an emitter follower may be added between R_b and T_1 base, observing the precautions previously mentioned (Fig. 7.7).

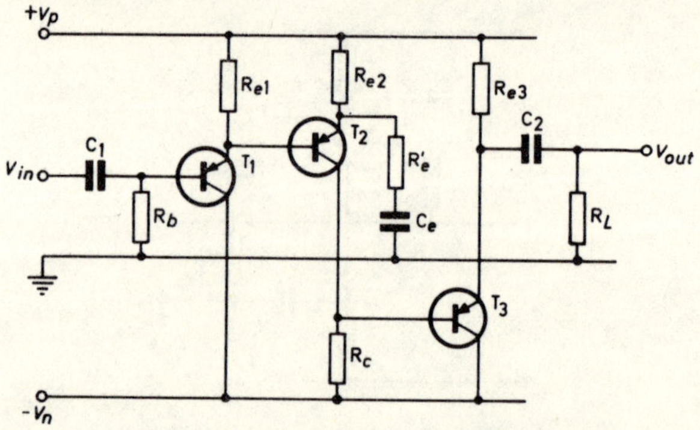

FIG. 7.7 Amplifier with input and output buffer stages

It often happens that R_L is variable (sometimes to infinity) and yet V_{out}/V_{in} is to be constant. In this case R_L must be isolated from the collector circuit and an emitter follower may again be used; it can usually be directly coupled to the collector as shown in Fig. 7.7. It is now T_3 which requires a standing current capable of supplying the load peak current, and T_2 current will normally be decided by R_C. The loading of T_3 (including R_L) on T_2 collector is approximately $\beta_3(R_{e3}//R_L + 1/g_{m3})$ and since the idea is to avoid changes in T_2 gain when R_L varies, R_C is chosen so that $R_C \ll \beta_3(R_{e3}//R_L + 1/g_{m3})$. T_2 current may now be designed as if the only load were R_C. As in previous circuits, any or all of the transistors may be *n–p–n* or *p–n–p*.

Practical Example

An amplifier is required to have a gain of 10, frequency response of 50 Hz to 50 kHz (−3-dB frequencies), load of 1·5 kΩ, $\hat{V}_{out}$ of

3 V, input impedance of $\geqslant$ 10 kΩ. Removal of the load must cause < 10 per cent rise in output. Supply lines ±15 V.

The load conditions immediately dictate the use of buffering (or an amplifier with overall negative feedback, *see* Chapter 8); the need for an input emitter follower is not yet known.

Referring to Fig. 7.8, T_2 collector potential must be known before T_3 emitter current can be determined. If loading on T_2 collector (other than R_C) is to be negligible, the standing requirement for T_2 is that T_2 collector potential must stand at least 3 V positive from $-V_n$ (i.e. > −12 V) and at least 4 V negative from T_2 base (i.e. < −4 V). A tentative value could therefore be −8 V.

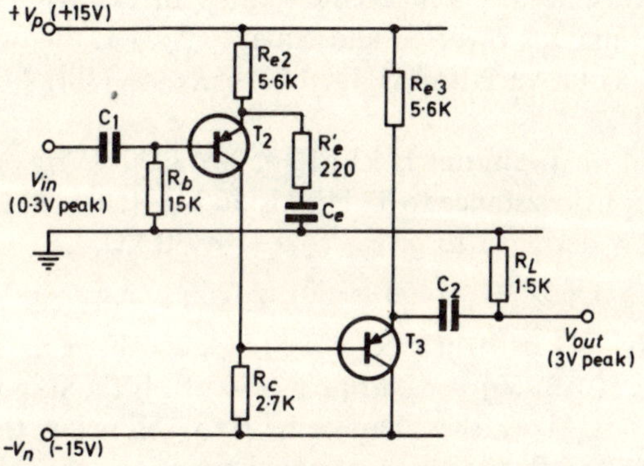

FIG. 7.8 Practical amplifier (*see text*)

The voltage at T_3 base at $+\hat{V}_{out}$ is therefore −5 V and the emitter (if a silicon type is used) is about −4·3 V. At this level the current in R_{e3} must exceed $\hat{V}_{out}/1{\cdot}5$ kΩ = 2 mA. The voltage across R_{e3} is 19·3 V and a value of 5·6 kΩ gives 3·4 mA nominal. This is likely to exceed the necessary 2 mA for all normal tolerances, but this must be checked finally. Standing current is 22·3/5·6 = 4 mA. Total load on R_3 emitter is now 5·6//1·5 = 1·2 kΩ, which itself represents a load on R_2 collector of $\beta_3(1{\cdot}2\text{ k}\Omega + 1/g_m) \approx 60$ kΩ if $\beta_{3(min.)}$ is 50. When this loading is removed, the output from T_3 (and therefore T_2) must change by less than 10 per cent, so that $R_C \leqslant 6$ kΩ. This makes no allowance for loss in T_3, the gain of which is $1{\cdot}2\text{ k}\Omega/[1{\cdot}2\text{ k}\Omega + 1/g_{m3}]$ and an $R_C = 2{\cdot}7$ kΩ is proposed.

This fixes T_2 collector current at (15 − 8)/2·7 = 2·6 mA, giving

$R_{e2} \approx 14{\cdot}3/2{\cdot}6 \approx 5{\cdot}6$ kΩ. For a gain of 10, $R_e' + 1/g_m = 2{\cdot}7$ kΩ/10 = 270 Ω. Now, $1/g_m \approx 20$ at 2·6 mA (to be confirmed for the chosen transistor), so that $R_e' \approx 250$ Ω, e.g. 220 Ω.

Now, maximum value of T_2 base current = 2·6/50 (assuming $\beta_{(min.)} = 50$) = 52 μA; at high temperatures and with high β, I_{cbo} may predominate and give T_2 base current = 4 μA inwards, giving total I_b variation of 56 μA. If T_2 base is allowed to vary 1 V due to I_b, R_b must be $\leqslant$ (1/56) MΩ, e.g. 15 kΩ.

Input impedance ≈ 15 kΩ//$\beta_1 R_e'$ = 15 kΩ//50 × 250 = 6·8 kΩ, which is too low.

An extra transistor (as T_1 in Fig. 7.7) is therefore required and its emitter current must be $> I_{cbo2}$, e.g. 0·5 mA, giving $R_{e1} = 33$ kΩ. T_1 base current has a maximum value of $(V_p/R_e + I_{b2})/\beta_{min.}$ = 10 μA and may be reversed and equal to I_{cbo2} at the other extreme, i.e. 4 μA. Total variation is 14 μA, so $R_L \leqslant$ (1/14) MΩ = 70 kΩ, e.g. 56 kΩ.

The load on T_2 emitter is 33 kΩ//[50 × (220 + $1/g_m$)] ≈ 9 kΩ so that the input resistance to T_1 base is 56 kΩ//[50 × 9] kΩ ≈ 50 kΩ.

C_1 is now designed to give $1/(\omega_L C_1) \ll 50$ kΩ, i.e.

$$C_1 \gg 10^3/(2\pi 50 \times 50) \approx 0{\cdot}06\ \mu\text{, e.g. } 2\ \mu\text{F.}$$

Similarly, $C_2 \gg 10^3/(2\pi 50 \times 1{\cdot}5) = 2{\cdot}2$ μF, e.g. 50 μF and $C_e \gg 1/(2\pi 50 \times 260) = 12$ μF, e.g. 250 μF. If the response is to fall by 3 dB at 50 Hz (rather than be level to well below this frequency as designed here), then one of these equations should become an equality, e.g. let $C_1 = 0{\cdot}06$ μF.

Since $R_e' \approx 5(1/g_m)$, frequency response for T_2 stage is given (*see* Appendix 3) by

$$f_{3db} = \frac{f_o}{\beta_o}\left(\frac{R_e + 1/g_{mo}}{1/g_{mo}}\right) \approx f_o \frac{6}{\beta_o} = \frac{1}{8} f_o$$

The transistor type must therefore have an f_o of about 400 kHz to give an upper h.f. 3-dB point of 50 kHz. Further reduction in h.f. response is caused by collector capacitance of T_2 and T_3 in parallel with R_C effectively reducing R_C and, hence, the gain at h.f. If this effect causes the gain to be 3 dB down at 50 kHz, then

$$(C_{c2} + C_{c3}) = \frac{1}{2\pi 50 \times 10^3 \times 2{\cdot}7 \times 10^3} = 1200\ \text{pF}$$

In practice the transistor capacitances would be much less (e.g. 6 pF each) so this effect may be ignored.

At this point the transistors may be specified as $\beta_{min.} = 50$, $I_{cbo} \leqslant 4\ \mu A$ at maximum junction temperature, collector voltage ratings $\geqslant 15$ V, power ratings $\geqslant 50$ mW, $f_o \geqslant 400$ kHz. The 2N3703 is suitable, as are many other small-signal transistors. The use of very-high-gain *n–p–n* planar silicon types (e.g. 2N930) would have greatly simplified design and led to the omission of T_1, but it is more instructive to overcome the difficulties of low gain and high I_{cbo}.

USING POWER TRANSISTORS

The design of linear higher-power amplifiers follows the principles just described, but transistor power dissipation is usually more significant. The instantaneous power dissipated in a transistor is the product of collector voltage and collector current plus the product of base–emitter voltage and base current, the latter product usually being negligible. This causes temperature rise in the junction which is dissipated by conduction to the transistor case and from there to the surrounding air via a heat-sink if fitted.

The thermal time lag before heat at the junction reaches the case depends on the construction of the transistor, but is usually of the order of 20 msec. The designer must therefore ensure that the average dissipation over a time of 20 msec is within the transistor rating. Within this limit, peak dissipation may be several times the rating. For example, a transistor rated at 100 mW can safely dissipate 1 W for 1 msec and then zero for 11 msec; it could not, however, dissipate 1 W for a month and then zero for a year, as destruction would have occurred within the first 20 msec.

Use of a heat-sink can increase greatly the mean power rating, and manufacturers usually quote the junction to case thermal resistance θ_M in degC/W and also the maximum permissible junction temperature $T_{j(max.)}$. If the case is assumed to be mounted on a perfect (i.e. infinite mass) heat-sink, then for an ambient temperature $T_{amb.}$, the maximum permissible dissipation P_M is that which raises the junction temperature to $T_{j(max.)}$, i.e. $P_M\theta_M = T_{j(max.)} - T_{amb.}$.

If the heat-sink is not perfect, it has also a thermal resistance (heat-sink to ambient) of θ_H; there is also an imperfect thermal connection between transistor and heat-sink represented by θ_I. In this case

$$P_M(\theta_M + \theta_H + \theta_I) = T_{j(max.)} - T_{amb.}$$

Some data sheets omit θ_M and give instead a graph of permissible dissipation versus ambient temperature in free air (i.e. without heat-sink) and with infinite heat-sink.

The only special precautions to be observed in these calculations are the duration of any peaks of high power and the use of worst case figures for thermal resistance (i.e. highest values of θ_M).

Chapter 8

Negative feedback

Complete understanding of negative feedback systems is difficult, but Bode and Nyquist (p. 292) have succeeded in explaining feedback behaviour in practical terms by interpretation of the results of analysis.

In this chapter only the simpler consequences of the application of feedback will be dealt with in order to give a basic understanding of the problems involved. The methods discussed here for preventing instability in feedback amplifiers are easy to apply and usually adequate, but better overall performance can sometimes be obtained by using the more sophisticated techniques described by Bode and Nyquist.

BENEFITS OF NEGATIVE FEEDBACK

The reasons for using feedback are best illustrated by a simple example (Fig. 8.1), where an inverting amplifier of gain $-A$ and

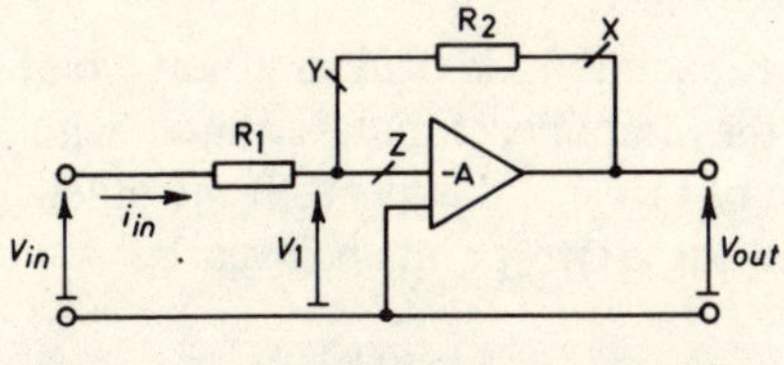

FIG. 8.1 Simple feedback amplifier

infinite input impedance has added to it the input resistor R_1 and the feedback resistor R_2. It is shown in Appendix 4 that the gain v_{out}/v_{in} is given by $-R_2/R_1$ if A is very large.

This result can be obtained without complete analysis by the following argument. Suppose that an input is present and that the resulting output is within the linear range of the amplifier. Under these conditions the input voltage of the amplifier itself, v_1, must equal $-v_{out}/A$, and if A is very large the implication is that v_1 is very small.

If v_1 is much smaller than v_{in}, then the input signal current i_{in} is v_{in}/R_1; since Z_{in} for the amplifier is infinite, all of this current must flow into R_2, giving a voltage $-v_{in}R_2/R_1$ across R_2. As v_1 is very small, the voltage across R_2 is equal to v_{out}. Therefore $v_{out} = -v_{in}R_2/R_1$.

This result implies that variations in A have no effect on the overall gain provided A is always sufficiently large, so negative feedback enables an amplifier of accurately known gain to be obtained from an amplifier of high but badly defined gain. Since accurate gain is essential in most designs, the use of feedback is very common.

As calculated in Appendix 4, feedback in the form shown has other advantages also: the input impedance is known to be R_1; the output impedance is reduced; the bandwidth is increased. Feedback does not improve signal/noise performance, nor does it reduce zero drift in a d.c. amplifier.

LOOP GAIN

The gain required in a negative feedback amplifier to obtain a good approximation to the above results is given in Appendix 4 as a mathematical expression

$$\frac{AR_1}{R_1 + R_2} \gg 1$$

Now, if the feedback loop is broken at any point and a signal injected towards the amplifier input terminal with v_{in} earthed, an amplified version of this signal appears at the other side of the break. The gain between these two points is given by

$$-\frac{AR_1}{R_1 + R_2}$$

and the magnitude of this is known as the 'loop gain'. If, for example, the loop is broken at X, and signal e is applied to R_2 at X, with v_{in} earthed, v_1 is given by $eR_1/(R_1 + R_2)$ and v_{out} is $-AeR_1/(R_1 + R_2)$, which appears at X. Loop gain is therefore $AR_1/(R_1 + R_2)$, as stated above. A similar result is given by breaking the loop at Y or Z or in the middle of the amplifier circuit; in fact, anywhere in the loop, but not for instance in R_1, which is outside the loop.

Note that it is the *loop gain* which must greatly exceed unity. This applies to all negative feedback systems regardless of circuit details; the loop is opened as described above and the gain round the loop must be large. If this condition is not obeyed, the advantages of feedback are reduced, and the inverse of loop gain is the fraction by which performance departs from the ideal.

Thus, in the present example a loop gain of 10 means that the overall gain v_{out}/v_{in} will be $-R_2/[R_1(1 + \frac{1}{10})]$ instead of $-R_2/R_1$. If tolerance in A is such that loop gain varies from 10 to 20, overall gain will vary from $-R_2/[R_1(1 + \frac{1}{10})]$ to $-R_2/[R_1(1 + \frac{1}{20})]$. To be certain of obtaining an overall gain which is constant to within 5 per cent, the loop gain must therefore be at least 20 so that A will be 20 times bigger than in a non-feedback circuit of similar overall gain.

It can be seen from Appendix 4 that the loop gain appears in all the important expressions: input impedance, output impedance, bandwidth, and gain.

VIRTUAL EARTH

This term is applied to a point such as v_1 in Fig. 8.1, where, although it is a vital part of the signal circuit, the voltage amplitude present is very much less than signal voltages in the rest of the circuit. For calculating signal currents it can be regarded as being at zero potential or virtually earthed, and the error which results is negligible (e.g. $i_{in} = (v_{in} - v_1)R_1 \approx v_{in}/R_1$, since $v_1 \approx 0$).

Effect of Amplifier Input Impedance

It is shown in Appendix 4 that the presence of a resistance R from v_1 to earth changes circuit performance but not so much as in a non-feedback circuit with R_2 open-circuit.

This is obvious if the feedback system is understood, because v_1 is known to be much less than v_{in}. Any resistance R to earth causes a current of only v_1/R to flow, which is much less than if the same resistor were placed across v_{in}. Therefore a small fraction of the signal current i_{in} flows into R and the rest flows through R_2 to v_{out}. Gain is therefore virtually unchanged.

The above argument is plausible but can lead to wrong conclusions if the significance of R is not appreciated. If for example $R = R_1$, then overall gain is hardly affected but the *loop gain* has been reduced (to half its former value if $R_2 \gg R_1$). This means that definition of overall gain, bandwidth improvement, etc. are worse by the ratio by which the loop gain has fallen. The loop gain can always be worked out as before by breaking the loop at X, Y, or Z.

PROBLEMS IN NEGATIVE FEEDBACK LOOPS

Desirable as negative feedback is, there is always a practical limit to the amount of loop gain which may be applied. This is due to inevitable phase shifts within the loop which at some frequency, either the l.f. or h.f. end of the band or both, add up to an extra 180 degrees. Feedback at this frequency is therefore positive and if the loop gain

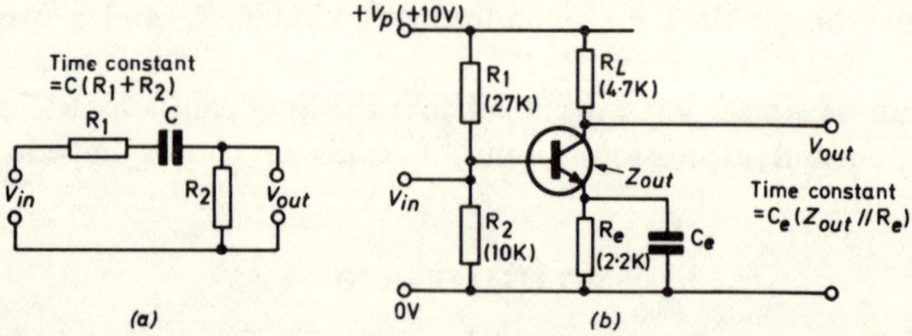

FIG. 8.2 Low-frequency time constants: (*a*) coupling, (*b*) decoupling

(which always falls as the phase angle increases) is still unity or more, oscillation results at that frequency. If the gain is close to but less than unity, the system is stable, but application of a step input causes a burst of sine waves at the critical frequency which die away exponentially in the same way as in a damped resonant circuit. This effect is known as 'ringing'. A sine-wave input test over the frequency spectrum will show a peak in the gain at the critical frequency.

These forms of complete or partial instability are caused at the l.f. end of the amplifier response band by the same components which cause the gain to fall. If no coupling or decoupling capacitors, transformers, or chokes are used and the response is constant to zero frequency, as in directly coupled amplifiers, no instability can result in this region, since 180 degrees phase reversal is impossible. In fact the presence of one simple CR network in the whole amplifier in the typical circuits of Fig. 8.2 is also guaranteed to be 'safe' as the maximum phase shift obtainable from such a network is 90 degrees. In theory two networks cannot produce 180 degrees phase shift but even though oscillation is impossible, ringing is likely if the angle exceeds 150 degrees with a loop gain of unity.

At the high-frequency end of the response band the components which cause the fall in gain and the accompanying phase shift are much more numerous. Transistor α, which falls in accordance with the equation

$$\alpha = \frac{\alpha_0}{1 + jf/f_0}$$

behaves like a *CR* network. Collector capacitance acts in the same way in conjunction with the collector load in an earthed emitter stage. Wiring strays and lead inductance complicate the issue still more.

For this reason overall feedback is rarely used in applications where the limits of frequency response for the transistors are being approached: there are so many elements contributing phase shift that after taking measures to prevent oscillation the response is greatly reduced. The designer thus generally tolerates relatively badly defined gain in the interest of simplicity and higher frequency response.

A situation which is identical from the point of view of feedback theory exists in many low-frequency amplifiers, such as chopper amplifiers or voltage stabilizers, where the response falls as the frequency increases, because of networks deliberately added. This is still a phase shift and gain drop at h.f. in that the phase lags increase and gain falls as frequency rises. In these cases the number and magnitude of such phase-shift networks is known, and the transistor effects and stray effects may be ignored if their effective time constants are very much smaller than those of the networks. As before, such a loop containing one simple *CR* network is free from oscillation, but two or more can ring or oscillate and the loop response must be examined.

Preventing L.F. Instability

The first step when designing the circuit detail of a feedback amplifier is to avoid all unnecessary l.f. coupling or decoupling circuits, which usually means avoiding capacitors, transformers and chokes. Direct coupling should be used where possible, and the aim is to reduce the *number* of phase-shift-producing networks. Making capacitor values so large that phase shift is negligible within the signal band does not help, since there is always some frequency, however low, where phase shift is appreciable. (It is little consolation that the resulting oscillations are lower in frequency than the input signal band!)

If the number of phase shift networks at low frequency, more

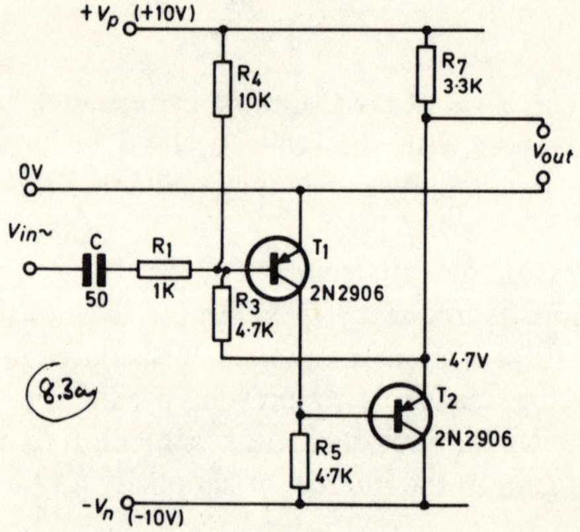

(*a*)

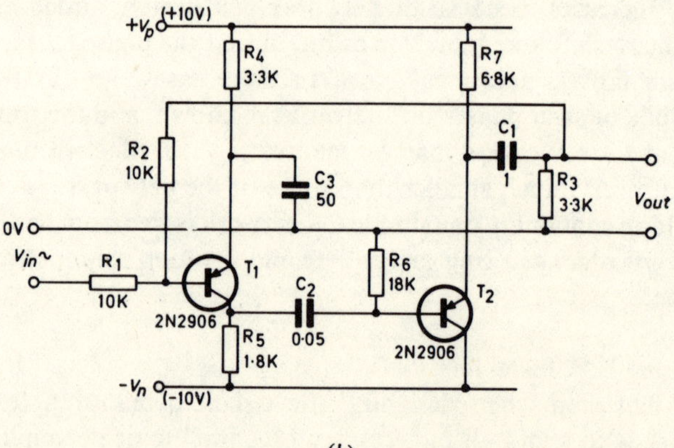

(*b*)

FIG. 8.3 Feedback amplifiers: (*a*) stable, (*b*) unstable

conveniently referred to as 'l.f. time constants', can be reduced to one, any amount of feedback may be applied and no l.f. oscillation can occur.

An example is given in Fig. 8.3 (*a*) and (*b*), where performance without feedback of the two amplifiers is identical over the audio band, provided that C, C_1, C_2, and C_3, are large enough, $R_6 \gg R_5$, and operating currents and voltages are correctly designed.

When feedback is added as shown, the circuit of Fig. 8.3 (*b*) is likely to oscillate, whereas Fig. 8.3 (*a*) is guaranteed to be free from l.f. oscillation.

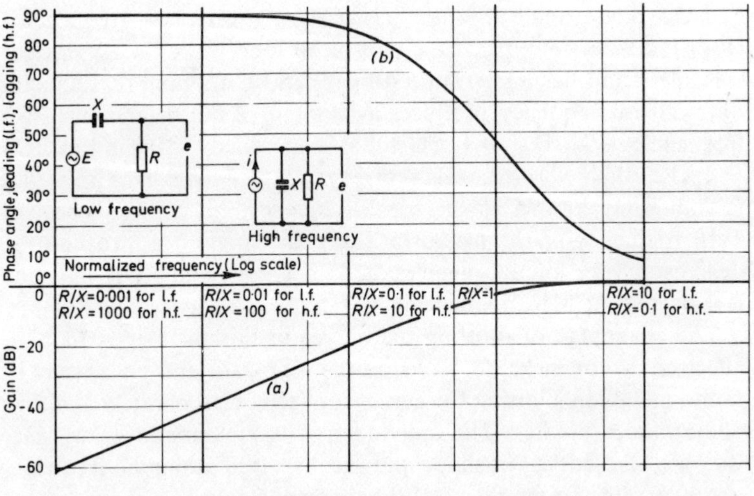

FIG. 8.4 Single *CR* network

If operating levels cannot be adjusted to allow direct coupling and two or more a.c. couplings have to be used, the important parameters are the ratios between the various time constants involved, as explained below.

Effect of several time constants

The normalized gain and phase response plots for a single *CR* network are given in Fig. 8.4 (*a*) and (*b*), where the frequency scale is logarithmic and expressed in multiples of $1/CR$. The gain of the network is plotted on a logarithmic scale in the form $G = 20 \log_{10} (V_{out}/V_{in})$, commonly called decibels (dB or db). (The use of the term 'decibels' is strictly incorrect, since it should be used for voltage ratios only if the impedances associated with the voltages are identical. However, the use of decibels for any voltage ratio is almost universal.)

The curve is obtained by calculating

$$V_{out}/V_{in} = \frac{j\omega CR}{1 + j\omega CR}$$

so that

$$|V_{out}/V_{in}| = \frac{\omega CR}{\sqrt{(1 + \omega^2C^2R^2)}}$$

and the phase angle $\theta = \tan^{-1}(1/\omega CR)$. At very low frequencies, $|V_{out}/V_{in}| \to \omega CR$, so that the plot of $20 \log_{10} V_{out}/V_{in}$ against log ω is linear and has a slope of 6 dB per octave, meaning that a factor of 2 fall in frequency produces a factor of 2 fall in gain. At high frequencies $|V_{out}/V_{in}| \to 1$, which has been called 0 dB; and at $\omega CR = 1$, $|V_{out}/V_{in}| = 1/\sqrt{2}$, which corresponds to -3 dB. This frequency is commonly known as the break-frequency, cut-off frequency, or 3 dB frequency of the network. The phase angle θ approaches 90 degrees at very low frequencies, is 45 degrees when $\omega CR = 1$, and is 0 degrees at high frequencies.

The advantage of plotting the curves in this manner is that the effect of two or more networks on the loop gain and phase may be found merely by plotting the curves for each one separately and then adding the ordinates. The individual plots are easy to draw, since they are all identical in shape and are displaced along the frequency axis according to the ratios of the time constants.

Returning to the circuit of Fig. 8.3 (*b*), there are three networks causing l.f. phase shift, and it will be assumed that $C_1(R_3//R_2)$ is 3 msec, $C_2(R_5 + R_6)$ is 1 msec, $C_3(R_{in\ e}//R_4)$ is 5 msec, and $R_2 \gg R_3$, $R_6 \gg R_5$, $R_4 \gg R_{in\ e}$. The plot corresponding to the 5 msec time constant is shown in curves (1) and (2) in Fig. 8.5: the gain is 3 dB down at $\omega = 1/(5 \times 10^{-3})$ with $\theta = 45$ degrees, and curves (1) and (2) are identical with the curves in Fig. 8.4.

The second largest time constant (3 msec) produces similar-shaped curves, but the 3 dB frequency (and the entire frequency scale) will be shifted in the ratio 5/3, as shown in curves (3) and (4), Fig. 8.5.

The last time constant of 1 msec is shifted in frequency by a factor of 5 from the first (curves (5) and (6)).

The combined effect of all three is shown in curves (7) and (8), which are obtained by direct addition of curves (1), (3), (5), and (2), (4) (6), respectively.

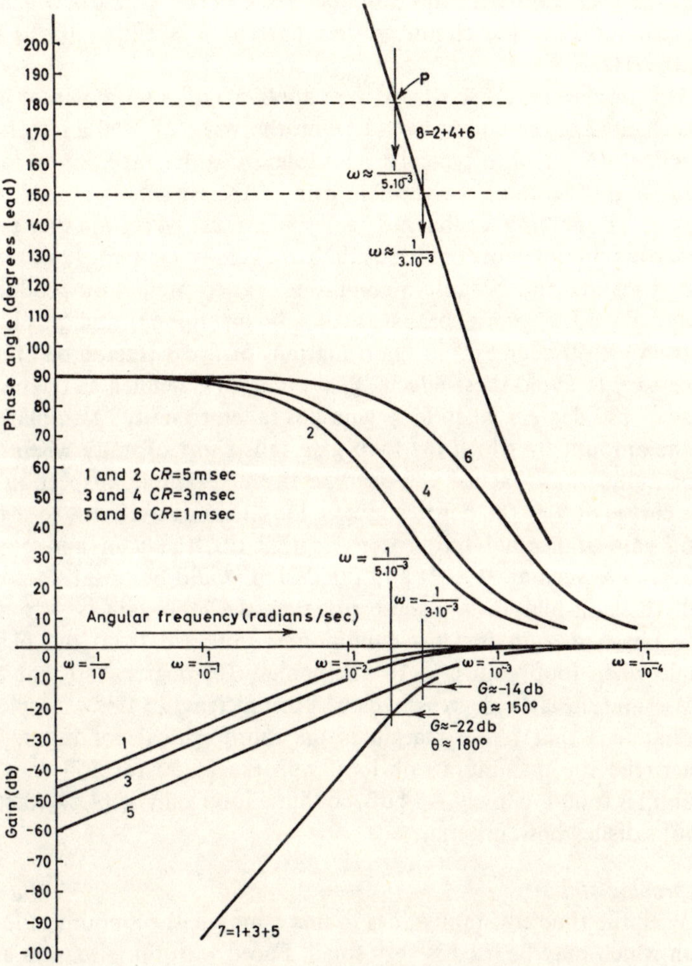

FIG. 8.5 Low-frequency response for unstable feedback amplifier (Fig 8.3*b*)

Note that at point P the phase angle has reached 180 degrees and continues to rise as the frequency falls, finally approaching 270 degrees at an infinitely low frequency (90 degrees for each network).

At P the frequency is given by $\omega = 1/(5 \times 10^{-3})$* and the gain is -22 dB compared with its medium-frequency value, where the networks have little effect.

If the loop gain at medium frequencies exceeds $+22$ dB, which is a factor of 12·6, the circuit will oscillate at $\omega = 1/(5 \times 10^{-3})$, i.e. at 31·8 Hz.

This implies that if feedback were applied with a loop gain of just less than 12·6, the circuit would be on the verge of oscillation. In a practical design, with typical circuit tolerances the variation in loop gain is considerable: the variable gain of the circuits is indeed one reason for applying feedback. The designer must be certain that even under maximum gain conditions the loop gain is less than 12·6. Even this is insufficient, because a condition near to oscillation produces 'ringing' and response peaks as already mentioned, and a 'phase margin' of 30 degrees and 'gain margin' of 6 dB (factor of 2) are necessary to avoid these effects. Phase margin is defined as the nearness to 180 degrees when loop gain has fallen to unity; gain margin is the amount by which the loop gain falls short of unity when the phase angle has reached 180 degrees. In our example, inspection of the curves at $\theta = 180$ degrees gives a loop gain of -22 dB, so if the loop gain at normal frequencies is $+22$ dB, the gain and phase margin are zero at 31·8 Hz and the design would be unsatisfactory.

If the gain margin is made 6 dB, then the loop gain is $+16$ dB. The phase margin for this condition is obtained from the phase angle when loop gain is -16 dB, namely 163 degrees, so that the phase margin is 17 degrees, which is insufficient. In this example it so happens that phase margin is the more critical condition. To determine the maximum safe loop gain the figure for 150 degrees margin is found, namely -14 dB, so that a loop gain of 14 dB (factor of 5) satisfies both criteria.

Improving stability

With the time constant values in this example, the amount of loop gain which may be used is very small. Faced with this situation and a requirement that the loop gain is to be at least, for example, 10 (i.e. 20 dB) in a particular application, there are three courses open to the designer. The first is to reduce the number of time constants, as already stated; a momentary glance at the curves shows the value of removing the second (3 msec) time constant. The second is to alter

* It is pure coincidence that this corresponds to the largest time constant.

the relative magnitudes of the time constants, the effect of which is to be described. The third is to add networks which favourably change the phase response of the system.

Assuming that the first method can be pursued no further, the second should be adopted; in all but the most critical applications this will succeed. Suppose that the smallest time constant is made 10 times smaller, i.e. 0·1 msec. The new plots of gain and phase are given in Fig. 8.6 (*a*), and the summed curves are labelled (7) and (8). This time the value of ω at 180 degrees is $2{\cdot}6/10^{-2}$ and the gain is -38 dB, so that in the limit $+38$ dB loop gain could be applied. For a gain margin of 6 dB, $+32$ dB is permissible, but the corresponding phase margin is then only 25 degrees. Again, the phase margin criterion is the more critical, and for 30 degrees margin (150 degrees phase angle) the gain is -30 dB. The safe limit for loop gain is therefore $+30$ dB (a factor of 31·6) and the desired loop gain of 20 dB is quite safe.

The reason for this great improvement in stability from a permissible loop gain of 5 to 31·6 is obvious on examining the curves: the greater the ratio of the smallest two time constants, the lower has the gain fallen when the phase reaches 180 degrees. The absolute values of the time constants do not affect permissible loop gain but merely change the time-scale and the overall bandwidth of the feedback system.

The designer should therefore make the *ratio of the two smallest l.f. time constants as large as possible*. Having done this, a safe rule is that *the loop gain may be as large as the ratio of the two smallest time constants*. (For a more precise statement *see* Littauer, op. cit.)

The validity of this rule can be checked in the two cases considered. In the first the ratio in question was 3 and the permissible loop gain 5. In the second the ratio was 30 and the permissible loop gain 31·6. The rule is not empirical and has a sound mathematical basis, but the proof is involved and will not be attempted here.

There is a certain disadvantage in the second method which is usually tolerable. If for any reason the smallest time constant cannot be less than 1 msec in the above example, and for good definition of gain a loop gain of 10 or more is required, then the second method demands the increase of the other two time constants to 10 msec or more. This is sometimes impossible, in which case the third method should be adopted, but only after the first and second have been carried as far as possible.

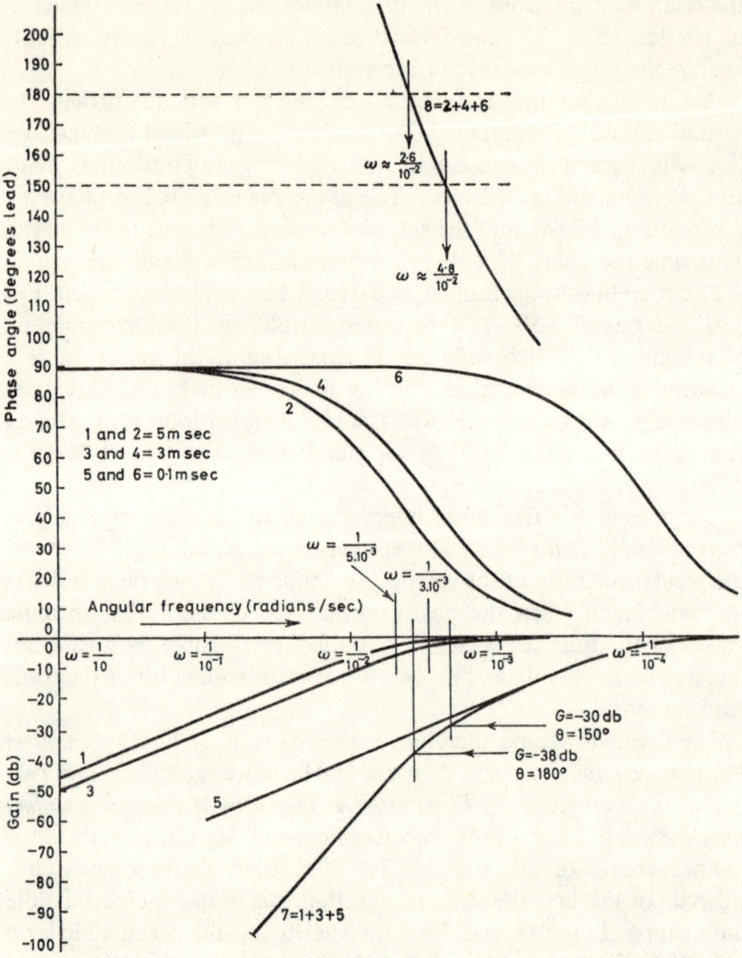

FIG. 8.6 (*a*) Stabilized version of Fig. 8.5 (second method)

The third method involves the addition of components to one or more of the phase-shifting networks and in its simplest form consists of a parallel *CR* combination in series with the capacitor of the main network (*see* Fig. 8.7). The effect of the extra components depends on the ratio of the two capacitors and of the two resistors; Figs. 8.9–8.19 show the gain and phase response curves for several ratios. The curves show that although the phase plot can be changed dramatic-

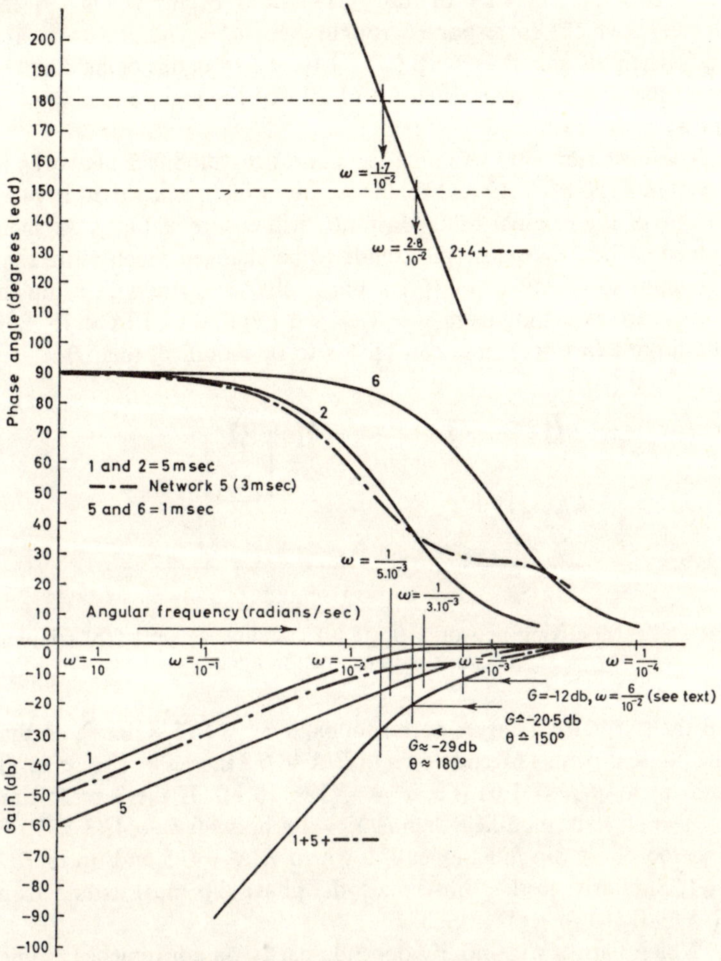

FIG. 8.6 (*b*) Stabilized version of Fig. 8.5 (third method)

ally to give a greatly undulating phase angle, the gain curve is changed much less.

The use of these networks is best illustrated by its application to the original example with time constants of 5, 3, and 1 msec giving the curves of Fig. 8.5, where the maximum permissible loop gain is 14 dB ($\times 5$) for a phase margin of 30 degrees. If it is required to use a loop gain of $+20$ dB ($\times 10$) it is evident that the dangerous region from the point of view of loop oscillations begins when $\theta \approx 150$ degrees, which corresponds roughly to $\omega = 1/(3 \times 10^{-3})$, and ends when the gain is -26 dB ($\omega = 1/(6 \times 10^{-3})$), this being deduced from the required phase margin of 30 degrees and gain margin of 6 dB.

The next step is to examine the curves for modified networks as given in Figs. 8.9–8.19 and to choose one which, when used in place of one of the original time constants, will ensure stability. As mentioned earlier, the phase plot tends to be changed much more than the gain, so initially, look for a phase characteristic with a dip towards zero extending from $\omega = 1/(3 \times 10^{-3})$ to $\omega = 1/(6 \times 10^{-3})$. If the largest (5 msec) time constant is to be modified, then $R/X = 1$

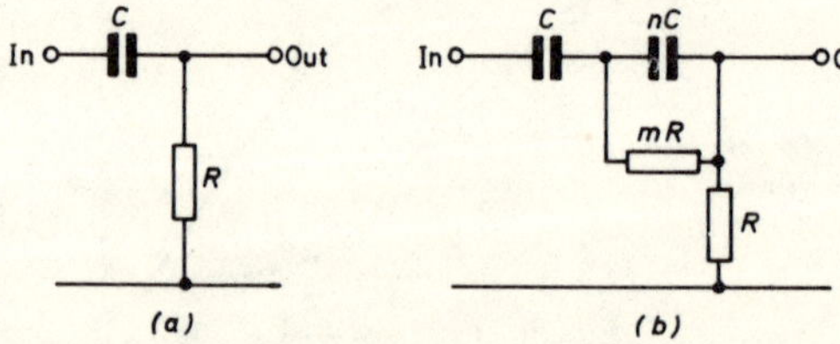

FIG. 8.7 Modifying networks (l.f.): (*a*) coupling network, (*b*) coupling network with correction added

on the normalized curves corresponds to $\omega = 1/(5 \times 10^{-3})$, so that the phase dip must extend down to $R/X = 0{\cdot}8$ (i.e. $\omega = 1/(6 \times 10^{-3})$) and up to $R/X = 1{\cdot}66$ (i.e. $\omega = 1/(3 \times 10^{-3})$). If the 3 msec time constant is to be modified, then $R/X = 1$ represents $\omega = 1/(3 \times 10^{-3})$ and the phase dip must extend down to $R/X = 0{\cdot}5$ and up to $R/X = 1$. Similarly, in the 1 msec case the phase dip must extend from $R/X = 0{\cdot}166$ to $R/X = 0{\cdot}33$.

Which network to modify depends partly on convenience; sometimes the required additional components would require to be very large accurate capacitors which are expensive or unavailable, and, on the other hand, it may be impossible to stabilize the system by operating on the network the designer would prefer to modify. In some designs more than one network may require to be changed.

Having decided tentatively on which network to alter—for example, the 3 msec time constant—the designer picks out a selection of curves which have the desired shape of phase response. In our example, curves (5), (6), (9), (10), and (13) look promising and if superimposed on Fig. 8.5 such that the $R/X = 1$ line coincides with $\omega = 1/(3 \times 10^{-3})$ the phase angle is reduced in the critical region.

Deciding which curve to use can be difficult: the general rule is that the less violent the change in phase, the better will be the overall l.f. response of the system. This rule must not, however, take precedence over the need for complete stability with all component tolerances. In the present case the author's choice was curve (5) (Fig. 8.13), and this is shown in dotted lines in Fig. 8.5. The curves are now summed as before, remembering that the new curves completely replace curves (5) and (6).

The resulting phase curve now reaches 180 degrees at $\omega = 1{\cdot}7/10^{-2}$ at which frequency the gain totals -29 dB, so that if the loop gain is $+20$ dB, there is a gain margin of 9 dB. When the gain is -20 dB the phase angle is 149 degrees, which gives a phase margin of 31 degrees.

The use of this network requires the insertion, in series with C_1 in Fig. 8.3 (*b*), of the parallel combination of a capacitor $C_1/\sqrt{10}$ and a resistor $(R_2//R_3)$.

As indicated above, this is not necessarily the optimum network to use, but the gain and phase margins are not excessive. A safer network is curve (10) (Fig. 8.18), which gives very large margins but has a worse effect on the overall response. Again, it may be preferable to treat one of the other networks instead (e.g. curve (11) applied to the 5 msec. time constant).

Overall response

The totalled curves obtained after deciding the correcting networks now represent the open-loop response of the amplifier, and the overall response with feedback may be calculated. In the example given, the gain curve (Fig. 8.6 (*b*)) shows a loss of 12 dB at $\omega = 6/10^{-2}$. At this frequency the loop gain is (20 – 12), i.e. 8 dB (ratio of 2·5) and this implies a 3 dB loss in overall gain (overall gain $= G/[1 + (1/2{\cdot}5)] = 0{\cdot}7G = G - 3$ dB).

Had the network modification been unnecessary, the 3dB frequency for the overall system with 20 dB loop gain would have been $\omega = 4/10^{-2}$.

Preventing High-frequency Instability

h.f. instability is caused and cured in the same way as l.f. instability. It is, however, much more difficult to deal with in practical transistor circuits. The number of h.f. time constants is often outside the designer's control; there is no equivalent to the use of direct coupling which removes l.f. time constants. The time constants are often unpredictable; transistor current gain cut-off frequency often has a spread of 3 or 4 to 1. The time constants often interact and cannot then be considered separately.

Overall feedback is consequently rarely applied to amplifiers intended to work near the transistor frequency limit. The problem more usually facing the designer is how best to restrict the bandwidth in, say, an audio amplifier so that it does not oscillate in the megahertz region.

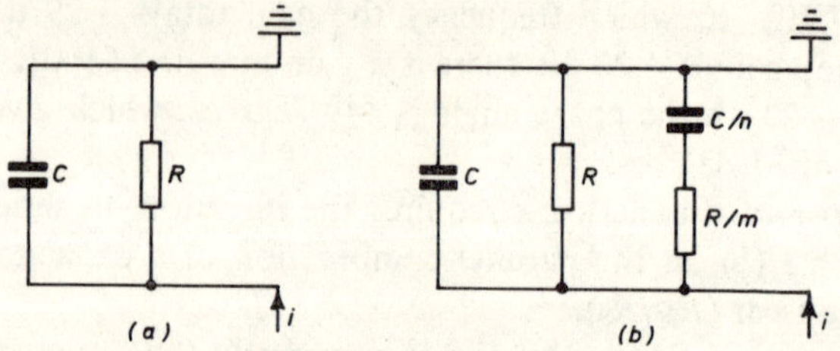

FIG. 8.8 Modifying networks (h.f.): (*a*) collector load, (*b*) collector load with correction added

The simplest solution is to follow 'method 2' and add just one h.f. time constant which is larger than any other in the circuit by a factor at least equal to the mid-frequency loop gain. The best solution is obtained if the time constant known to be the largest is further increased. This is known as the 'dominant lag' technique.

As indicated earlier in the chapter, similar h.f. problems can occur in servo systems, low-frequency amplifiers, or control loops where smoothing circuits, inductance of motors, and other predictable components produce a fall in gain and a phase lag as frequency rises. In such applications all of the procedures used for l.f. instability can be adopted.

There are some practical differences in their application. For instance, in the second method the important ratio is now between the *largest* and next largest time constant. The normalized curves are plotted in the opposite direction (which is made clear in Fig. 8.4) and the correction networks are the 'dual' equivalents of the l.f. circuits (Fig. 8.8).

Other Causes of Instability

It is an all too frequent occurrence that, having carefully designed a feedback loop to be free from instability, the designer finds that in practice the circuit oscillates.

Assuming that no gross error has been made either in estimating the time constants or in drawing and adding the curves, there are two common reasons for oscillation. The most usual is that the power supply rails and many other connections which are assumed to be zero impedance in most design calculations often have considerable impedance at both very high and very low frequencies. These introduce either extra feedback paths or add more time constants to the system. Much trouble can be avoided by the methods described in Chapter 16, and where the supply lines are suspected of introducing feedback the addition of extra capacitance between the lines will prove the point by changing the oscillation frequency. If this proves to be the case, then isolation between stages must be provided, preferably by means of a simple Zener diode stabilizer (Chapter 1) to supply the lower-powered stage.

The second common cause of oscillation is an unsuspected feedback loop coupled through stray capacitances. These can occur in wiring looms where high- and low-level signals may be carried on closely coupled wires and by mutual inductance between chokes or transformers. The effects can be reduced by improving cable routing and by mounting transformer and choke cores at right-angles.

If there is doubt as to whether the negative feedback loop is the cause of oscillation, reduce the gain of the negative feedback loop; if the loop is causing the trouble, any l.f. oscillation will become lower in frequency and h.f. oscillation will become higher in frequency, and the violence of oscillation will decrease; if the loop is not responsible, oscillations will usually increase and will tend to move further in frequency from the bandwidth limits of the amplifier.

SUMMARY

The benefits of negative feedback should be already well known to the student and most of this chapter has therefore been devoted to the methods of preventing instability. The use of separate gain phase plots rather than the composite gain-phase Nyquist diagram has the advantage that the influence of each network can be clearly seen. Corrective measures can then be taken as described and, finally, a Nyquist plot made if considered desirable.

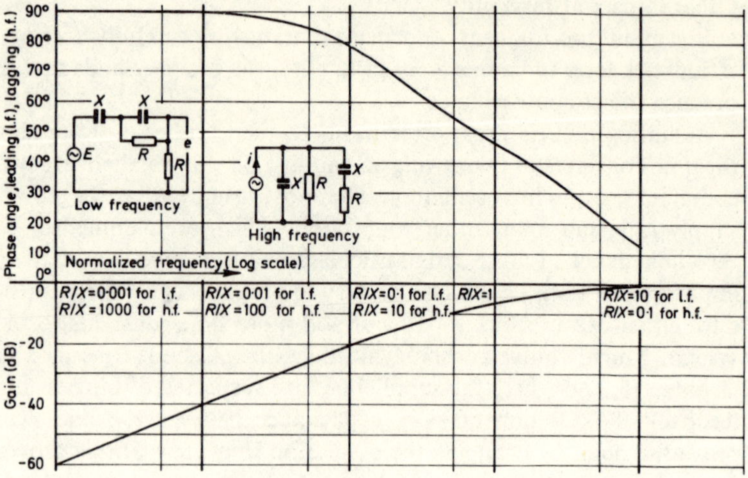

FIG. 8.9 Network 1

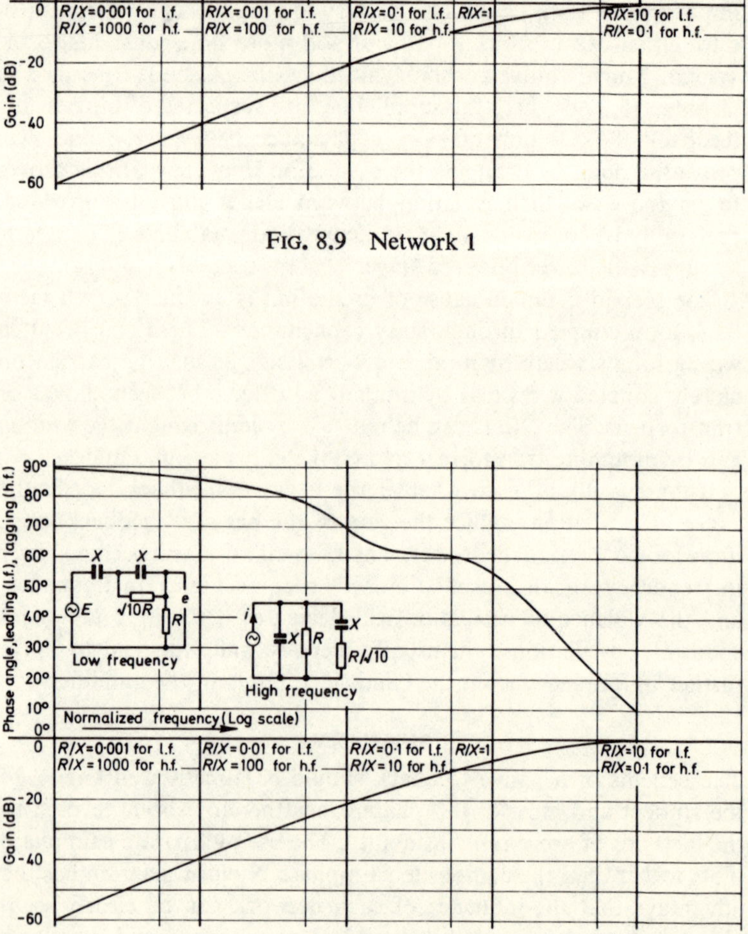

FIG. 8.10 Network 2

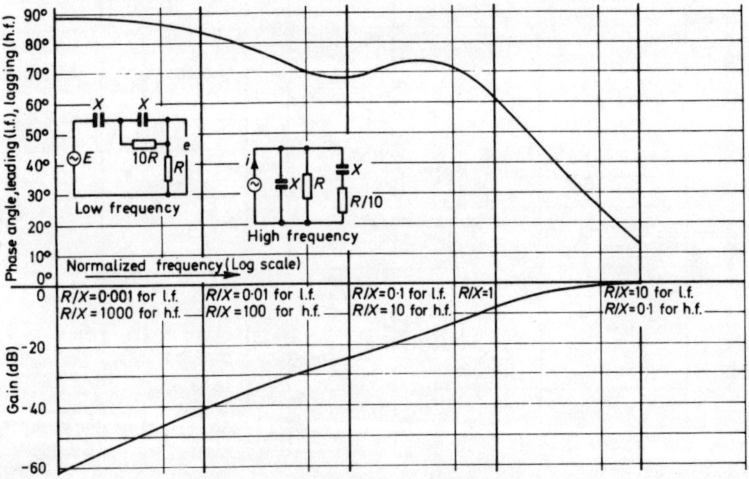

FIG. 8.11 Network 3

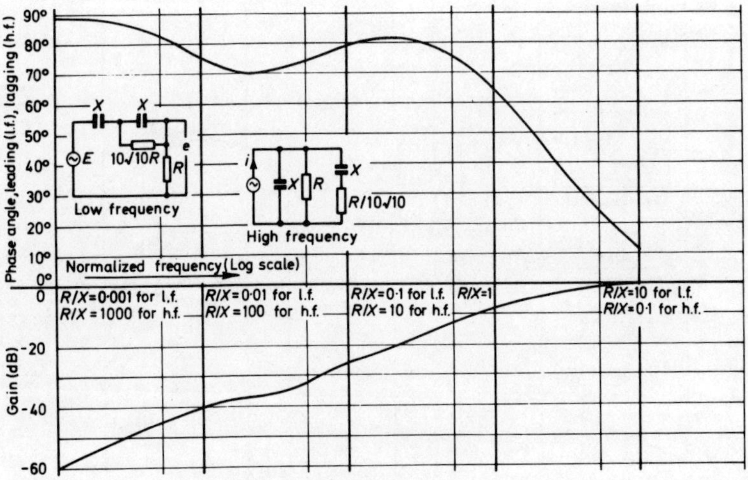

FIG. 8.12 Network 4

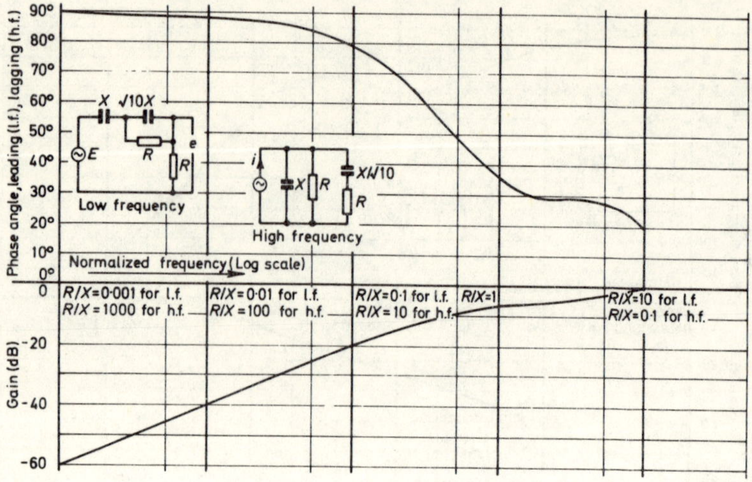

FIG. 8.13 Network 5

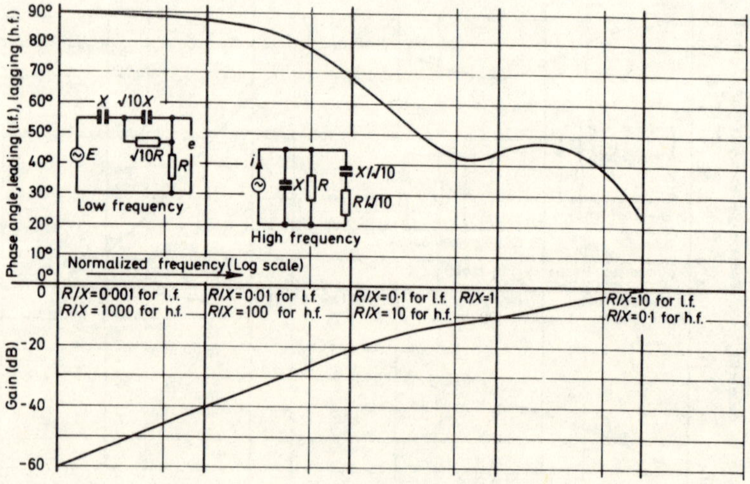

FIG. 8.14 Network 6

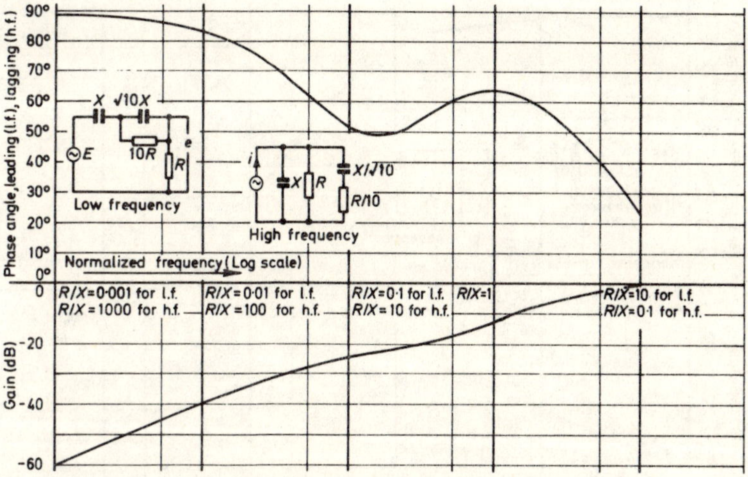

FIG. 8.15 Network 7

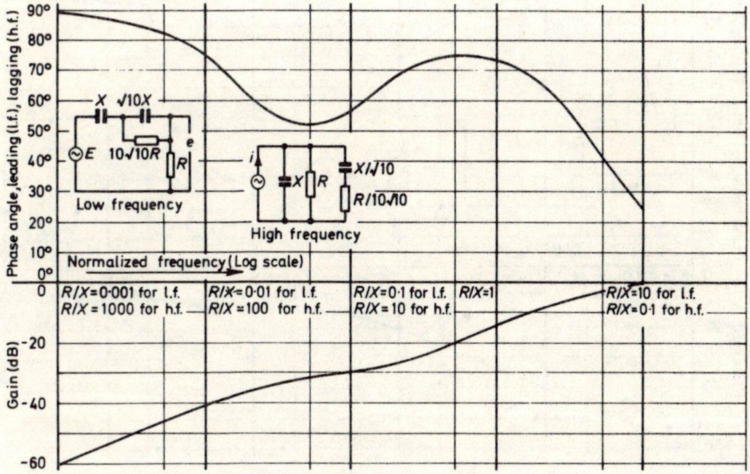

FIG. 8.16 Network 8

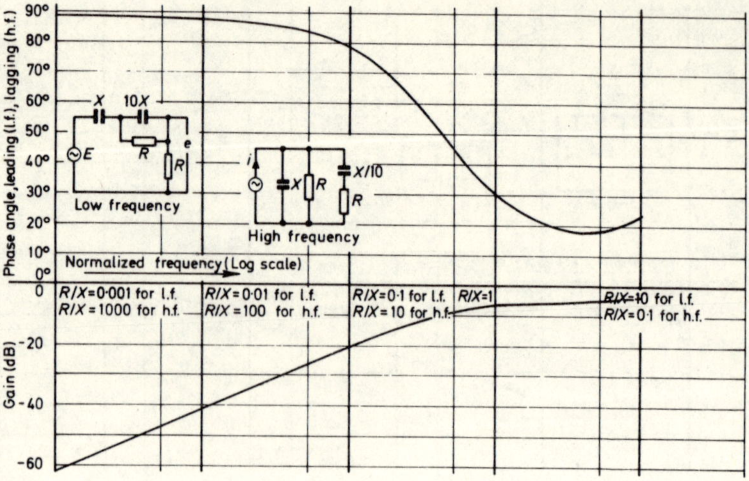

FIG. 8.17 Network 9

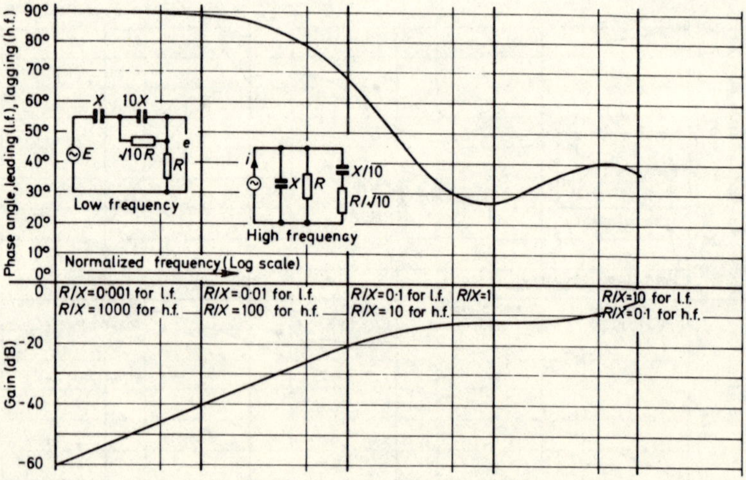

FIG. 8.18 Network 10

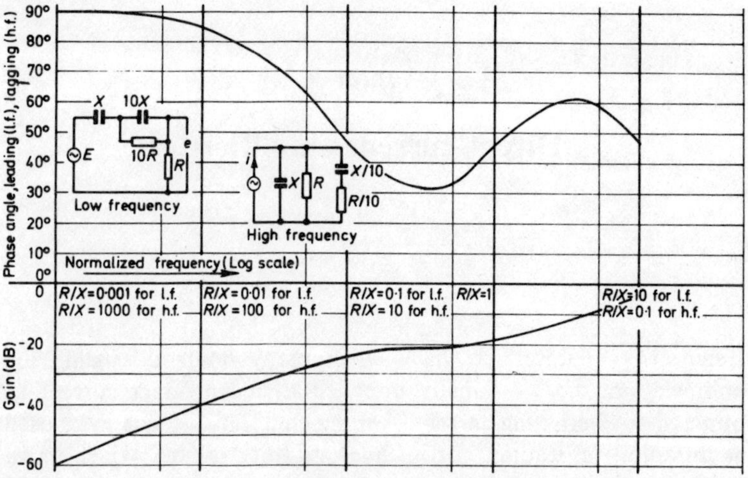

FIG. 8.19 Network 11

Chapter 9

Direct-current amplifiers

Engineering philosophers have spent many hours discussing the abbreviation 'd.c.', originally intended to mean direct current as opposed to alternating current. Among engineers it tends to be used as an adjective, leading, for instance, to 'd.c. voltage' which to an engineer means a zero-frequency signal.

When applied to an amplifier the term 'd.c.' could be taken to mean 'directly coupled' (i.e. no capacitor or transformer couplings) or 'direct current' (i.e. 1 μA input gives several milliamps output).

In the present context it is intended to mean an amplifier which operates as a voltage amplifier down to infinitely low frequencies: it is not necessarily direct-coupled nor does it necessarily amplify the input current.

DIRECT-COUPLED AMPLIFIERS

The simplest approach to d.c. amplifiers is to use the same circuit configurations as in Chapter 7. The practical difficulties involved are different in character, because changes of operating levels in the transistors caused by ambient temperature variations affect the output in the same way as if the signal were changing. Correct design of operating levels is therefore even more important.

For example, a simple earthed emitter amplifier has a gain of

$$\frac{R_L}{1/g_m + R_e + R_s/\beta}$$

and for maximum a.c. amplification R_e is by-passed, giving a gain of $g_m R_L$ for a zero resistance source. For d.c. amplification the by-pass capacitor must be replaced by a resistor or Zener diode or some other element which operates down to zero frequency, while at the same time the operating current must remain correct.

The base operating potential will be determined by the input signal, and the collector voltage, although it represents in amplified form the input signal, has a standing voltage level which may be undesirable.

Suppose, for example, that a signal of 0 to -50 mV from a source resistance of 100 Ω is to be amplified by 10, and the output level is to be 0 to $+500$ mV into a 1 kΩ load.

The direct attempt shown in Fig. 9.1 fails in at least two respects. First, the collector voltage has to be negative by 0·5 V or more for correct transistor operating, yet this is also the load voltage, which has to be 0 to $+500$ mV.

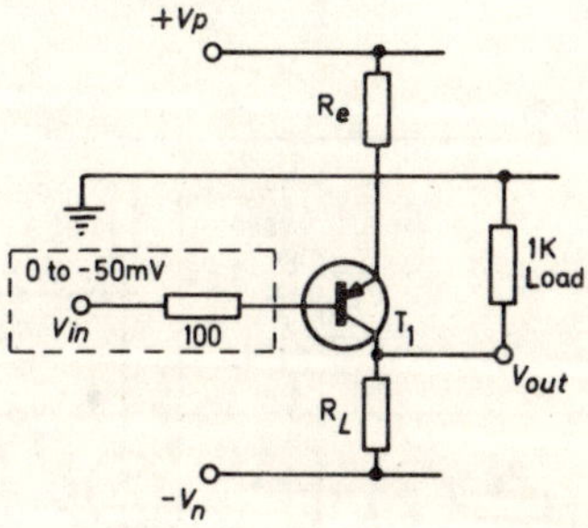

FIG. 9.1 Simple direct-coupled amplifier

Secondly, the voltage gain has to be 10, so that even if R_L were infinite, $R_e + 1/g_m + R_s/\beta$ would have to total 100 Ω since the gain is

$$\frac{R_L//1000}{R_e + 1/g_m + R_s/\beta}$$

This implies R_e of about 30 Ω and $1/g_m$ of the same order, which requires I_e of 2 mA and a positive supply of $V_p = (V_{eb} + 60 \text{ mV})$.

This leads to very poor temperature-stability and great variation if the transistor is changed for another of the same type; in fact, on V_{eb}-tolerance alone, some transistors would conduct heavily and others cut-off. There remains also the problem of collector voltage.

The above disastrous example illustrates that when d.c. gain is achieved, stability of levels is much more difficult to maintain; this is only to be expected, since the main technique for good stability is low d.c. gain.

Persevering with the simple circuit of Fig. 9.1, some of the snags can be overcome. The supply voltage requirement of (V_{eb} + 60 mV) for V_p can be obtained as shown in Fig. 9.2 from a more readily available supply line; provision can be made for adjustment for different V_{eb}, which could vary from 0·4 to over 1 V for a silicon transistor, and a change of level can be added in the load circuit (V_Z).

The use of setting-up potentiometer (RV_1) is often required in d.c. amplifiers, because even when temperature stability is good, some means have to be provided to offset resistor and V_{eb} tolerances. In Fig. 9.2 RV_1 is adjusted so that with zero input voltage V_{out} is zero. The voltage gain cannot be predicted accurately, as the setting of

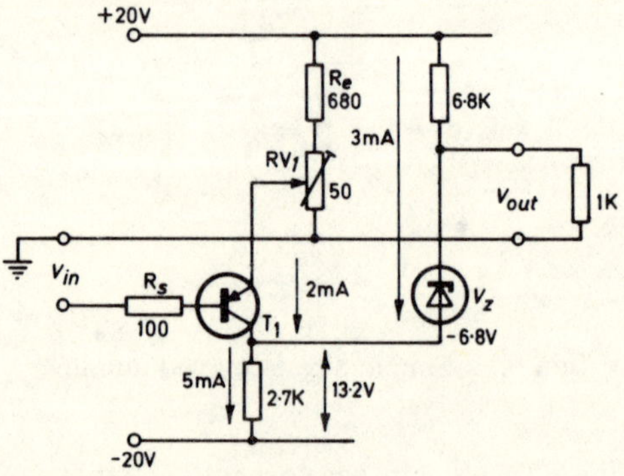

FIG. 9.2 Practical form of simple direct-coupled amplifier (Fig. 9.1)

RV_1 directly affects the gain and $1/g_m$ is also subject to wide variation. The first problem could be overcome by making RV_1 much lower in value (e.g. 10 Ω), but the standing current through it (≈ 100 mA) then represents a considerable drain on the supply.

Temperature-stability is poor, since a 10 degC rise causes V_{eb} to change 20–25 mV, having the same effect on the output as an input signal change of the same amount.

By modifying the circuit as shown in Fig. 9.3, the above difficulties are relieved, with the exception of gain variation due to $1/g_m$-tolerance, which is unchanged.

In this emitter-coupled pair the emitter circuit of the second transistor acts as the emitter load of the first and the setting-up control has much less influence on the gain, since its apparent value in the emitter circuit is reduced by a factor β. Temperature changes tend to cause equal V_{eb} variation in T_1 and T_2, and if these match exactly the output change is only $\delta V_{eb}(R_L/R_e)$ rather than $\delta V_{eb}(g_m R_L/2)$.

The signal behaviour of this circuit is identical to the a.c. performance of the emitter-coupled pair when the free base is decoupled, and is dealt with in Chapter 4.

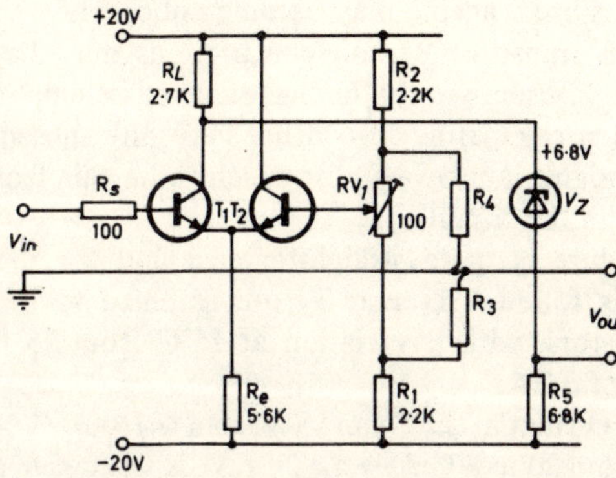

FIG. 9.3 Improved version of Fig. 9.2

Drift in d.c. Amplifiers

d.c. amplifiers suffer from two main defects, 'zero drift' and 'gain drift'.

Zero drift

Ideally, when the input signal is zero the output should remain constant (not necessarily at zero), but in practice temperature and voltage supply variations cause the output to drift.

The importance of this drift depends on how much signal change would have been required to produce the same effect at the output. It is customary therefore to quote the zero drift of a d.c. amplifier in terms of the equivalent input signal, and this is known as 'referring the drift to the input'.

Gain drift

When the input changes, the output changes by a larger amount such that $V_{out} = GV_{in}$. 'Gain drift' refers to the change in G due to temperature, supply variation, and ageing effects.

This effect is usually not so important as zero drift, since change of G always represents the same percentage error in signal. Zero drift is independent of signal and therefore represents an infinite percentage error if this signal is infinitely small.

Calculation of drift

Unless the amplifier is kept in a constant-temperature enclosure, the drift contributions to be considered are power supply and temperature variations, ageing effects normally being negligible. When an enclosure is used, ageing may become noticeable.

The various transistor parameter variations must be considered as in Chapter 2; usually only the first stage of an amplifier need be examined when calculating zero drift, since any subsequent variations are less significant by a factor which is the gain from the input to that point in the circuit.

The procedure is quite straightforward and for the circuit of Fig. 9.3 is as follows. T_1 and T_2 are assumed to be low-power silicon transistors, with β variation at 25°C from 25 to 100, and I_{cbo} at 50°C of 3 μA.

Input base current at 25°C can vary from $(I_c/100) - I_{cbo}$ to $(I_c/25) - I_{cbo}$, i.e. from 20 to 80 μA $- I_{cbo}$ if RV_1 is set to give about 2 mA in T_1 and T_2. At this temperature I_{cbo} may be 0·5 μA, giving a range of I_b from 19·5 to 79·5 μA.

At 50°C, for example, β variation will be from 150 to 37·5, assuming 2 per cent per degree C rise, and I_{cbo} may be 3 μA, giving a range of I_b from 10·5 to 50 μA.

Possible base current limits are therefore 10·5 to 79·5 μA, but this represents initial transistor β tolerances as well as temperature effects and indicates the voltage adjustment required from RV_1 because of β and I_{cbo}.

For a given circuit, after setting up RV_1, drift is worst for a low-β transistor and I_b can then vary from 79·5 μA at 25°C to 50 μA at 50°C, a drift of 29·5 μA. This causes the input base to move by $29{\cdot}5 \times R_s$ μV = 2·95 mV (25 to 50°C).

The base of T_2 has a similar influence on the output as the base of T_1 but opposite in sign. Its maximum variation is the same as above, except that R_s is replaced by the impedance between the base and ground. This is highest when RV_1 is central and is then 75 Ω [(50 + 100)//(50 + 100)] if R_1 and R_2 are ignored, giving a maximum shift of +3·9 mV, which is equivalent to −3·9 mV at T_1 base.

Unfortunately, these drift voltages cannot be added, because it can well happen that one transistor has a high β with no temperature coefficient and zero I_{cbo}. If this applies to T_2, then T_2 base variation with temperature is zero.

Input drift from β and I_{cbo} can therefore total 2·95 mV for a 25 to 50 degC rise.

Changes of V_{be} with temperature tend to cancel, since if both drift exactly together at the maximum value of $-2{\cdot}5$ mV/degC rise, the effect is merely to increase the drop across R_e by 62·5 mV for a 25 to 50 degC rise. This is an increase in ($T_1 + T_2$) emitter current of $62{\cdot}5/5{\cdot}6 \approx 11$ μA, i.e. 5·5 μA in each transistor. The input voltage which would be required to increase I_{c2} by this amount is given by $2(5{\cdot}5/g_m)$ μV, or approximately 0·5 mV. It is clear that the larger R_e can be, the less is the error due to equal V_{be} changes; replacement of R_e by a constant-current device (Chapter 6) virtually eliminates this error, but of course drift in the device has to be considered.

When the two drifts are unequal, the equivalent input drift is the difference between the two. For dissimilar transistor types the drift is therefore $\pm(2{\cdot}5 - 2)$ mV, i.e. $\pm 0{\cdot}5$ mV/degC, giving a total drift of $\pm 12{\cdot}5$ mV from 25 to 50°C.

It is therefore advantageous to use similar transistor types, when this figure will be reduced to about ± 5 mV, and if dual transistors can be used a further reduction to ± 1 mV or even less will be obtained.

Other sources of temperature drift are the resistors and ZD_1. If the coefficient of ZD_1 is $+0{\cdot}07$ per cent per degree C, then over a range from 25 to 50°C this is equivalent to an input drift of 70 mV$/G$, i.e. about -7 mV. Resistor drifts of up to $\pm 0{\cdot}02$ per cent per deg C can be calculated similarly and referred to the input.

Power supply variations also contribute to drift and the effect of the variation can be calculated by considering separately each entry point to the circuit.

For instance, if the -20 line changes by ± 10 per cent, the direct effect on the emitter circuit is to change the drop across R_e by ∓ 2 V, giving a change in each emitter current of $\mp(1/2)(2/5{\cdot}6)$ mA, i.e. 0·18 mA. This is equivalent to an input change of $\mp 2 \times (0{\cdot}18/g_m)$ mV $= \mp 18$ mV.

The -20 line affects T_2 base to an extent slightly dependent upon the setting of RV_1. If RV_1 is central, the change in base potential for ± 10 per cent in the -20 line is about ∓ 40 mV giving an equivalent input of ± 40 mV.

In this case a cancelling effect can be taken into account, since R_e and R_1 are attached to the same -20V line, the result being a drift of ± 22 mV referred to the input.

Variations in Directly Coupled Amplifiers

Because of the inherent cancellation of V_{be}, the differential amplifier is the basis of most low-drift directly coupled amplifiers.

Many refinements are used to increase input impedance, to reduce the effect of changing supply voltages, and to define the gain accurately. To reduce zero drift the only approach is to use the lowest possible base currents in the transistors directly connected to the source, and to balance V_{be} accurately in each half of the amplifier.

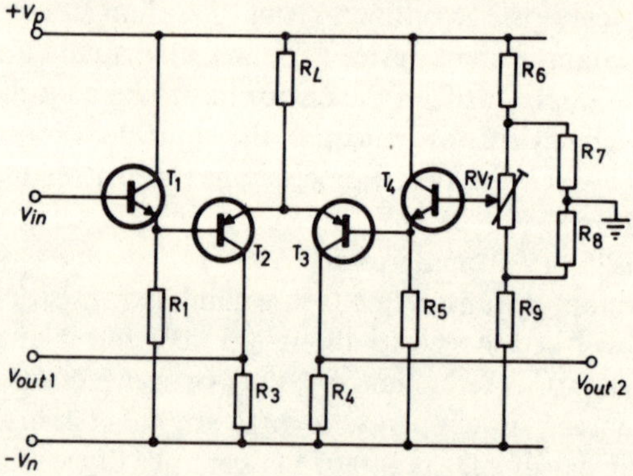

FIG. 9.4 Use of emitter follower

To this end emitter followers, preferably using planar epitaxial types of high gain, either *p–n–p* or *n–p–n*, may be added as shown in Fig. 9.4. An even better method, due to Bénéteau (Fig. 9.5), is to replace each transistor of the differential amplifier by a complementary pair (*see* Chapter 10). The advantage is that, as in the two-transistor differential amplifier, only two V_{be}s need to be balanced, yet the base currents are as low as the emitter followers in Fig. 9.4.

Supply voltage has its biggest effect in coupling to the second base of the input pair. One method to reduce this at the cost of 'rejection ratio' and gain is to earth this base and adjust for zero balance by an emitter potentiometer (Fig. 9.6). To be effective the drop across R_e' when the slider is at one end must exceed the possible V_{be} differential when half the emitter current in R_e flows through R_e'.

Bootstrapping, described in Chapter 15, may be added to increase the input resistance, but it must always be remembered that the input drift of an emitter-coupled pair will be V_{be} differential drift plus the

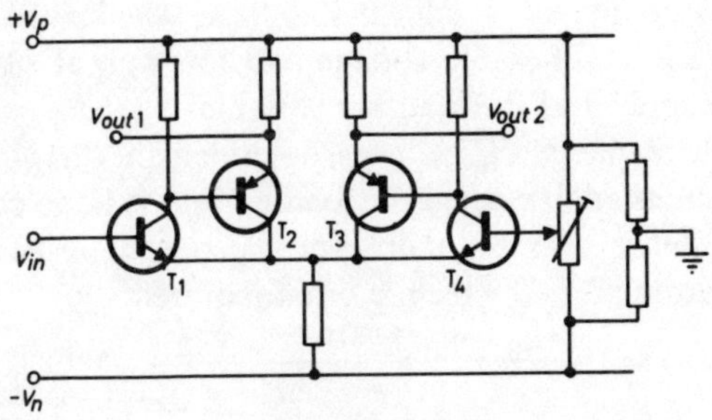

FIG. 9.5 Bénéteau's circuit

voltage produced when I_{b1} flows into the resistance seen by the base. If by bootstrapping or any other method the input resistance is made infinite, then the drift due to I_b is $I_b R_s$ where R_s is the source resistance. Removing the source then gives 'infinite' drift, i.e. the amplifier drifts until some limiting occurs.

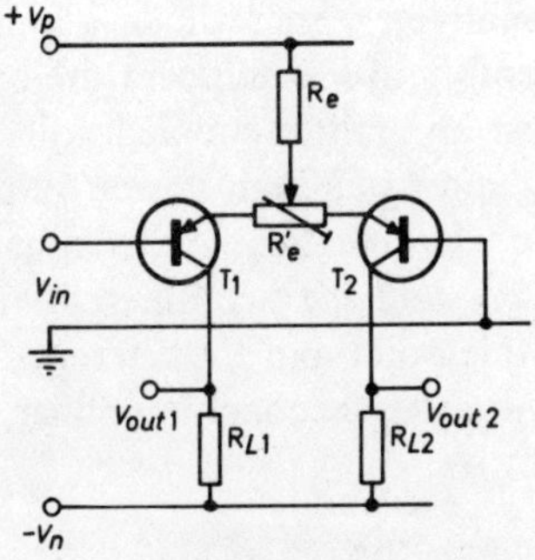

FIG. 9.6 Alternative zero set

In assessing a design it is therefore important to know its input voltage drift V_D with $R_s = 0$ and its current drift I_D which causes a further drift of $I_D R_s$. For example, an amplifier with $V_D = 1$ mV and $I_D = 1$ μA/degC will drift 1 mV + 1 V/degC when driven from a 1 MΩ source.

Typical Application: Voltage Stabilizer

By far the most common use for a directly coupled amplifier is the reference amplifier of a voltage stabilizer. The stabilizer is used to provide a source of stable voltage and low output resistance when the main supply and the load are variable.

A simple Zener stabilizer was described in Chapter 1, but this becomes undesignable when the load current is large compared with the rated Zener current; moreover the output resistance is much greater than the 0·1 Ω which is often required.

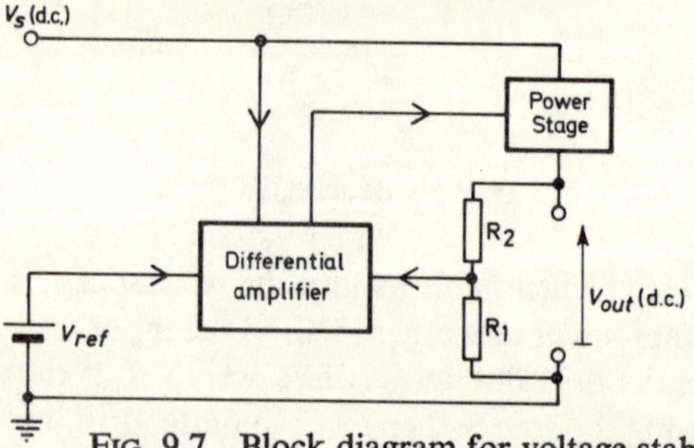

FIG. 9.7 Block diagram for voltage stabilizer

The principle of operation is shown in Fig. 9.7, where the 'reference' voltage is usually a Zener diode supplied as described in Chapter 1 and preferably of low temperature coefficient. A differential amplifier such as an emitter-coupled pair is supplied from the main d.c. supply V_s and has a high-power output stage from which the stabilized output V_{out} is taken. An attenuated version of V_{out} is coupled by R_1 and R_2, and the two differential amplifier inputs are connected between this point and the reference voltage V_{ref}.

The system is a negative feedback amplifier, and if correctly designed causes V_{out} to be

$$\frac{R_2 + R_1}{R_1} V_{ref}$$

regardless of V_s variations. The connections are such that if V_{out} is too small, the differential amplifier changes its current in the correct direction to increase V_{out}, and vice versa.

In practical stabilizers the power stage is usually a simple emitter follower. If it is arranged so that any output load current tends to increase the emitter follower current the circuit is called a 'series stabilizer' and the emitter follower is known as the 'series transistor' or 'series element'. If load current tends to decrease emitter follower current, the terms are 'shunt stabilizer', etc.

Typical Design

A practical example of the above system is given in Fig. 9.8, where T_1 and T_2 form the differential amplifier and T_3 and T_4 enable currents of about 1 A to be delivered to the load. Phasing can be checked by imagining the loop broken between R_6 and T_4 and levels adjusted to produce normal transistor operating conditions.

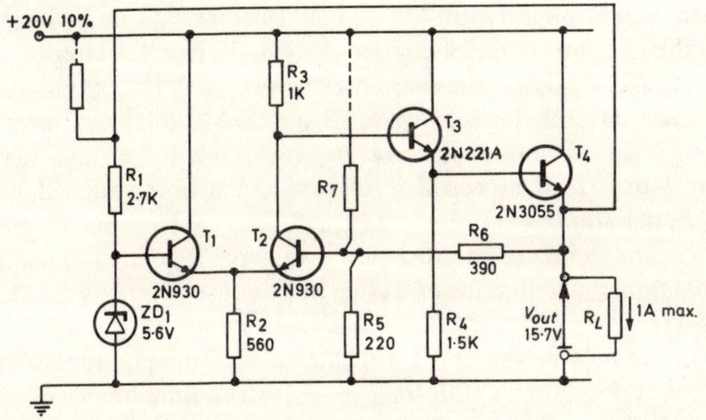

FIG. 9.8 Practical voltage stabilizer

Then move the free end of R_6 negatively. This causes T_2 to pass less current, so that its collector moves positive. T_3 emitter and T_4 emitter therefore move positive, and when R_6 is reconnected this will tend to move R_ϵ positive, thus resisting its negative movement.

To design such a circuit, assume that all voltage levels are at the desired level. In this case the output required is taken to be about 15 V with a current 0–1 A for $V_s = 20 \pm 10$ per cent, temperature 0–100°C. To make the potential of base 2 of the differential amplifier correct (5·6 V), a ratio of about 2·8 to 1 is required for $(R_5 + R_6)/R_5$. For reasons of drift and loop gain, to be discussed later, the values of R_5 and R_6 should be as low as possible as long as they do not drain a large proportion of the output current capability of T_4. $R_5 = 220\ \Omega$ and $R_6 = 390\ \Omega$ are suitable; lower values in similar proportions may also be used.

Since the output transistor will carry 1 A and have a V_{ce} of 20 − 15, i.e. 5 V, it must have a power rating of at least 5 W at the maximum operating temperature. If this is 100°C, then the 2N3055 in conjunction with a small heat-sink (of about 5 degC/W) is satisfactory and since its guaranteed minimum β_L at 1 A is 50, I_{b4} is 20 mA maximum. 'Minimum' current occurs at no-load (leaving only R_5 and R_6 as loads), when only I_{cbo} flows. At 100°C this is about $1\frac{1}{4}$ mA maximum, but if the full load has been applied, raising the junction temperature to 130°C, and is then removed, I_{cbo} may be 10 mA. The base current of T_4 can therefore be 20 mA flowing outwards, at one limit, or 10 mA inwards, at the other.

This is of extreme importance, as it shows the need for R_4; the base current requirement for T_4 could be inwards, and without R_4, T_3 then has no source of emitter current. T_4 base falls until $I_{b4} = 0$, giving $I_{e4} = \beta_4 I_{cbo4}$; the loop loses control and the output remains at some unstable level between 15 and 20 V. Stabilizers have often been designed without R_4, and they invariably fail at high temperature with no load, especially if full load had just been applied, making T_4 hotter still.

The only safe design procedure is to make R_4 current equal to the maximum possible value of $T_4 I_{cbo}$, in this case 10 mA, so that R_4 = 1·5 kΩ.

T_3 can now be chosen; it will carry a maximum current of 10 mA (from R_4) + 20 mA (full load for T_4 with minimum β_L of 50 and zero I_{cbo}), i.e. 30 mA at 5 V, a power rating of 150 mW at 50°C. A 2N2221A is suitable and has a minimum β of 25, giving a maximum I_{b3} of 1·2 mA.

R_3 must now be designed so that when I_{b3} is maximum the potential across R_3 is such that its more negative end never goes more negative than $(15 + V_{eb}T_3 + V_{eb}T_4)$ V, i.e. +16 V. (No great accuracy is required here so long as a large margin is allowed in the value of R_3.) For the first time in the design the more positive limit of V_s is important, since the loop will fail completely as soon as the drop across R_3 is too large. This positive V_s limit is +18 V, so that 2 V drop in R_3 can be allowed in the limit when I_{b3} is maximum. This gives a limit value for R_3 of (2/1·2) kΩ, i.e. 1·6 kΩ, and to ensure a large margin for tolerance and to avoid T_2 approaching cut-off under I_{b3} maximum conditions, a value of 1 kΩ is selected.

Under the opposite set of conditions, where $V_s = +22$ and $I_{b3} \approx 0$, which can occur if T_3 and T_4 have high β and high I_{cbo}, then to bring the output to +15 a current of $(22 - 15)/R_3 = 7$ mA must be supplied to R_3. (This time no allowance for V_{be} was made, since in the worst case this approaches zero.) This current must therefore be available from R_2, the value of which must be $(|V_{ref.(min.)}| - |V_{eb1}|)/7 = 4{\cdot}3/7 = 610\ \Omega$, if $V_{ref.} = 5{\cdot}6 \pm 10$ per cent and $V_{eb1} = 0{\cdot}7$ V. R_2 is chosen as 560 Ω.

T_1 and T_2 have to carry a maximum current of (5·6 V + 10 per cent)/R_2, i.e. 11 mA, and T_1 may at the same time have 16 V, giving a dissipation of 176 mW. This may be reduced by connecting T_1 collector to T_4 emitter, since this is just as effective a supply line as V_s. Dissipation is then 121 mW maximum in T_1 and T_2, and the 2N930 or BC108 are suitable.

R_1 may now be decided to suit Zener diode ZD_1, which may be of low power rating, e.g. 300 mW, with an optimum current of 5 mA. T_1 base current may reach 1/3 mA, which has little effect, and the value of R_1 may be $(20 - 5{\cdot}6)/5 \approx 2{\cdot}7$ kΩ.

Refinements and their dangers

As in the case of T_1 collector, there appears to be no reason why R_1 should not be connected to T_4 emitter, with a change in value to $(15 - 5{\cdot}6)/5 = 1{\cdot}8$ kΩ, since this is a more stable supply than V_s and will result in less variation in the reference when V_s changes. In this particular design this is a good idea and should be adopted. In general this 'gimmick' always needs careful examination, because the result may be that the circuit never 'starts', the output remaining at zero. (If this should happen in a stabilizer, a momentary resistive link from V_s to the reference Zener will 'start' the stabilizer and the link may then be removed.)

The only reason for its success in the present circuit is that R_3 is pulled positive by V_s and this pulls the output positive, so turning ZD_1 on until full stabilization is reached. An example where non-starting results is shown in Fig. 9.9, in which T_3 is an earthed emitter stage instead of an emitter follower. To overcome the inversion of signal thus produced, T_3 is supplied from T_1 collector instead of from T_2. If ZD_1 is supplied from the output as shown and V_s switched on, R_A and R_B are pulled positive, but unless the I_{cbo} of T_4 pulls R_C positive enough to turn on T_4 and ZD_1, the output never reaches stabilizing level. Momentary connection of R_D or R_E starts the circuit.

Compensation for line changes

Returning to Fig. 9.8, it is clear that variations in V_s are coupled into the system by R3, thus undesirably varying the output; the percentage variation is reduced by the loop gain (which is typically 10 in this simple circuit), but this still amounts to ± 1 per cent.

One way to reduce this is to couple some of the V_s variation into T2 base at such an amplitude that it cancels the V_s change seen at the collector. This requires a resistor R7 of such a value that

$$V_s \frac{R_5//R_6//R_{in2}}{R_5//R_6//R_{in2} + R_7} R_3 g_{m2}/2 = V_s$$

R_7 therefore depends on g_{m2} so that it cannot be given an exact value and will require adjustment for the particular transistor. This is not

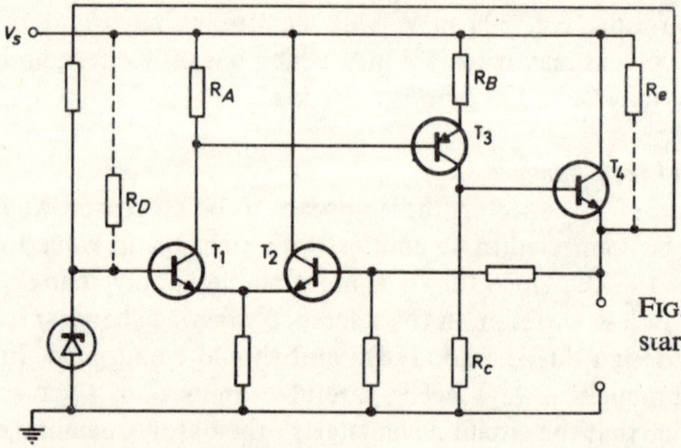

FIG. 9.9 A 'non-starter' (*see text*)

a welcome situation, but even without exact setting a marked improvement is given and this applies to any form of V_s variation, including ripple. In Fig. 9.8 R_7 should be about 10 kΩ. Note that in Fig. 9.9 R_A and R_B move together when V_s varies, so that if the output resistance of T1 ($\approx r_c/\beta_1$) is much higher than R_A, no change of current results in T3; thus, no change appears at the output.

Positive feedback to give zero output resistance

The use of positive feedback within a negative feedback loop has a surprising effect. If the positive feedback is applied to one stage of the loop amplifier to the critical point where the positive loop would normally oscillate, the negative loop behaves as if it had infinite gain. Oscillation does not occur (except for other unconnected reasons), and the output impedance of the negative feedback system is zero. If positive feedback is increased beyond what would normally cause oscillation, the system is still stable and the output impedance becomes negative.

Although this is a useful result, it is by no means the cure for all stabilizer or feedback problems. The main difficulties are, first, that the value of positive feedback for $Z_{out} = 0$ is highly critical and cannot be held with great accuracy, so that the risk always exists that Z_{out} may become negative, causing oscillations with certain types of load; and, secondly, that the bandwidth over which the critical condition holds is restricted to a fraction of the normal system bandwidth.

Both these considerations make this idea useful only in narrow-band systems where negative Z_{out} is tolerable—namely stabilizer circuits.

Here the idea is practical provided the designer does not attempt to convert a poor performance into perfection solely by its means.

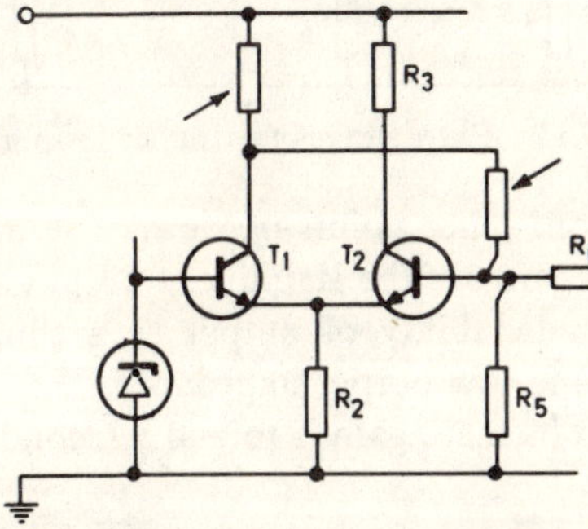

FIG. 9.10 Adding positive feedback to practical voltage stabilizer (Fig. 9.8)

If, for example, Z_{out} is 1 Ω and the use of positive feedback is used to reduce Z_{out} to 0, then normal tolerance effect will result in $Z_{out} = 0 \pm 0{\cdot}1$ Ω, not $Z_{out} = 0 \pm 1$ mΩ.

Fortunately, the incorporation of this scheme is very simple and can be done by adding a collector load in T_1 (so designed that T_1 cannot saturate on $I_{c1(max.)} = 10$ mA) and a resistor from T_1 collector to T_2 base. The value of the latter component should ideally be such that the gain round the T_{1c}–T_{2b}–T_{1c} loop is just unity. The extra components affect T_2 base, and some change in R_5 or R_6 is required to reset V_{out} correctly. These conditions are easily calculated: what is not easy is to estimate the long-term drift in the positive loop gain. As stated above, this 'trick' should be regarded as a final touch to an already adequate design (Fig. 9.10).

Extra loop gain

Using the original circuit of Fig. 9.8, the loop gain of about 10 is often insufficient, although attention to some of the above refinements will often remove the need for more. (Only isolation, not loop gain, is required to improve performance against changes in V_s; although more gain also improves this, it is bad design to use more

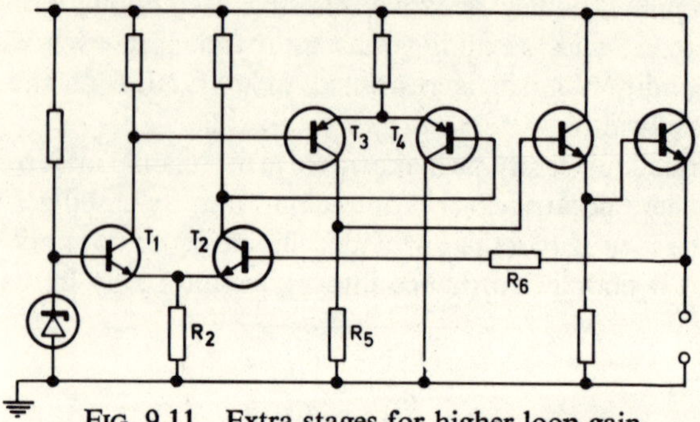

FIG. 9.11 Extra stages for higher loop gain

than is required from other points of view.) The important advantages of more loop gain are the stability of output against amplifier gain variations; predictability of output as a simple ratio of the reference voltage; and low output impedance.

The usual way to increase gain is to add a second coupled pair of

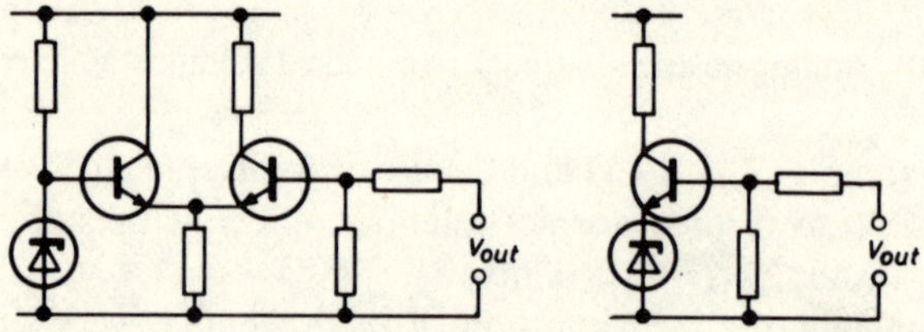

FIG. 9.12 Alternative forms of reference amplifier

the opposite type (*see* Fig. 9.11); another method is to use an operational amplifier in the loop or to use a ready-made integrated circuit stabilizer.

Combined reference amplifier

An alternative to the Zener diode/emitter-coupled pair combination is the use of a single transistor with the reference diode in its emitter circuit (Fig. 9.12).

This has certain conveniences counterbalanced by slightly worse performance.

The advantages are: the saving of one transistor; and the omission of the Zener feed resistor if the transistor emitter current is suitable.

Disadvantages are: temperature performance depends on V_{be} matching with V_Z (e.g. 6·8 or 8·2 V Zener tends to match a transistor V_{be} in temperature coefficient), whereas in the first circuit matched transistors and low-coefficient Zener can be obtained; change of output current varies output transistor V_{be} and results in some change of Zener current; variable supplies are difficult because the Zener cannot be tapped down, as a very-low-resistance potentiometer would be needed and if the feedback resistors are varied, minimum output is V_Z, not zero; positive feedback is more difficult to add.

The first snag has been tackled by some manufacturers, who supply a matched Zener-transistor reference amplifier.

On the whole, the single-transistor version tends to be used in fixed supplies unless performance is to be exceptional.

Loop stability

Since a stabilizer is a feedback circuit the problem of loop stability exists. The principles of controlling loop response were given in Chapter 8.

The designer must assume that the output load may be capacitive to any degree and so the only safe method to cure loop oscillation is to add output capacitance to make the output load the major time constant. Additional load capacity will then improve stability. In many applications the a.c. performance of the loop is unimportant and the simplest techniques may be used. If the loop has to have good h.f. response it must be treated as a wide-band feedback amplifier as in Chapter 8.

SUMMARY

Directly coupled amplifiers are designed by assuming the desired conditions have been achieved and the values chosen accordingly. Feedback circuits incorporating such amplifiers are equally simple to design, the example chosen being the voltage stabilizer.

CHOPPER AMPLIFIERS

When amplifying a slowly moving signal such as the output of a thermocouple by means of a conventional direct-coupled amplifier, slow drift in the amplifier caused by ageing or ambient temperature changes are indistinguishable from the signal. Although precautions can be taken to keep amplifier drift small, some drift inevitably takes place and this sets a lower limit to the input signal level which can be measured to a given accuracy.

If, however, the input signal is treated before entering the amplifier in such a way that it can always be distinguished from other signals not so treated, then amplifier drift becomes unimportant. (Strictly, it is the 'zero drift' which becomes unimportant; 'gain drift' is still as significant as before, but this causes a percentage rather than a fixed error so that it imposes no lower limit to the useful signal level.)

One system using this principle is the 'chopper amplifier', which in its simplest form consists of a vibrating switch S_1 which alternately connects and disconnects a short-circuit across the signal; often a series resistor R_1 is added where a direct short-circuit could cause excessive current flow (*see* Fig. 9.13).

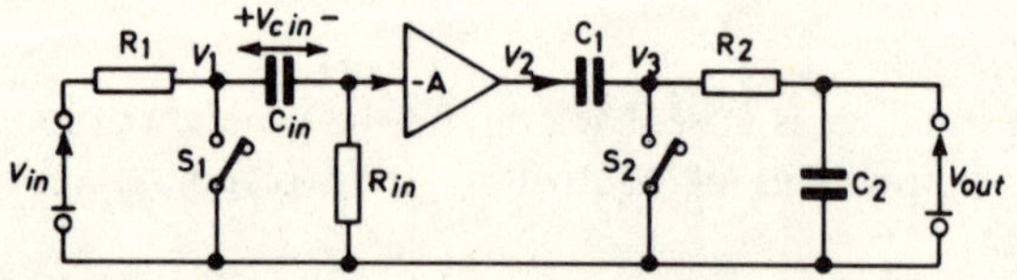

FIG. 9.13 Chopper amplifier block diagram

In this way the input V_{in}, which is assumed to move very slowly compared with the switch period ($T_1 + T_2$), appears at v_1 in the form of a square wave which moves between zero and V_{in} alternately. If v_1 is now a.c. coupled to an a.c. amplifier, the signal is always recognizable as the peak-to-peak amplitude of the square wave, any slow drift of the mean level in the amplifier being insignificant.

After amplification to a level $AV_{in(pp)}$ sufficient for direct monitoring, the signal can now be recovered in its original form (but amplified) by simple diode peak rectification which produces an output of $(1/2)AV_{in}$. This method has the fault that input signals of either polarity always produce the same output polarity, which depends only on the polarity of the rectifier diode connections.

Usually it is desirable to reproduce the input polarity, in which case a 'synchronous rectifier' is used. This is S_2 in Fig. 9.13 and is driven in synchronism with S_1, so that with the positive V_{in} assumed in Fig. 9.13, S_2 closes during the more positive 'half'-cycle of V_2. C_1 therefore acquires an extra charge corresponding to AV_{in} (the left terminal being the more positive), compared with the level it would carry if S_2 did not exist. When S_2 opens, V_2 simultaneously falls by AV_{in}, giving the waveform V_3 shown in Fig. 9.14.

Hence, V_3 swings from zero to $-AV_{in}$ and the average value of V_3, obtained by the smoothing circuit C_2R_2, is

$$-A\frac{T_1}{T_1+T_2}V_{in}$$

Knowing T_1/T_2 and A, V_{in} is therefore known in magnitude and sign.

Note that if S_2 is operated in the opposite sense to S_1 (i.e. closed

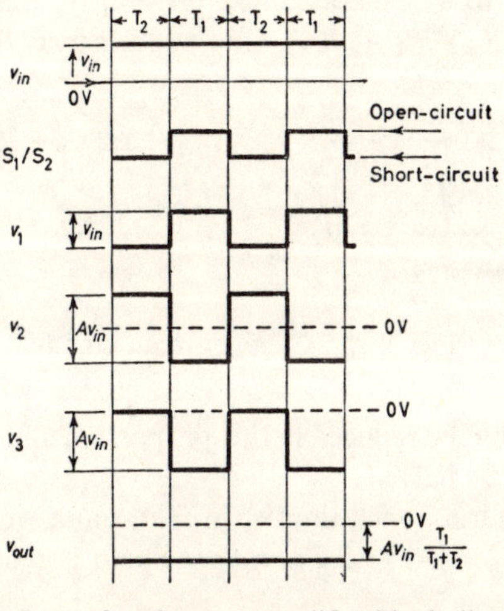

FIG. 9.14 Waveforms for chopper amplifier block diagram (Fig. 9.13)

during T_1), the output polarity reverses and is positive for positive inputs; the same result applies if the amplifier has a gain $+A$ instead of $-A$.

The synchronous rectifier possesses the valuable property of being insensitive to signals which are unrelated in frequency to its own switching frequency. The easiest way to understand this is to assume that no normal signal is present and that only random noise is coupled to S_2 through C_{in}. When S_2 closes, C_{in} is being charged and discharged randomly, so that when S_2 re-opens the output begins at zero and follows the amplifier variations until S_2 closes again. Since the noise has zero mean level, the integration of the output over many openings of S_2 yields zero output from the smoothing circuit. This applies however large the noise is, provided the amplifier does not change the form of the noise, e.g. by cutting off, and that S_2 works as a perfect switch.

Design Problems

Apart from the switches S_1 and S_2, which will be dealt with later, a number of problems arise which are of a somewhat unusual nature.

The first of these concerns the input impedance of the a.c. amplifier, which is represented in Fig. 9.13 as a resistor R_{in}. It is obvious that this causes a reduction in overall gain, because after S_1 opens V_1 cannot rise to V_{in}, as assumed previously, but only, it would appear, to $(V_{in}R_{in})/(R_1 + R_{in})$. One might expect therefore that the

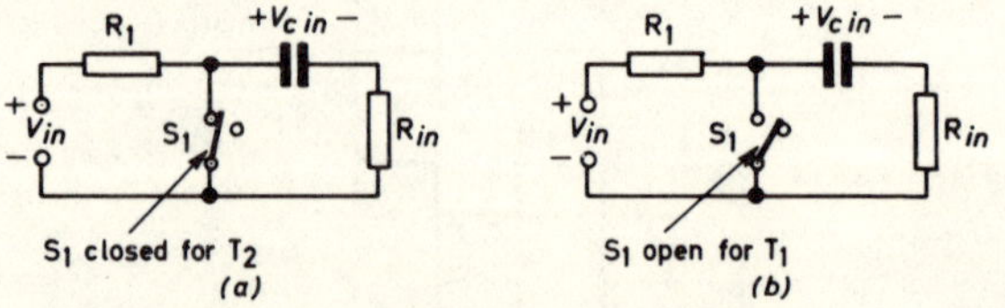

FIG. 9.15 Chopper input circuit

overall gain would be reduced in this proportion and that if $R_{in} = R_1$ the gain would be halved.

This very natural conclusion is, in fact, quite wrong and, as will be shown, if $R_{in} = R_1$ and $T_1 = T_2$ the gain is reduced by a factor of $\frac{2}{3}$, not $\frac{1}{2}$.

The method of calculation follows the principle given in Chapter 1: the charge and discharge of C_{in} per cycle are calculated and assumed equal, which must be so when equilibrium is reached. C_{in} is assumed very large ($C_{in}R_{in} \gg T_1 + T_2$), so that the a.c. waveform across C_{in} is negligible and it is assumed to have, at equilibrium, a steady voltage V_{Cin} between its plates.

When S_1 is closed, C_{in} discharges through R_{in} and since V_{Cin} changes negligibly during T_2 (i.e. no a.c. waveform on C_{in}) the current, assumed flowing from left to right, is constant at $-V_{Cin}/R_{in}$ (*see* Fig. 9.15 (*a*)). Charge flow in the direction shown is therefore $(-V_{Cin}/R_{in})T_2$.

When S_1 is open, the current is given by $(V_{in} - V_{Cin})/(R_1 + R_{in})$

in the direction shown (*see* Fig. 9.15 (*b*)) and charge flow is therefore $T_1(V_{in} - V_{Cin})/(R_1 + R_{in})$.

Since both $i_{C(T_1)}$ and $i_{C(T_2)}$ were assumed to be in the same direction, the sum of these charges must be zero.

Hence,

$$(-V_{Cin}/R_{in})T_2 + T_1(V_{in} - V_{Cin})/(R_1 + R_{in}) = 0$$

i.e.

$$V_{Cin} = \frac{V_{in}}{1 + (T_2/T_1)[1 + R_1/R_{in}]}$$

To calculate the peak-to-peak output across R_{in}, Kirchhoff's second law is used to find V_{Rin} in each position of S_1, regarding C_{in} as a battery of e.m.f. V_{Cin} in the direction shown. When S_1 is closed (i.e. during T_2) the voltage across R_{in} is $-V_{Cin}$ and when S_1 is open (during T_1) the voltage across R_{in} is

$$\frac{(V_{in} - V_{Cin})R_{in}}{R_1 + R_{in}}$$

The peak-to-peak voltage across R_{in} is therefore

$$V_{Cin} + \frac{(V_{in} - V_{Cin})R_{in},}{R_1 + R_{in}}$$

which reduces to

$$V_{in}\frac{1 + T_2/T_1}{1 + T_2/T_1(1 + R_1/R_{in})}$$

In the special case where $T_1 = T_2$,

$$V_{Rin} = \frac{2}{2 + R_1/R_{in}} V_{in}$$

so that if $R_1 = R_{in}$, $V_{Rin} = \frac{2}{3} V_{in}$.

Note also that if $T_1 \gg T_2$, $V_{Rin} \approx V_{in}$ provided $R_1 \not\gg R_{in}$, and if $T_1 < T_2$, $V_{Rin} \approx V_{in}/(1 + R_1/R_{in})$, an attenuation equivalent to a direct loading of R_{in} on R_1.

A second difficulty appears in connection with the synchronous rectifier. In the description of its action it was assumed that the capacitor, when connected to earth by S_2, would become charged to the signal level at the amplifier output within the closure time of S_2. In practice this does not always occur, since the current required to charge C_1 in this time may be more than the amplifier can supply. In this case C_1 becomes only partly charged and on successive cycles receives more charge until finally the correct level is reached. This represents a delay between the application of an input signal and the obtaining of the final output and is additional to the delay in the smoothing circuit which follows S_2.

In designing the output stage it is advisable to avoid this effect. The delay in itself may be tolerable and is calculable if the maximum current capability $I_{max.}$ of the output stage is known (from $I_{max.} = C_1 dV/dt$), but in most cases the value of $I_{max.}$ differs according to the direction in which C_1 is being charged. This gives unequal response times for rising and falling input signals which can cause instability if used in a servo loop (a common use for a d.c. amplifier). A more subtle consequence is that when the output stage is near cut-off, just before $I_{max.}$ is reached, any spurious ripple or noise associated with the signal becomes rectified and gives an output error, thus nullifying one of the best features of a synchronous rather than a diode rectifier.

An alternative arrangement which makes the design of the output stage simpler consists in adding resistor R_3 in series with C_1. The advantage of avoiding high current demand from the amplifier almost always outweighs the longer delay for a given value of R_2C_1 and C_2. In practice the longer delay is slight, is easy to calculate and is independent of output direction, provided the amplifier is designed to remain linear when the load of R_3 is switched in by S_2.

Chopper and synchronous rectifier switches

S_1 and S_2 may be either mechanical or electronic switches. Generally, the former behave more like the ideal switch in giving virtually perfect open-circuits and short-circuits, but the life and switching frequency are limited. Electronic switches can be transistors which are alternately cut-off and saturated, or may be photoconductive cells which change from low to high resistance when illuminated by a flashing light source.

Mechanical choppers

When amplifying input signals in the microvolt region from a high impedance source of several megohms, the present-day semiconductor chopper is unsuitable and a mechanical chopper has to be used. These have been highly developed and are now of much greater reliability than the standard astable relay which was often used in the past.

Even so, these devices should be used only when dictated by the circuit specification, since they are expensive, have short life, and usually fail to operate under military vibration conditions. Usually the contact arrangement consists of one changeover switch so that to work as S_1 and S_2 the moving contact has to be earthed.

The power consumed by the driving coil depends on whether a resonant structure is used. For a resonant vibrator the drive is of relatively low power (100 mW) but the 'over and back' time is fixed regardless of drive period: this results in non-unity mark/space of chopping if the drive period differs from twice the resonance period or if temperature changes affect the resonance. The non-resonant type is free from this difficulty and can be driven at any frequency up to its top limit, but requires several times the drive power (2 W) and is more expensive.

Other design problems when using mechanical choppers are the need to protect the chopper unit from mechanical shock and the need to screen the input chopper connections (which are close to the high-level synchronous rectifier leads) and eliminate earth currents in the lead to the common contact. It is also highly desirable to operate the chopper at a frequency unrelated to any mains fields which might enter the a.c. amplifier and be treated as a chopped signal, so that the use of the mains supply for the drive is questionable.

In some applications it is important that the signal earth should be 'floating' relative to the amplifier earth, and this can be done by using an input transformer. Specialists in mechanical choppers will supply complete chopper/transformer units, doubly screened to avoid undue stray pick-up.

Transistor choppers

As explained in Chapter 3, the alloy or planar epitaxial transistor can be operated as a switch by alternate cut-off and saturation. The performance of a transistor switch is not perfect, and can be represented approximately as a battery $V_{ec(sat.)}$ when saturated (Fig. 9.16 (*b*)), and as a source of current I_{ebo} when cut off (Fig. 9.16 (*c*)), if used in the optimum configuration (Fig. 9.16 (*a*)).

Assuming that $V_{ec(sat.)}$ may double for a 50°C rise from perhaps 3 to 6 mV, and that I_{ebo} may be negligible at 0°C and 1 μA at 50°C, the errors caused by these imperfections are easily calculated. If E_{in} were zero, e_{out} ought to be zero also, but will be alternately $V_{ec(sat.)}$

and R_sI_{ebo}. The sign of $V_{ec(sat.)}$ is negative and R_sI_{ebo} is positive, so that the resulting waveform has an amplitude of $(V_{ce(sat.)} + R_sI_{ebo})$ peak-to-peak. This is equivalent to an unwanted input of $(V_{ce(sat.)} + R_sI_{ebo})$ and this is therefore the drift caused by the chopper referred to the input. It is often referred to as the 'pedestal' of the chopper, since the input signal sits on top of it.

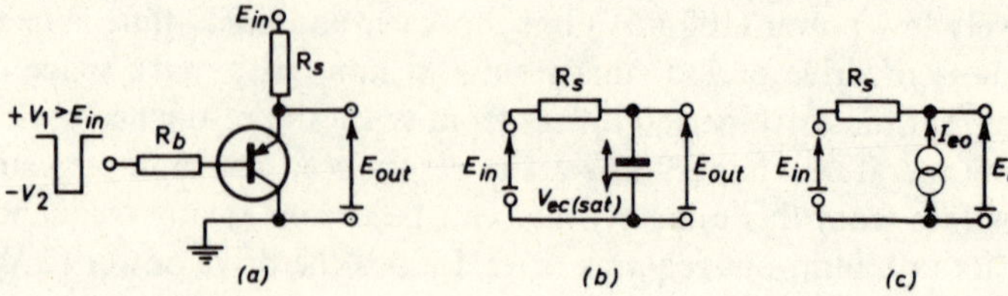

FIG. 9.16 Transistor chopper equivalent circuits

To minimize this figure, R_s must be as small as possible and a transistor with low I_{ebo} specified. When *only* low-level input signals will be present, the effect of I_{ebo} can be reduced, it would appear, to zero by driving the base only to collector potential (Fig. 9.17).

In this circuit D_1 cuts off when the drive goes positive, so that T_1 base remains at zero potential. Provided E_{in} is no larger than a few tens of millivolts in the positive direction, T_1 is cut off (or at least

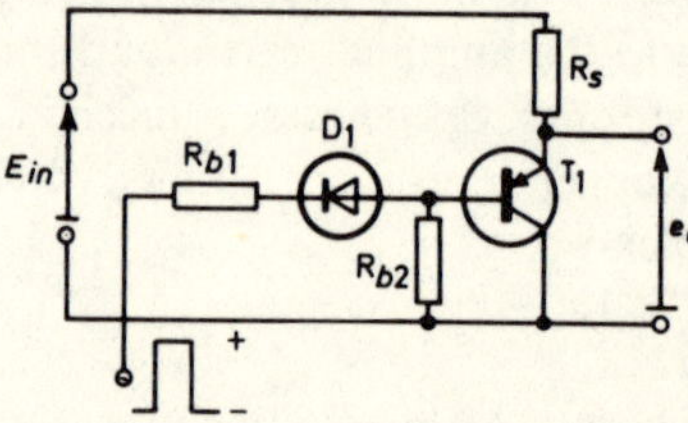

FIG. 9.17 Circuit to reduce effect of I_{eo}

presents a high resistance); and since there is no reverse bias from base to emitter, no I_{ebo} can flow. This is an over-simplification, because transients caused by hole storage in T_1 and D_1, and by transistor and diode capacitance, momentarily reverse T_1 base–emitter during which time I_{ebo} flows.

One idea to reduce the effect of $V_{ce(sat.)}$ is to introduce some of the drive waveform into the collector circuit of the chopper transistor.

This can be done by transformer coupling or by resistive injection from an inverted version of the drive. These methods do not help temperature drift of $V_{ce(sat.)}$, so that where temperature changes of 50 degC are likely, the total zero offset at one temperature extreme would normally be halved. A more useful compensation method for use under these conditions is to back-off the $V_{ce(sat.)}$ of two identical

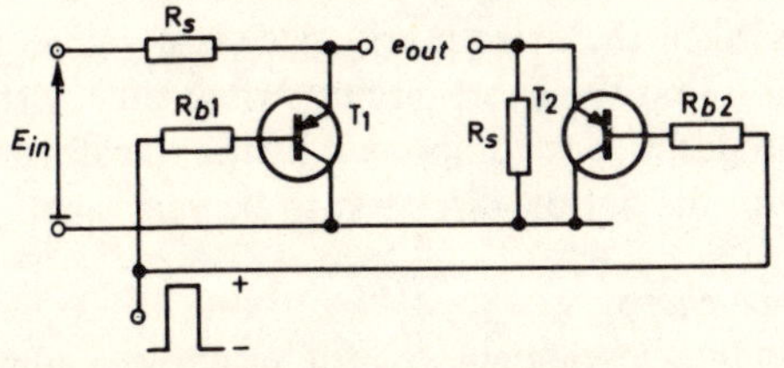

FIG. 9.18 Balancing $V_{ec(sat.)}$

transistors driven simultaneously (Fig. 9.18). In the circuit illustrated, a differential amplifier such as an emitter-coupled pair must be used to amplify the difference signal from T_1 and T_2 emitter. This may be overcome by transformer-coupling the drive (Fig. 9.19); if the transistors are matched for β and $V_{ce(sat.)}$, the emitter 1 to emitter 2 voltage at saturation can be less than 100 μV over a 50 degC temperature range. Suitably matched transistors are available from several

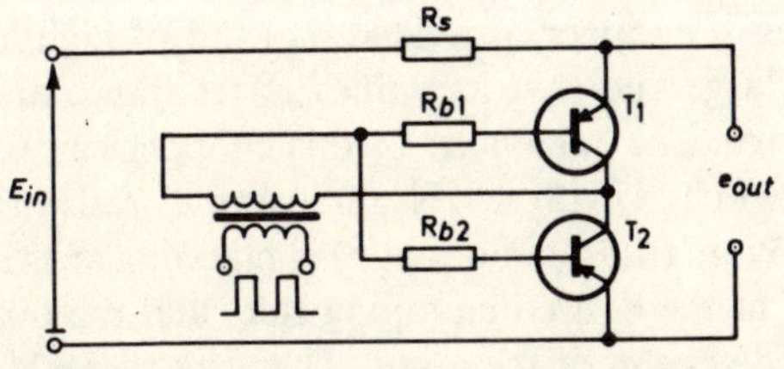

FIG. 9.19 Balancing $V_{ec(sat.)}$—transformer coupling

manufacturers, and in some cases they are encapsulated with a transformer and called solid-state choppers.

The circuit of Fig. 9.19 has further useful properties. Consider the state condition when no base drive is applied, and assume T_2 emitter is earthed and a potential is applied to T_1 emitter. If this signal is very small (a few millivolts) then no forward current or reverse current flows. If the signal is large and positive, T_1 conducts, causing T_2 base and collector to move positively. This cuts off T_2, since its emitter–base junction is reverse-biased and its collector and base are at the same potential, so that I_{ebo2} flows from the signal source. Conversely, if the signal is negative, T_1 cuts off and I_{ebo1} flows into the source.

The chopper is therefore open-circuit in the static condition for all signal voltages up to the V_{eb} reverse breakdown level. This can be advantageous when the drive mark/space is designed to be small in such a direction that the short-circuit switch time is short, since the removal of zero frequency component by the transformer then causes the excursion in the cut-off direction to be very small.

Photoconductive choppers

A photoconductive element of, for example, cadmium sulphide has the property of increasing its conductance when illuminated,

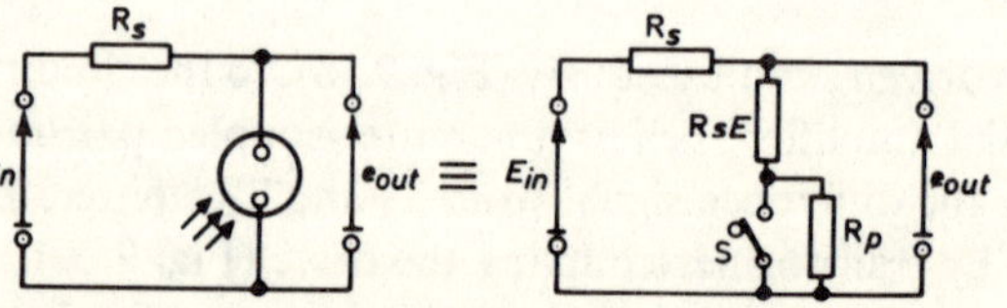

FIG. 9.20 Photoconductive cell used as a chopper

without generating internal voltages or currents. Such an element may be used as a chopper, provided its ratio of high/low resistance is sufficiently large and that its values of resistance are suitable for the signal source. The behaviour of this chopper can be represented as a perfect switch S with series and parallel resistors R_{se} and R_p (Fig. 9.20). A lamp (tungsten or neon) is placed near the element and made to flash at the desired chopping rate, and the output signal is then a chopped version of the input. The peak-to-peak output is less than the input signal by an easily calculated amount which depends on source resistance R_s, R_{se}, and R_p. This quantity varies considerably with ambient temperature and with lamp drive voltage but represents only a gain drift, not zero drift, so that overall feedback (as described later) masks the drift to any desired extent.

Zero drift would be non-existent if there really were no voltage generation in the element. Any asymmetry in its construction does, in fact, lead to some e.m.f., although this is only in the microvolt region even for quite inexpensive elements.

Practical difficulties in the use of photo-choppers. Two of the best features of transistor choppers over mechanical and photoconductive types are that they may be operated at a few kilocycles per second before transient effects ruin the performance. This implies that the overall bandwidth of the chopper amplifier can be a few tens of cycles after smoothing the output. The disadvantage of transistor chopping is the generation of unwanted e.m.f.s and currents.

The photo-chopper is more or less free from these defects but only the expensive materials have a fast response, cadmium sulphide being limited to about 100 Hz. As the drive frequency is increased, the element is unable to reach its high- and low-resistance values within the half-cycle, so that peak-to-peak chopper output falls and, since the waveform becomes non-rectangular, further loss occurs in the synchronous rectification.

Unless expensive elements (e.g. lead telluride) are used, the overall response is limited therefore to a few cycles per second. A further difficulty, directly related to speed of response, is that the chopped waveform lags several milliseconds behind the drive waveform. This occurs also in mechanical choppers but is much smaller (a few microseconds) in transistor choppers. If no correction is applied and the synchronous rectifier is driven by the same drive waveform, reduction of output results. Since the phase shift is not predictable and varies with lamp intensity (which changes by ageing), any compensation by extra phase shift in the drive is approximate only.

Another problem is the flashing light source, which may be a constantly lit source separated from the photo-element by a motor-driven rotating shutter, or may be a lamp driven on and off. The motor method has been used successfully in a range of commercial instruments. The flashing tungsten lamp has a limited life and a neon lamp (although spectrally suitable) requires higher voltage drive than is readily available in a transistor instrument.

Transformer coupling from the drive generator to obtain a high voltage requires a bulky transformer owing to the low drive frequency. One solution, devised by A. Errington (formerly of B.A.C., Stevenage), is to use the drive waveform to start and stop a high-frequency oscillator (100 kHz) the output of which is transformed up in voltage, rectified, and coupled to the neon lamp.

This last arrangement appears to be the best driving system, and although apparently complicated the cost and bulk of components is low and no special power supply is required.

Field effect transistor choppers

Field effect transistors behave in a similar manner to photo-conductive cells when used as choppers. The source-drain channel resistance switches from its normal value (between 10 Ω and 5 kΩ) to a very high value (several megohms) when the gate voltage is driven to cut-off. No e.m.f. is generated and gate-source leakage is very low.

Unlike the photo device, the isolation between drive and switch electrodes is not perfect due to gate-source and gate-drain capacitance which couples the drive waveform to the switch connections during each transition. Similarly 'back-wash' is coupled from the switching electrodes back to the gate. (In a photo-conductive cell it is naturally impossible for voltages present across the cell to modulate noticeably the brightness of the lamp.) Even though the isolation seems at first sight adequate (only a few picofarads) its presence gives rise to spike problems as in the transistor chopper. This is due to the large gate voltage swing needed to effect cut-off—usually 10 V more negative than both gate and source potentials for an *n*-channel depletion type such as the 2N3824.

Because of the spike problem the FET is sometimes used as a modulator rather than a switch. This is achieved by driving the gate with a sine wave superimposed on a d.c. level in such a way that neither cut-off nor full conduction is reached. This attractive-sounding solution is not ideal since a special drive waveform is required, having a.c. and d.c. levels suitable for the particular field effect transistor sample. Another snag is that the resulting amplitude of the modulated input d.c. signal is very much less than that obtained by using the field effect transistor as a true switch. A more subtle problem is that the field effect transistor gate-source capacitance causes a greatly attenuated version of the gate waveform to appear across the source-drain terminals in quadrature to the main signal (in quadrature because it is coupled by a very small capacitance). If the output chopper drive is exactly in phase with the input chopper drive and if no phase shift occurs within the amplifier then this quadrature signal produces no output error from the synchronous rectifier, unless it is so large as to cause limiting. If, however, a phase error does exist then the quadrature 'leakage' contributes to the output in the same way as if an input d.c. were present. This point has been emphasized because generally a slight phase error has no significance in a chopper amplifier, merely resulting in a slight loss of gain. Consequently great care must be taken to check this point if it is proposed to modify an existing chopper amplifier by using a field effect transistor modulator.

For most applications the field effect transistor, used as a switch, should be considered if a transistor is inadequate due to its $V_{ce(sat.)}$. It is easier to drive than a photo-conductive element but is less convenient to drive than a transistor.

Synchronous rectifier switches

The requirements here are less difficult to meet than in the input chopper. Signals levels are much higher, so that a switch 'pedestal' as high as several tens of millivolts is normally tolerable. The only extra problems arise in the use of a transistor switch where the base drive must exceed, in the cutting-off direction, the largest signal peaks, so that the series diode method for reducing I_{ebo} cannot be used.

If spurious signals are present which the synchronous rectifier is intended to ignore, then the base potential for cut-off must exceed the total possible signal peak (i.e. the sum of wanted signal and spurious signal); if not, the spurious signal becomes rectified and the resulting output is equivalent to an input error.

Overall Feedback in Chopper Amplifiers

Zero errors and zero drift in the input chopper cannot be improved by signal feedback, since no sensing element following the input chopping can distinguish between a true input with a perfect switch and zero input with an imperfect switch.

Feedback can be used with advantage in the reduction of the variations of gain, which occur when the mark/space changes, and also, in photoconductive choppers, if illumination changes. These changes apply also to the synchronous rectifier, and further overall gain changes occur due to variation in the gain of the a.c. amplifier and in its input impedance.

Whenever gain accuracy of better than a few per cent is required, it is therefore advisable to add overall feedback as shown in Fig. 9.21. As in any feedback system, the improvement obtained depends on the amount of loop gain, i.e. gain 'thrown away' in feedback. The designer must therefore assess the likely change in gain without feedback and deduce the amount of feedback required; then add the extra gain and apply the feedback remembering that the extra gain may also change.

Where the amplifier is to handle relatively large signals such as a few volts, but is to have very high input impedance (hundreds of megohms) an alternative feedback connection can be made (Fig. 9.22). Here the output signal after rectification smoothing and attenuation by n to 1, is connected to the input chopper in such a way that the input switch oscillates between input and attenuated output.

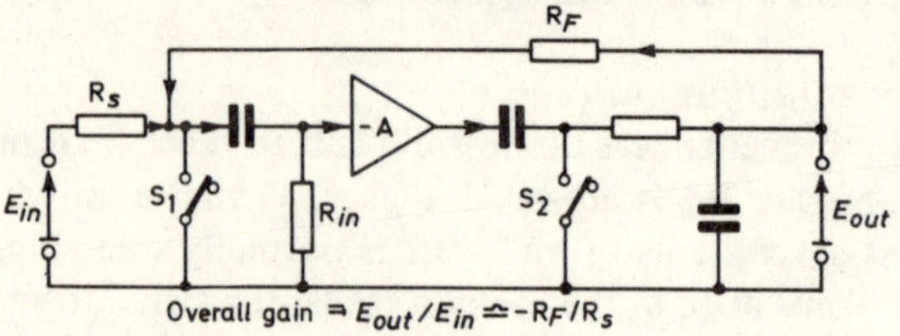

FIG. 9.21 Chopper system with overall feedback

With correct phasing the system settles in the state where the chopper switch output is very small and the output is n times the input to a degree of accuracy which again depends on the loop gain. If the gain from input chopper to smoothed output is A, then the loop gain is

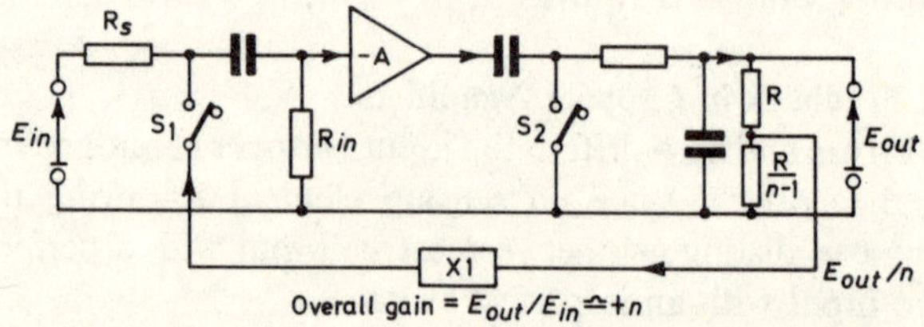

FIG. 9.22 Use of negative feedback for high input resistance

A/n and the input impedance is of the order of $R_{in}A/n$. Since R_{in} can easily be 1 MΩ either by bootstrapping (Chapter 15) or by the use of low-current planar transistors, and A/n can be 1000, an overall input resistance of 1000 MΩ with a gain accuracy of 0·1 per cent may be achieved. Zero drift is still present, so that the overall gain equation is of the form $V_{out} = nV_{in}(1 \pm 0{\cdot}1) \pm nV_0$ where V_0 is the zero drift referred to the input.

This feedback system is particularly effective with photochopping, since V_0 is then negligible.

Detailed Design of Chopper Amplifiers

The problems described above are mainly concerned with the chopper system rather than individual circuits: the details of circuit design follow normal considerations.

To take a practical example, suppose the output of a thermocouple of 100 mV maximum is to be amplified by 20 with an accuracy of ±10 per cent, with permissible drift of 10 mV referred to the input. (This zero drift must always be specified, as well as percentage accuracy.) Source resistance is 100 Ω and the source may be short-circuited without damage. Ambient temperature range is 0–50°C.

Choice of switches. The possibility of using a single-transistor chopper should be considered first, since this is the least expensive arrangement. The zero drift of a transistor chopper in the earthed collector configuration (Fig. 9.16) is given by $V_{ce(sat.)} + I_{ebo}R_s$, where $V_{ce(sat.)}$ is the value of V_{ce} for heavy saturation with zero emitter current, and I_{ebo} is the reverse emitter leakage at 50°C.

Manufacturers seldom quote the value of $V_{ce(sat.)}$ under the conditions mentioned above: they quote a value of perhaps 0·1 V for 10 mA I_c and 1 mA I_b in the earthed emitter configuration. This is useful in designing multivibrators and binary circuits but gives no indication of chopper performance. Fortunately, a very simple static

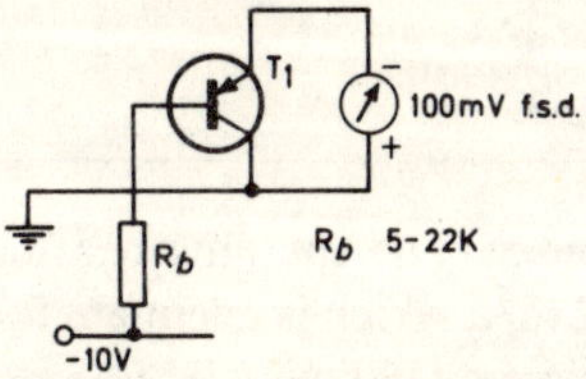

FIG. 9.23 Direct measurement for $V_{ce(sat.)}$

test can be made to convince the reader that the quoted 5 mV figure is correct. Using the circuit of Fig. 9.23, $V_{ce(sat.)}$ can be measured directly at several base currents, showing that I_b is not critical for a $V_{ce(sat.)}$ of 5 mV. Generally, a high-β transistor has lower $V_{ce(sat.)}$, provided it is of epitaxial or alloy construction.

Returning to the example, $I_{ebo}R_s$ is only 3 mV, so that a normal-drive waveform may be used, i.e. no series diode need be added (Fig. 9.17). The synchronous rectifier switch can be another transistor operated in the same mode. It is likely that the resistance in its emitter circuit will be higher than 100 Ω, so that its I_{ebo} will cause much more than 3 mV drift; however, the significance of any drift will be reduced by a factor of 20, so the rectifier may contribute 60 mV or more to the drift and the design will still remain within specification. This point must be confirmed when the design is complete.

Drive waveform. The simplest arrangement is a conventional multivibrator of unity mark/space, and if the two switches are of the same type (e.g. *p–n–p*) and are driven from opposite collectors of the multivibrator, a signal inversion in the a.c. amplifier will give an overall gain inversion.

A possible circuit is given in Fig. 9.24, where a 'standard' symmetrical free-running multivibrator is operated between +5 and −10 V supplies, giving a total period of $1{\cdot}4CR = 1{\cdot}4 \times 0{\cdot}1 \times$

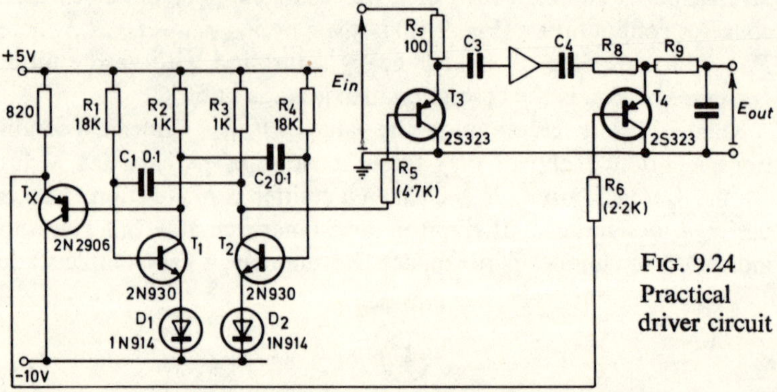

FIG. 9.24
Practical driver circuit

$18 \times 10^{-3} \approx 2{\cdot}5$ msec. The coupling resistor to T_3 of 4·7 kΩ gives a base current of 2 mA, which is adequate for good saturation at the maximum input current (100 mV/100 Ω, i.e. 1 mA). The value of R_6 which supplies $S_2(T_4)$ cannot be determined until R_8 and R_9 are fixed.

Synchronous rectifier components. These are so chosen that R_9 does not unduly load R_8, giving direct loss of gain; at the same time, R_9 cannot be so large that significant error is caused by the leakage of any emitter follower which may be added at the output; also, R_8 cannot be so low that closure of T_4 causes the amplifier to limit. The last condition is usually the most important, so it will be assumed that 10 mA is the maximum current which the amplifier can supply when T_4 closes. Since maximum peak signal is at least 2 V (to produce 2 V output), T_4 would have to pass $2/R_8$ A if the input jumped to maximum within one switching period. This leads to a value for R_8 of about 200 Ω, and to allow for output losses which increase the amplifier signal, R_8 should be about 270 Ω.

Note that in this application this is an extremely conservative value, since in the steady state (i.e. fixed input) T_4 would pass only enough current to make up for the output drain V_{out}/R_L. This current would be $2V_{out}/R_L$ and is much less than $2/R_8$. One point in favour of using the conservative figure is that no further protection against excessive amplifier current from the destructive point of view need be taken and sudden application of maximum signal by a switched input connection cannot cause damage.

C_4 can now be designed on the basis that it must charge to the correct voltage within the closure time of S_2. The conservative value is given by assuming that to charge fully in 1·25 msec. could occupy

a time of $4C_4R_8$, i.e. $C_4 = \dfrac{1{\cdot}25}{4 \times 270} \times 10^{-3} \approx 1\ \mu\text{F}$.

During the open-circuit interval of S_2, C_4 must not discharge appreciably so that $(R_8 + R_9)C_4 \gg 1{\cdot}25$ msec., i.e. $R_8 + R_9 \gg 1{\cdot}25\ \text{k}\Omega$, e.g. 12·5 kΩ, giving $R_9 \approx 12\ \text{k}\Omega$.

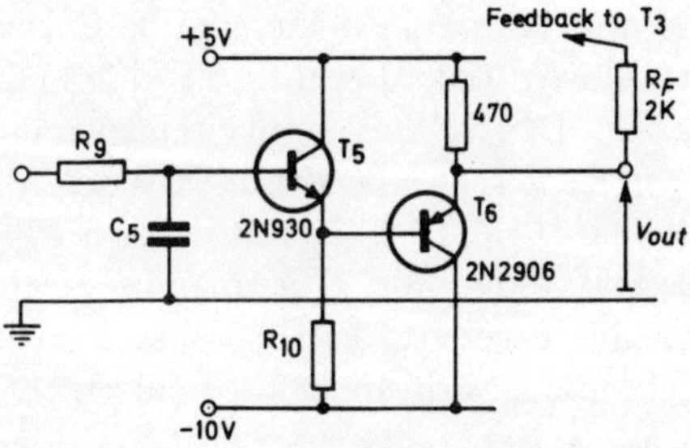

FIG. 9.25 Output buffer to drive heavy load or feedback resistor

If $R_L \geqslant 50\ \text{k}\Omega$, which is likely if the load is a pen-recorder or voltmeter, the attenuation $R_9{:}R_L$ is small. If $R_L \leqslant 10\ \text{k}\Omega$ then it is advisable to add a unity-gain amplifier between R_9 and R_L (*see* Fig. 9.25). One emitter follower could be used, but its V_{eb} variation of 125 mV from 0 to 50°C represents 6·25 mV zero drift and so two complementary transistors are preferable, as shown.

C_5 is now given by $C_5R_9 \gg$ switching period, i.e. $C_5R_9 \gg 2{\cdot}5 \times 10^{-3}$ or $C_5 \gg 0{\cdot}2\ \mu\text{F}$, e.g. 50 μF. This would have to be of the electrolytic type, but as its accuracy is unimportant and voltage rating less than 6 V, this is acceptable.

Returning to the base drive for T_4, the maximum emitter current for T_4 is 10 mA, so that a base drive of 2 mA would appear to be adequate, because this would cause near-saturation, though not to the millivolt level. When first turned on after an increase in signal from zero to maximum, T_4 would therefore saturate to perhaps 100 mV and on the next closure the emitter current would be only 100 mV/270 or 0·4 mA. On this closure, saturation to a few millivolts would now occur.

This reasoning is, however, incorrect, as it fails to take into account the reversed emitter current which flows if the input signal is suddenly disconnected. In this situation 10 mA reverse current flows and if β_{Lr} is < 5, S_2 does not close to saturation level. This gives an overall delay in amplifier response which differs according to whether the input is increasing or decreasing, an effect which R_8 was intended to prevent.

Two solutions are possible; either T_4 is a symmetrical transistor ($\beta_{Lr} \approx \beta \geqslant 20$) or the base drive is increased to $10/2 = 5$ mA. In either case the 'normal' saturation level is worse than for a very asymmetrical type with very small forward emitter current. In this design a 2S323 will be specified and the base drive increased by adding emitter follower T_x and putting $R_6 = 2{\cdot}2$ kΩ.

Temperature drift. Drift in the output circuitry is caused by T_5 base current flowing in R_9 so that $I_b R_9/20$ must not exceed a few millivolts in either polarity; if $|I_b| R_9/20 \leqslant 2 \times 10^{-3}$, then $|I_b| \leqslant 3{\cdot}3$ μA. The leakage component alone for a germanium transistor at 50°C (30 μA) would give excessive drift, so that a silicon planar type 2N930 should be used, with an emitter current of no more than 100 μA, giving $R_{10} \approx 100$ kΩ. T_6 may be silicon alloy or planar type, e.g. OC202, 2S323, or 2N2906, etc., with emitter current of no more than 1 mA (so that base current $\leqslant 30$ μA), giving $R_{11} = 5{\cdot}6$ kΩ.

Drift caused by the synchronous rectifier T_4 consists of its $V_{ce(sat.)}$ and its I_{ebo}. $V_{ce(sat.)}$ when divided by the gain of 20 will be negligible for any transistor suitable for chopping. I_{ebo} is more difficult to assess: when S_2 opens, I_{ebo} flows into R_9 in parallel with C_4 and R_8. If C_4 were infinite and R_8 were zero, no error would be caused, but in fact an immediate step of $I_{ebo}(R_8/R_9)$ occurs followed by a positive rise as I_{ebo} charges C_4.

If I_{ebo} is 1 μA, then the step error is $(12\ \text{k}\Omega//270) \times 10^{-6} = 0{\cdot}16$ mV, or 8 μV referred to the input. By the end of the open period for S_2, C_4 has charged by $(I_{ebo}T/2)/C_4$ which is

$$\frac{10^{-6} \times 2{\cdot}5 \times 10^{-3}}{10^{-6} \times 2} = 1{\cdot}2 \text{ mV}$$

or 60 μV referred to the input. The sum of errors is tolerable, so that T_4 may be a 2S323 transistor.

a.c. Amplifier. The circuits which are peculiar to the chopper have been dealt with and it only remains to specify the a.c. amplifier, which may then be designed using conventional techniques.

The required a.c. gain depends on whether overall feedback is to be used and this depends on the conditions of operation and installation. There are several causes of error: for example, mark/space of chopping, where the circuit is reasonably stable against ageing and temperature variation but where the absolute conditions have a wide spread. In such cases adjustment could be provided (such as one base resistor in the drive multivibrator or a gain control within the amplifier) to set this right in the factory. On the other hand, this involves extra labour and may be more costly than incorporating feedback which, especially in the rather wide gain tolerance permitted here (±10 per cent) removes the need for any adjustment.

If feedback is to be incorporated, the expected maximum errors which would occur without feedback must be assessed. Experience helps to obtain the right order of tolerance and if in this example a variation of ±50 per cent in the overall gain, including the amplifier, is assumed, this will be a conservative estimate. This assumes that components which directly affect the gain are of ±5 per cent tolerance: this includes multivibrator resistors, R_8 and R_9, and presupposes that the a.c. input resistance of the amplifier is to be much greater than R_8.

Given this expected variation, a gain of 5 must be 'thrown away' in feedback to convert the ±50 per cent into ±10 per cent. Amplifier gain must therefore be 20 × 5 × (chopper losses); chopper losses include a factor of 2 in synchronous rectification and further small losses such as $R_8:R_9$, loss in T_5 and T_6, input resistance of the amplifier loading R_8, and non-unity mark/space. These would amount to no more than 30 per cent with reasonable design, giving an amplifier gain requirement of 130.

Input resistance must be much greater than R_8, e.g. 1 kΩ, and the output resistance must be much less than R_8, e.g. 20 Ω, or alternatively included in R_8. Output current capability must be greater than 10 mA.

Bandwidth must be adequate to pass a square wave of 2·5 msec period without great deterioration, e.g. 50 Hz–3 kHz. Design can now proceed as described in Chapter 7 using two earthed emitter stages followed by one or two emitter followers, the final one having a standing current in excess of 10 mA. This gives adequate gain and no overall signal inversion. The lower 3 dB cut-off frequency should be about 50 Hz.

FEEDBACK CONSIDERATIONS

Negative feedback is added by connecting RF from final output to the input chopper S_1, the value being $20 \times R_8$, i.e. 2 kΩ. Since the loop gain is about 5, the system will be free from h.f. oscillations if the ratio of the two major lagging time constants is greater than 5. The main lags are R_9C_5, i.e. $12 \times 10^3 \times 50 \times 10^{-6} = 0{\cdot}6$ sec, and one period of chopping, i.e. 2·5 msec (since no mean output change could occur faster than this even if R_9C_5 did not exist). The ratio is 240 : 1, so the system is stable.

SUMMARY

Design of a chopper amplifier is straightforward provided the principles are understood; the system can then be divided into individual circuits each designable by techniques discussed in earlier chapters. There is a large amount of choice in the design of every circuit involved; this gives the designer a good opportunity to test his judgement.

Part Two

PRACTICAL CIRCUITS

Chapter 10

Complementary circuits

'Complementary circuits' is the name generally given to those circuits which employ both *n–p–n* and *p–n–p* transistors in such a way as to exploit their opposite bias polarities.

Very often in 'standard' circuits using two *n–p–n* or two *p–n–p* transistors, one transistor can be changed for its opposite type. The result is a circuit which in many examples is more economical in components, and often a performance improvement is also obtained.

Other complementary circuits are unique, not being derived from any 'normal' circuit.

The examples which follow are by no means exhaustive but are intended to assist the designer in inventing his own circuits and variations on the ones discussed.

It is often profitable when examining critically a newly completed design to consider the effect and possible improvement which would follow a change of any transistor in the circuit for its complementary version.

COMPLEMENTARY BISTABLE NO. 1

This circuit is derived from the two-state circuit given on page 69 and reproduced below.

If we replace T_2 in Fig. 10.1 by a *p–n–p* transistor, remembering to reverse the polarities applied to collector and base relative to emitter, the circuit of Fig. 10.2 is obtained. As in the original circuit, this arrangement has two stable states.

Assume, for example, T_1 is 'on' and saturated. Then its collector is near zero potential, so that the current in R_6 turns on T_2 to saturation (with correct design). The saturation of T_2 brings its collector potential to zero, so that current in R_1 maintains saturation of T_1.

This is therefore a stable state. Now assume T_1 is cut off, its collector potential being highly positive. With correct values the base of T_2 is now positive, cutting off T_2. Its collector now falls negative, thus cutting off T_1. This is also, therefore, a stable condition.

The derived circuit has two stable states, as did the original version; the changes in polarity, however, result in several important differences which are discussed below.

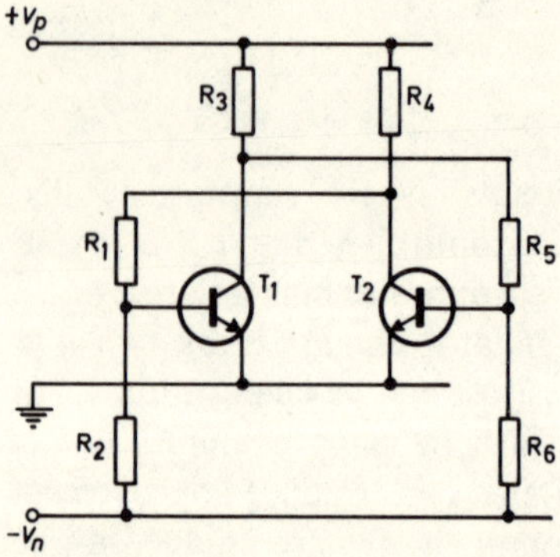

FIG. 10.1 Conventional two-state circuit

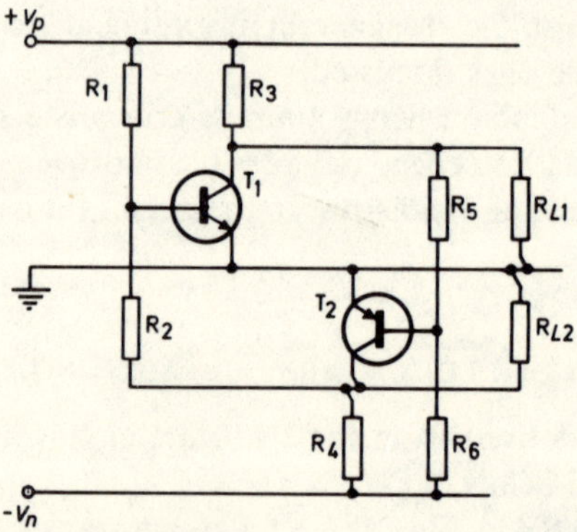

FIG. 10.2 Complementary bistable No. 1

Properties

First, instead of each transistor being 'off' while its partner is 'on', both are 'off' or 'on' together.

Secondly, instead of each output alternately and oppositely going from earth to positive (or negative if *p–n–p* types are used), both outputs are near earth in one state, and in the other state one is highly positive, the other highly negative.

Thirdly, the output waveforms of the two outputs of the original circuit each have a sharp fall and slow rise (the opposite for *p–n–p* types), whereas the complementary circuit has from T_1 a sharp fall,

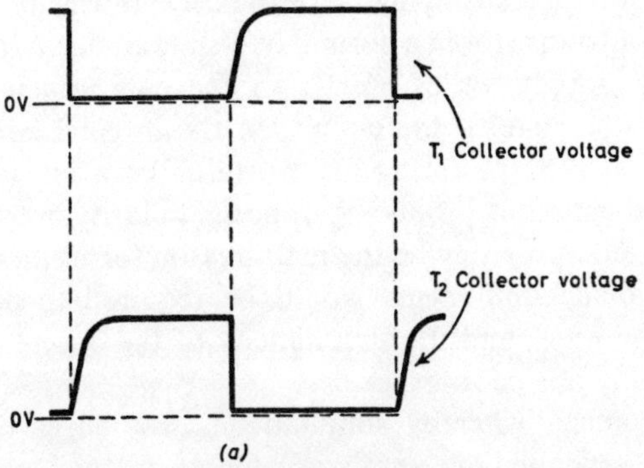

FIG. 10.3 (*a*) Waveforms for conventional two-state circuit (Fig. 10.1)

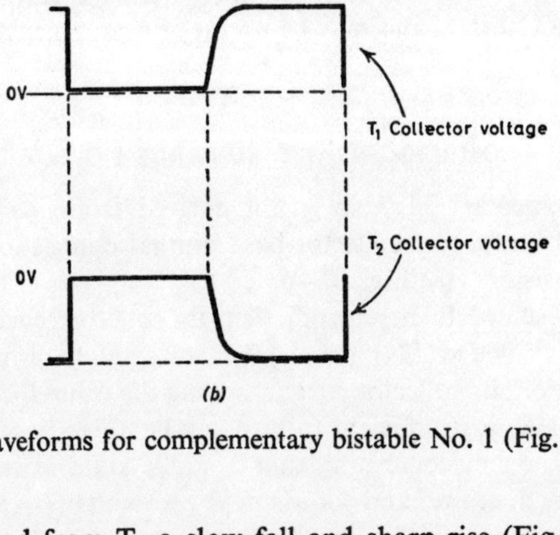

(*b*) waveforms for complementary bistable No. 1 (Fig. 10.2)

slow rise, and from T_2 a slow fall and sharp rise (Fig. 10.3).

Fourthly, the original circuit always requires a negative base drive to turn off either 'on' transistor, and the new circuit requires a negative drive on T_1 base to turn off T_1 or a positive drive on T_2 base to turn off T_2.

There are many consequences of the above properties.

The first result can be very useful where the circuit normally stands in its 'off state' with very low drain on the supply, yet, when turned on, delivers heavy current to two separate loads, from each saturated collector–emitter path. This same action also has its snags, since to turn off such a circuit two heavily saturated transistors have to be cut-off, which requires roughly twice the drive normally required to turn off the 'on' transistor in a normal two-state circuit. Moreover, the drain on the supply varies greatly in the complementary circuit according to the state of the circuit; in the original circuit supply drain is constant except during the transition between states.

The second action of producing opposite polarity outputs is most useful and would probably be the main reason for using this circuit. Many additional components would be required to achieve this result in other ways, and the simultaneity of the output waveforms would generally not be so exact.

The third effect, whereby simultaneous fast edges of opposite polarity are generated, is another valuable feature, enabling, by differentiation, fast positive- and negative-going pulses to be obtained, a frequent requirement in logic circuitry.

The fourth characteristic of differing turn-off drives is merely different from normal and may be useful or a nuisance according to circumstances.

COMPLEMENTARY BISTABLE NO. 2

This complementary bistable is not derived from a conventional bistable but uses direct collector-base mutual connections between the two transistors (*see* Fig. 10.4).

If T_1 is assumed to be cut off, then its collector current will be small (I_{cbo1}). Provided $(I_{cbo1} + I_{cbo2})R_2$ is too small to turn on T_2, then T_2 also is cut off, its collector current having the value I_{cbo2}. Provided $(I_{cbo2} + I_{cbo1})R_1$ is insufficient to turn on T_1, then T_1 is cut off, as originally assumed, confirming that a stable state exists where T_1 and T_2 are both cut off and V_{out} is near zero volts ($I_{cbo}R_{2L}$).

On the other hand, if T_1 is conducting such that $I_{c1}R_2$ turns on T_2, and if the current in T_2 collector is such that $I_{c2}R_1$ turns on T_1 further, equilibrium is reached when T_1 or T_2 saturates, i.e. a second stable state exists with both transistors conducting. V_{out} is now almost equal to V_p.

Properties

Since this circuit, like the previous example, has two transistors which turn off and on simultaneously, some of the points still apply; for instance, low quiescent current in the 'off' state, and the possibility of using either polarity of trigger for either changeover.

The main feature, however, is the ability of the circuit to supply heavy loads with low transistor dissipation. This comes about because at least one transistor is saturated and the other very nearly so, and within limits a heavier load current results in still heavier available drive from T_1, since T_1 itself receives more base drive from the load current passing through T_2. This useful effect applies until

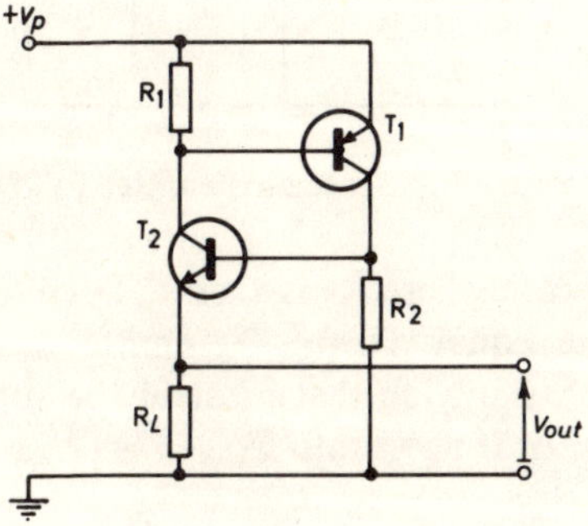

FIG. 10.4 Complementary bistable No. 2

eventually with very low values of R_L the load current is so great that the β_s of T_1 and T_2 fall. Ultimately, saturation is no longer maintained and transistor dissipation rises causing catastrophic failure.

COMPLEMENTARY EMITTER FOLLOWERS

The emitter follower is one of the most widely used circuit configurations, and is particularly helpful in isolating a voltage amplifier from its load.

The isolation produced by a single emitter follower is, however, very often insufficient, and it is common practice in such a case to use two emitter followers in cascade (*see* Fig. 10.5).

Although this circuit is usually quite satisfactory, there are certain conditions, for instance, in d.c. amplifiers, where the input-to-output d.c. voltage drop, now equal to two base–emitter voltage drops, constitutes a serious disadvantage. The amount of this combined voltage drop can be as high as 2 V and the associated temperature coefficient will lie between 4 and 5 mV/degC.

An alternative arrangement which does not have this fault, and has another important advantage, is shown in Fig. 10.6 (*a*) and (*b*).

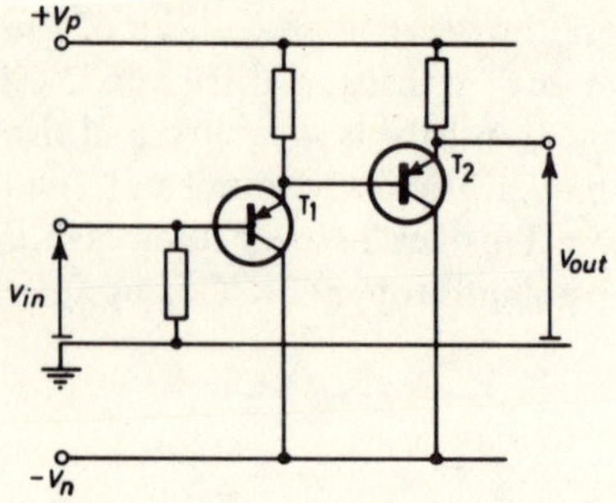

FIG. 10.5 Cascaded emitter followers

In the simple case where R_2 is absent, T_1 collector current and T_2 base current are equal. Hence, T_1 emitter current is $1/\beta_2$ of T_2 collector current, implying that most of the supply current I flows into T_2 and $1/\beta_2$ of it flows into T_1.

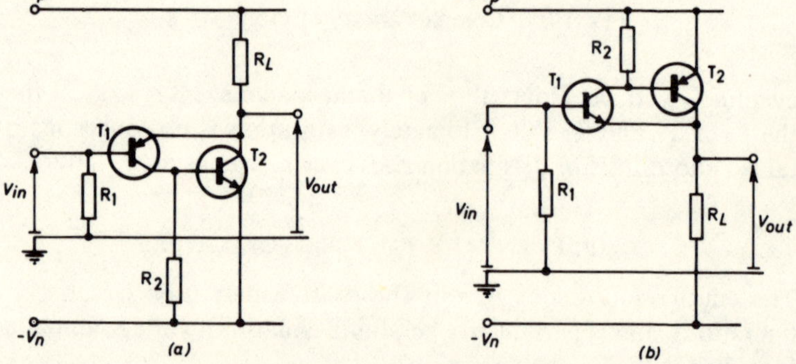

FIG. 10.6 (*a*) Complementary emitter follower (1); (*b*) complementary emitter follower (2)

The d.c. input-to-output drop consists of the V_{be} of T_1 only and calculation shows that in most respects the performance of the circuit is at least equal to that of Fig. 10.5.

The main disadvantage of the complementary version in practice is its liability to oscillation by the feedback loop T_1 emitter–collector, to T_2 base and collector, back to T_1 emitter. As in any feedback system, a total phase shift round the loop of 180 degrees will lead to oscillation if the loop gain then exceeds unity. It is shown in Chapter 8 that the most likely conditions to satisfy these criteria occur when the phase-shifting circuit elements each have similar phase-frequency characteristics.

Since T_1 operates, so far as the feedback loop is concerned, in grounded base, and T_2 is in grounded emitter, high-frequency oscillation is most likely to occur if the phase characteristic of α_1 approximates to that of β_2, i.e. f_{T_1} equal to f_{T_2}/β_2.

Using transistors of similar f_T is therefore marginally safe, but it is preferable if f_{T_1} is much greater than f_{T_2}.

As in many circuits where direct collector-base coupling is used, the addition of R_2 is desirable. R_2 is designed so that $(I_{co1} + I_{co2})R_2$ is insufficient to turn on T_2. It can often be omitted if T_1 and T_2 are silicon.

This circuit has been used extensively in audio output stages since, in the form shown in Fig. 10.6 (*b*), the circuit as a whole behaves like a *n–p–n* transistor (i.e. it tends to turn on when driven positive), yet the main output current passes through T_2, which is a *p–n–p* type. In effect, therefore, a *n–p–n* power transistor (in itself a rare and expensive device for commercial audio equipment) has been obtained from a low power *n–p–n* and a power *p–n–p* transistor (both less expensive).

In this application T_2 would often be a germanium power type and therefore R_2 would have to be very small to ensure correct high-temperature operation (I_{cbo} could be 10 mA at 50°C and V_{eb} for turn-on could be 0·05 V, giving $R_2 \leqslant 5\ \Omega$). Fortunately, high-temperature working is not normally important in these cases and a relaxation can be permitted. Some commercial designs are, however, questionable on this design point, and fail catastrophically even at 35°C.

The best arrangement is to return R_2 to a higher potential than T_2 emitter either by using an additional supply rail for R_2 or by adding diodes or a Zener diode in series with T_2 emitter (Fig. 10.7). The value for R_2 now becomes larger, since the voltage which $I_{cbo2}R_2$ must not exceed is not 0·05 V, but (0·05 + V_1) where V_1 can be typically 1 to 4 V, giving R_2 a value from 20 to 80 times its value in Fig. 10.6 (*b*).

Another difficulty which arises in the design of emitter followers occurs in stabilizer and power amplifier output circuits. Using the simple configuration of Fig. 10.5 to drive a heavy load requires the first transistor to carry $1/\beta_2$ of the load current, and a mean collector–emitter voltage which is the same as for T_2. Typical values for a

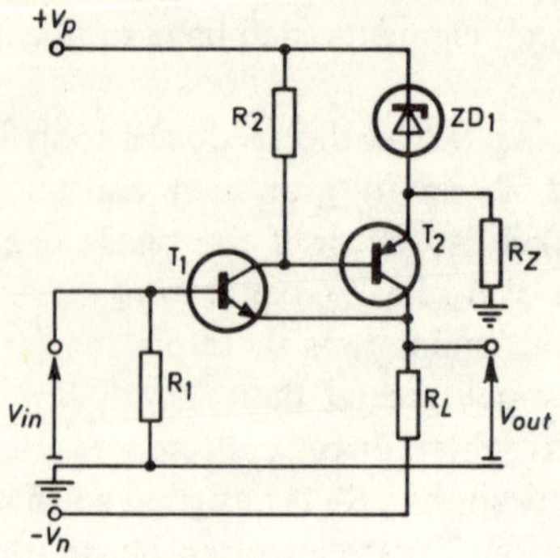

FIG. 10.7 Modified form of Fig. 10.6 for high-temperature operation

stabilizer are $I_{e2} = 2$ A, $V_{ec2} = 15$ V; T_1 carries $I_{e1} = 2/\beta$ A, e.g. 80 mA, at 15 V, so that the power dissipated by T_1 is 1·2 W.

In such a case T_1 must therefore be more than a small-signal transistor, and, in fact, power transistors are often used for both T_1 and T_2. Apart from the obvious disadvantages of size and cost, a

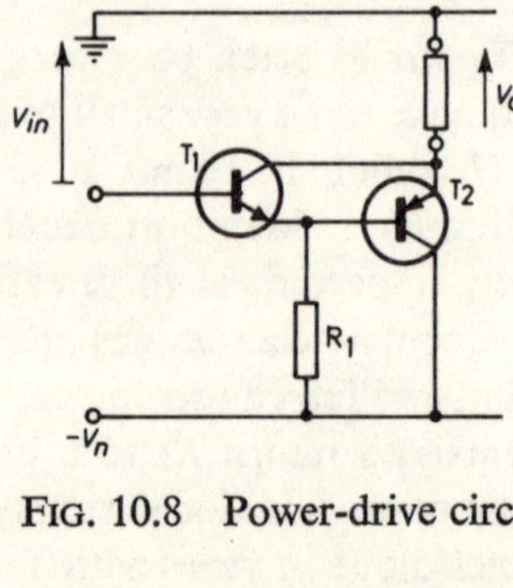

FIG. 10.8 Power-drive circuit

more subtle difficulty arises: the previous stage, probably the collector circuit of an amplifier, now has to drive T_1, the leakage current of which is likely to be 10 mA at an ambient of 50°C. Thus, yet another emitter follower may have to be used to drive T_1.

The circuit of Fig. 10.8 overcomes this problem. Here the current in T_1 is still I_{e2}/β_2, or 80 mA in the above example, but the V_{ec} for T_1 is only the base–emitter drop of T_2, which may be 0·3 V. The power dissipated in T_1 is now only 24 mW and a low-power type having a maximum leakage current of perhaps 100 μA at 50°C, if germanium, or 5 μA, if silicon, can be used. Hence, the two snags outlined above for the normal circuit are overcome.

The difficulties which have now been solved are replaced in part by the problem of designing R_1 in Fig. 10.8. In order that T_1 should never cut off, R_1 must always pass a current greater than I_{e2}/β_2, under any condition of supply voltage. If V_{out} is, for example, 40 V and V_n varies from 50 to 60 V (quite normal for a stabilizer output circuit), then for a load current of 2 A and β_2 of 25, the current in R_1 must exceed 2/25 A, i.e. 80 mA when V_n is at 50 V. R_1 must therefore

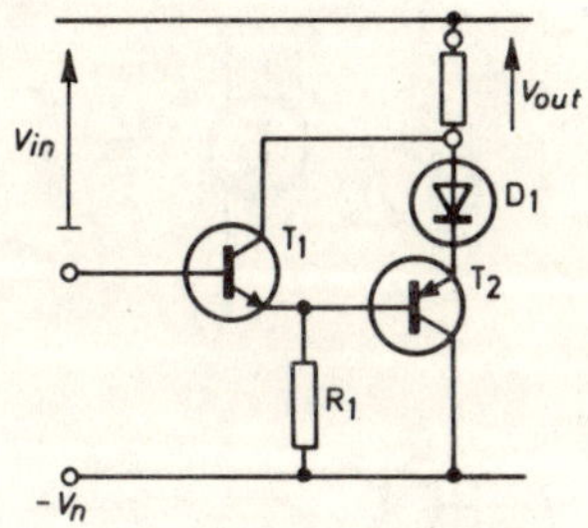

FIG. 10.9 Modified version of power-drive circuit (Fig. 10.8) to increase V_{ce} for T_1

be less than (50 – 40)/80 kΩ, i.e. 125 Ω. At the other limit of V_n (60 V) R_1 will therefore pass (60 – 40)/125 A, i.e. 160 mA.

This results in T_1 having to pass a maximum current of, in this example, twice the amount normally to be expected. Fortunately, power rather than collector current determines the physical size and therefore leakage currents for a transistor, so that in many instances this circuit is advantageous.

A second, less serious difficulty occurs when the emitter–base voltage drop of T_2 is insufficient for the type of transistor intended to be used for T_1. This is readily overcome by the addition of a diode in series with T_2 emitter, as shown in Fig. 10.9.

In compensation for these difficulties in fixing the operating conditions of T_1, the circuit has the property that temperature rise, which above all causes the I_{cbo} of T_2 to increase, tends to turn T_1 on rather than off.

COMPLEMENTARY AMPLIFIERS

Naturally, complete amplifiers may well contain a mixture of *p–n–p* and *n–p–n* transistors; the intention in this section is to deal with circuits where *p–n–p* – *n–p–n* pairs are used to obtain a special advantage.

The compound emitter follower configuration already described by Fig. 10.6 can also be used as a voltage amplifier by regarding the emitter of the second transistor as the collector of a new transistor, as illustrated in Fig. 10.10. The 'transistor' thus synthesized has

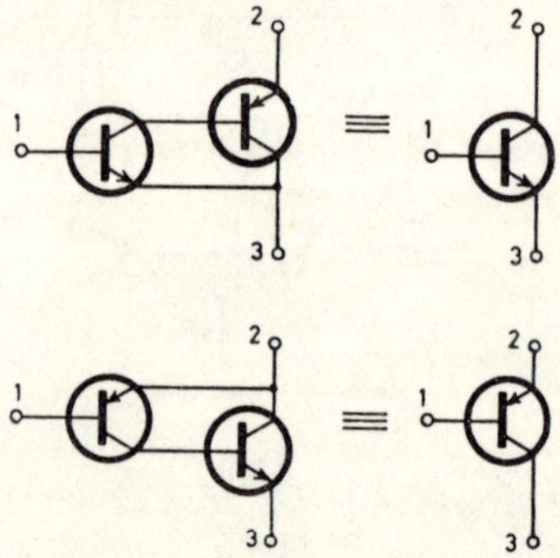

FIG. 10.10 Equivalence of complementary circuits to single transistors

approximately the same g_m as a single transistor run at the same current, but its apparent current gain is equal to $\beta_1\beta_2$.

This is the basis of a well-known differential amplifier first described by Bénéteau and reproduced in its simplest form in Fig. 10.11. As is evident from the above comments on the compound transistor of Fig. 10.10, this circuit behaves like a normal emitter-coupled pair with high input impedance, as high as if emitter follower drivers were added but without the inconvenient V_{be} drops which would then result. The further development of the circuit by replacing R_E with a constant-current device and by coupling into a further similar stage results in exceptional performance with regard to gain and drift.

A particularly useful feature of the compound transistor is the ease with which a single-transistor stage can be changed into the compound form. The need for this can arise if the input resistance of a single stage has to be increased using the minimum of extra com-

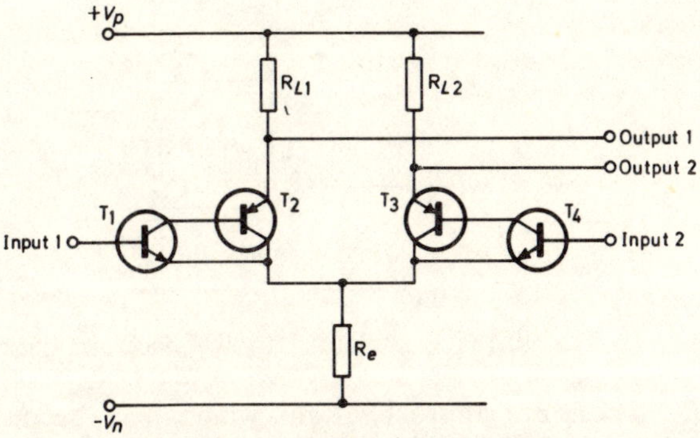

FIG. 10.11 Bénéteau's differential amplifier

ponents and a simple emitter follower cannot be added because of its additional V_{be} drop. This can occur following a modification to specification or from a previous design error. Fig. 10.12 illustrates the necessary reconnection.

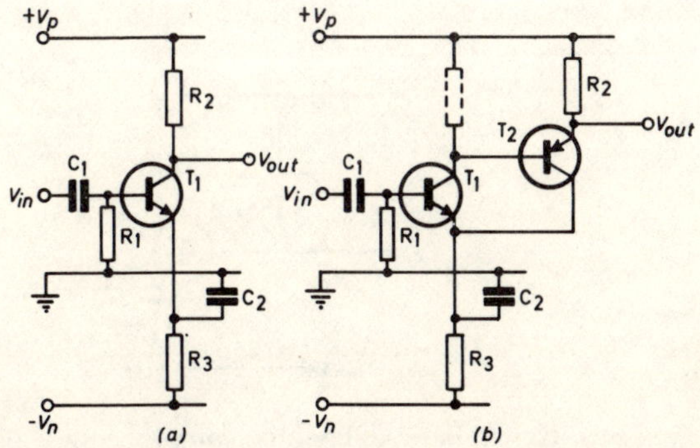

FIG. 10.12 Practical complementary modification: (*a*) before, (*b*) after

Another method which exploits the interconnection shown in Fig. 10.6 to form an amplifier is shown in Fig. 10.13. This circuit is easily proved to have a gain of approximately $(R_1 + R_2)/R_2$. Transistor parameters have little influence on the gain, so that although the input and output are in phase, the circuit behaves like a normal feedback amplifier having a loop gain of β_2.

Hence, the output impedance is R_1/β_2 and input impedance is

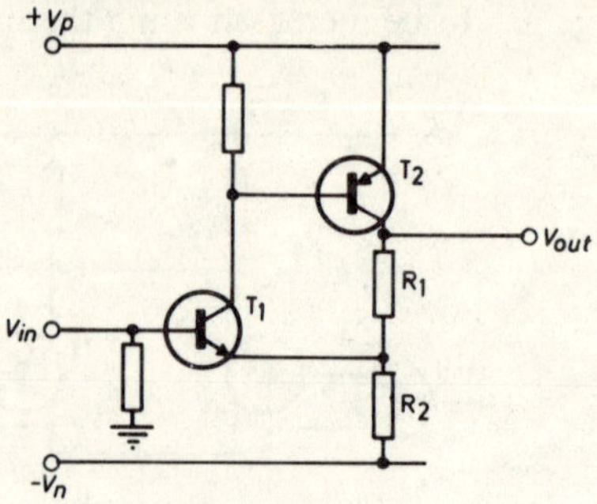

FIG. 10.13 Complementary feedback amplifier

$\beta_1\beta_2R_2$. These are approximations which may be deduced from equivalent-circuit analysis (*see* Chapter 4).

Complementary Differential Amplifier

If one transistor of the emitter-coupled amplifier described in Chapter 4 is replaced by a complementary type, the circuit of Fig. 10.14 results.

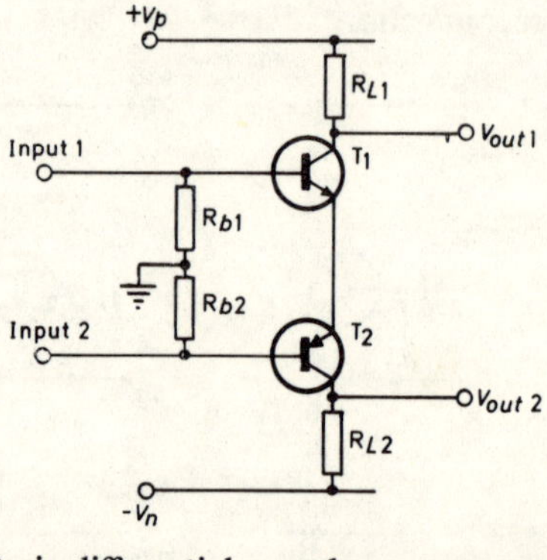

FIG. 10.14 Basic differential complementary amplifier

As it stands, this will not operate in a linear manner, since both transistors are cut off. Returning T_1 base to a more positive potential than T_2 base ensures current flow but the collector current is then undefined. A simple method of ensuring correct biasing is shown in Fig. 10.15.

Here V_{1b} and V_{2b} are defined by R_1, R_2, R_3, and V_p, and the resulting collector currents are each equal to $(V_{1b} - V_{2b})/R_e$, neglecting V_{be} drops.

The main difference between this circuit and the standard emitter-coupled pair is that there is now no need for a resistor to supply current to T_1 and T_2 from the zero or negative line.

It was shown in Chapter 4 that the presence of this resistor is the main cause of the undesirable 'push–push' gain of the normal circuit: for good 'push–push' rejection this resistor is usually replaced by a constant-current source.

Hence, one might expect the complementary circuit to have good 'push–push' rejection without additional circuitry. This proves to

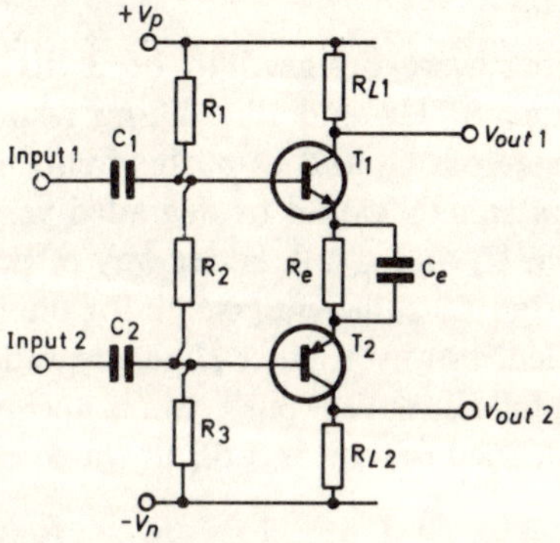

FIG. 10.15 Practical version of basic differential complementary amplifier (Fig. 10.14)

be true and the offsetting disadvantage is that inputs must be a.c. coupled unless the signal sources happen to sit at two suitable d.c. levels.

Design procedure is simple: $R_{1,2,3}$ are chosen so that the drop across R_2 is large compared with V_{be} but not so large that T_1 or T_2 bottoms on large output swings. R_e is chosen to give correct emitter current to supply the required load R_{L1} and R_{L2}. For maximum available output R_e, R_{L1}, and R_{L2} are chosen so that V_{C1} (V_{C2}) are halfway between $V_{b1(2)}$ and $V_{p(earth)}$. If only one output is required, the other load can be short-circuit without noticeably affecting 'push–push' rejection.

The choice of C_e is dictated by the same considerations as in a normal earthed emitter amplifier, but care must be taken in a wide-band amplifier that C_e is made no larger, physically, than is necessary. This is important because at high frequencies the stray capacitance from this large component to earth causes degradation in the 'push–push' rejection.

Another design point is that C_1 and C_2 must be so large at the lowest operating frequency that each input signal reaches its transistor base circuit without appreciable loss. Clearly, this criterion would be applied as a matter of course in order to avoid loss of gain, but the issue is more critical than this. If C_1 and C_2 are not large enough and if T_1 and T_2 are not identical (a likely condition, since *n–p–n* and *p–n–p* types rarely match precisely), then unequal signals will reach T_1 and T_2 base, even if identical inputs are applied to Input 1 and Input 2.

If this amplifier were designed to be 3 dB down (at the low-frequency end) at 50 Hz and this falling response were achieved by allowing loss across C_1 and C_2 at this frequency, then push–push rejection of hum signals would be degraded very severely owing to unequal losses in C_1 and C_2. A better way of producing the fall-off would be to design C_e sufficiently small for the purpose. This is still not recommended, however, since slight degradation is still caused (push–pull gain falls but push–push gain is unchanged), and the 3 dB point is badly defined (circuit is 3 dB down when

$$X_{c_e} = \left(\frac{1}{g_{m1}} + \frac{1}{g_{m2}} + \frac{R_{s1}}{\beta_1} + \frac{R_{s2}}{\beta_2}\right) // R_e$$

which depends greatly on transistor parameters).

The best procedure is to degrade the frequency response in a later stage where push–push rejection is no longer a problem. This is also the best procedure for the minimizing of circuit noise produced by T_1 and T_2.

The virtues of this circuit were brought to the author's attention by John Murray of Marconi Instruments Ltd. in connection with a low-level transformerless balanced amplifier.

Chapter 11

Wide-range voltage controlled oscillator

The frequency of either sine-wave or square-wave oscillators can be varied by changing the value of one of the frequency determining elements. When the frequency is to be varied over a wide range it is necessary in many circuits to vary two or more of these elements simultaneously in order, first, to achieve the required frequency swing and, secondly, to maintain constant amplitude at various frequencies.

When there is a requirement for a remotely controlled oscillator or where a feedback system calls for an oscillator whose frequency is to be controlled by a d.c. level (as in a frequency lock or a phase-lock loop) the normal methods of frequency control become unwieldy or impossible.

The voltage-controlled oscillator to be described is essentially a multivibrator using emitter circuit timing, so that the main output is a square wave, ideally suited to driving mixers. Because of the emitter circuit arrangement, however, a triangular waveform of constant amplitude is also available and this in turn can readily be shaped to a sine wave. The oscillator frequency is directly proportional to control-voltage over a wide range and the circuit has good temperature-stability.

BASIC CIRCUIT

This circuit has the unique property that an attempt to calculate component values without appreciating exactly how it operates usually results in a circuit which produces twice the expected output at all four output terminals for two quite different reasons.

To appreciate the operation of Fig. 11.1, assume that I_1 and I_2 are perfect constant-current sources and that T_1 is conducting and T_2 cut-off, i.e. T_2 emitter is negative with respect to T_2 base. Assume further that R_3 and R_4 are much smaller than $R_{1,2,5,6}$, and that voltage excursions are such that bottoming does not occur.

At this time T_2 collector current is zero, so that T_2 collector voltage is near zero and T_1 base is $V_pR_2/(R_1 + R_2)$ positive. Since T_1 is conducting, T_1 emitter potential is just above $V_pR_2/(R_1 + R_2)$.

Consider now the current in T_1 emitter. It is not merely I_1, but $(I_1 + I_2)$, since, T_2 being cut-off, I_2 must be charging C and adding to I_1 in T_1 emitter. Hence, T_1 collector current is $(I_1 + I_2)$ and T_1 collector potential is $(I_1 + I_2)R_3$ volts positive.

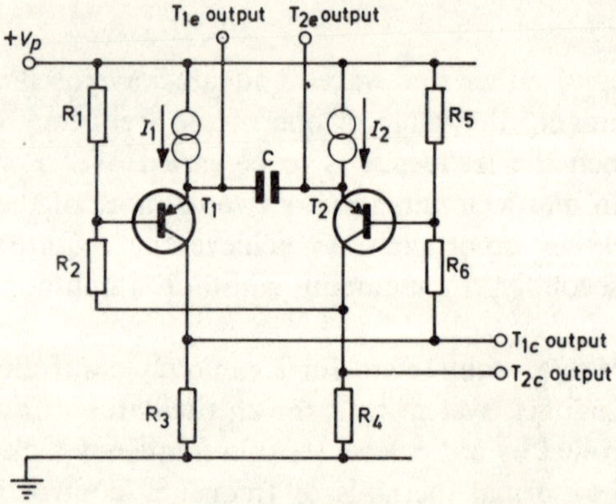

FIG. 11.1 Basic circuit of emitter-coupled multivibrator

A simple Ohm's law calculation shows that T_2 base potential is $V_pR_6/(R_5 + R_6) + (I_1 + I_2)R_3R_5/(R_5 + R_6)$ volts positive.

These are therefore the conditions which exist whenever T_2 is cut off, and similar expressions apply for T_1 cut-off (*see* Fig. 11.2).

Continuing the train of events, C is being charged by I_2 and since its T_1 emitter connection is fixed, just positive to T_1 base, T_2 emitter rises linearly at a rate of I_2/C volts/sec. (since $Q_C = I_2t = CV$). When T_2 emitter becomes just positive to T_2 base, T_2 begins to conduct, causing T_2 collector and, hence, T_1 base to rise sharply. Since T_2 is conducting, T_2 emitter cannot rise and, hence, T_1 emitter cannot follow the sharp rise of T_1 base, because C would have to charge instantly. T_1 therefore cuts off.

As soon as T_1 cuts off, its collector potential falls to zero, a change of $(I_1 + I_2)R_3$ volts, and T_2 base therefore falls by $(I_1 + I_2)R_3R_5/(R_5 + R_6)$, causing T_2 emitter and because of C, T_1 emitter to fall by the same amount.

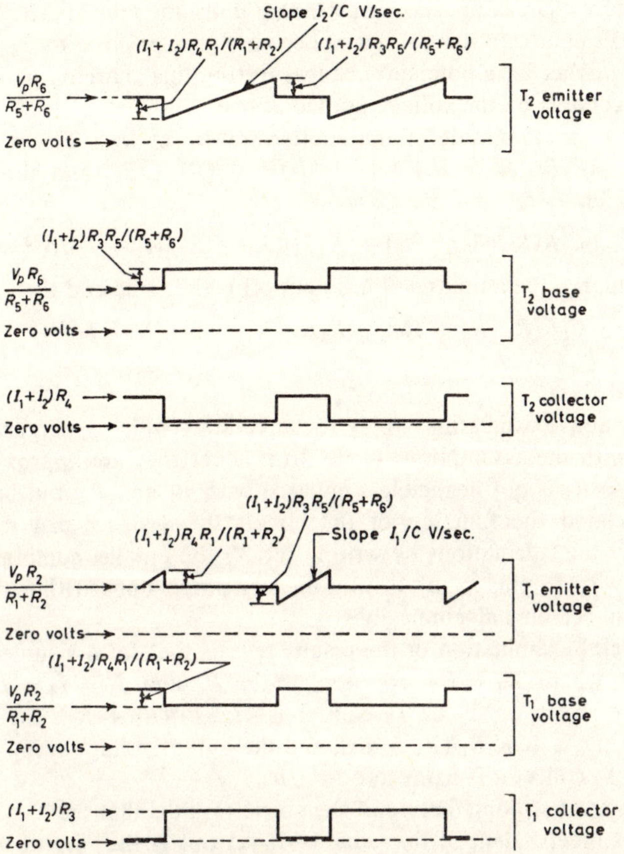

FIG. 11.2 Waveforms for circuit of emitter-coupled multivibrator (Fig. 11.1). Magnitudes are approximate and assume $R_3 \ll (R_5 + R_6)$, $R_4 \ll (R_1 + R_2)$

Also, when T_1 cuts off, the current $(I_1 + I_2)$ now flows into T_2, raising its collector by $(I_1 + I_2)R_4$ and therefore raising T_1 base by $(I_1 + I_2)R_4R_1/(R_1 + R_2)$.

Hence, immediately T_1 cuts off, its emitter and base potentials (which were virtually equal during the conduction of T_1) have moved in opposite directions by the amounts given and at the end of the transition differ by a voltage of

$$(I_1 + I_2)R_3R_5/(R_5 + R_6) + (I_1 + I_2)R_4R_1/(R_1 + R_2)$$

C now charges linearly from I_1 at a constant rate of I_1/C V/sec until T_1 emitter potential again becomes just positive to T_1 base. Since neither base potential changes during this charging period C has to charge by the voltage quoted above.

Thus the amplitude of the emitter waveform with T_1 cut-off is $(I_1 + I_2)R_3R_5/(R_5 + R_6) + (I_1 + I_2)R_4R_1/(R_1 + R_2)$ and the time for which T_1 remains cut-off is

$$C[R_3R_5/(R_5 + R_6) + R_4R_1/(R_1 + R_2)](I_1 + I_2)/I_1$$

Similarly, the amplitude for T_2 cut-off is the same and the time is

$$C[R_3R_5/(R_5 + R_6) + R_4R_1/(R_1 + R_2)](I_1 + I_2)/I_2$$

Special Case

The above expressions apply for any values which are in accordance with the assumptions made. In practice they are approximate because R_4 is not negligible compared with R_1 and R_2, but having appreciated the function of the circuit the designer can readily modify the calculations by writing for R_4 the parallel combination of R_4 with R_1 plus R_2, a substitution very simply done arithmetically but cumbersome algebraically.

Careful examination of the results reveals that for a symmetrical design where $R_3 = R_4$, $R_1 = R_5$, $R_2 = R_6$ and $I_1 = I_2 = I$, the voltage sweep on T_1 or T_2 emitter is $4IR_3R_1/(R_1 + R_2)$, the half-cycle time is $4CR_3R_1/(R_1 + R_2)$, and the output voltage swing from T_1 or T_2 collector is from zero to $2IR_3$.

The most striking feature of the above result is that the half-cycle time is independent of the value of I, yet this is the current which directly determines the charging rate of C. The reason is that I also determines the voltage by which C must charge, and since both this and charging rate are linear functions of I the charging time for C remains constant.

Two Common Difficulties

There are two difficult points in the action of the circuit which can lead if misunderstood to expecting half the output voltage actually obtained.

The first concerns the emitter current of the 'on' transistor, which, as pointed out, is $(I_1 + I_2)$, not merely I_1 or I_2. If this is difficult to accept, consider the moment when both transistors are conducting, a condition which exists transiently during the switching-over at the end of each period. Now, assume one or other transistor, e.g. T_2, cuts-off because its base is driven suddenly positive. It is clear that T_2 emitter cannot follow a sudden change, since an enormous current would be required to charge C rapidly and its T_1 end is anchored to T_1 emitter. Hence, T_2 emitter suddenly appears to become open-circuit. I_2 continues to flow and its only path is through C into T_1, adding to I_1. Even though C now charges, I_2 remains constant, and the current in T_1 remains equal to $(I_1 + I_2)$ until the end of that period.

If this point is not appreciated and I_1 equals I_2, the output voltage of $2I_1R_3$ is twice the (wrongly) expected I_1R_3.

The second point which can easily be missed even though the first is understood, arises when considering the emitter waveforms. Taking the case when T_2 is off because its emitter is more negative than its base, T_2 emitter is rising linearly because I_2 is charging C and reaches the level when T_2 conducts.

Two things now happen. First, because T_2 conducts, T_1 base potential rises rapidly to the level at which it will remain throughout the ensuing period; this leaves T_1 emitter at the potential T_1 base had previously remained at while T_2 had been 'off'. Secondly, because T_1 cuts off, T_2 base falls to the level at which it will now remain, pulling T_2 emitter and, because of C_1, T_1 emitter, more negative by the same amount.

The two sudden steps, the upward step of T_1 base and the downward step of T_2 base and emitter and T_1 emitter are equal if the circuit is symmetrical, and a common mistake is to suppose that the potential now existing between T_1 base and emitter is only the magnitude of one such step instead of two.

This leads to underestimating the voltage waveforms on T_1 and T_2 emitters and also the timings by a factor of 2.

If both the above errors are made, the timing calculation is wrong by a factor of 4.

Voltage Control for the Circuit of Fig. 11.1

The above discussion of the circuit of Fig. 11.1 has shown that although the charging rates for C are proportional to the values of I_1 and I_2, the cycle time tends to remain constant.

If, however, the voltage steps on T_1 and T_2 collector are made constant, the voltage excursions of C are fixed and timing can be varied by controlling I_1 and I_2.

This is readily achieved by catching the collectors as shown in Fig. 11.3. Provided $(I_1 + I_2)R_3$ and $(I_1 + I_2)R_4$ exceed the clipping voltage, the collector swings will now be fixed for all higher values of I_1 and I_2.

The symmetrical case is of special interest, since the unity mark/space square wave available from each collector is ideal for driving a mixer for use in a beat-frequency oscillator or in a frequency or phase-lock loop. In this case each base excursion is $(V_D + V_Z)R_1/(R_1 + R_2)$, where V_D is the forward drop of D_1 or D_2 giving emitter swings of $2(V_D + V_Z)R_1/(R_1 + R_2)$ and a total

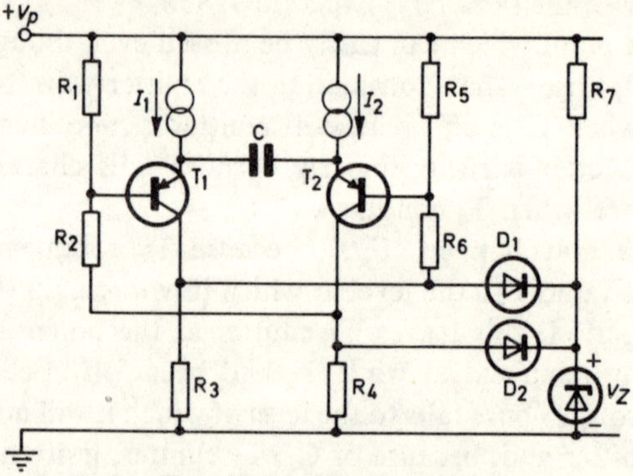

FIG. 11.3 Clipping circuit

cycle time of $4(C/I)(V_D + V_Z)R_1/(R_1 + R_2)$, or a frequency of $I(R_1 + R_2)/4C(V_D + V_Z)R_1$ Hz.

Frequency is therefore directly proportional to I, and it only remains to provide voltage-controlled high-impedance (i.e. 'constant-current') sources of current. Figure 11.4 shows a suitable circuit in which the collector currents of two transistors T_3 and T_4 are defined by the emitter resistors R_7 and R_8 (equal for symmetrical outputs) and the applied voltage V_C.

Provided V_C greatly exceeds V_{eb} for T_3 and T_4, the value of I is V_C/R_7 and the frequency is $V_C(R_1 + R_2)/4CV_ZR_1R_7$ Hz. The range of control begins at the value of V_C which produces sufficient current in R_3 and R_4 to reach the clipping level V_Z, and ends when either V_C is so large that T_3, T_4 bottom or the value of I produced by V_C is so large that transistor or clipping diode ratings are exceeded. Linearity of control will normally be lost before ratings are exceeded, since at high current levels the base currents of T_1 and T_2 will affect the voltages on the bases in an unpredictable manner (β has a wide tolerance).

Performance of the circuit of Fig. 11.4

The practical performance of this circuit is as predicted; the ratio of frequency limits depends on the linearity required, but typically 20:1 frequency swing with ±5 per cent linearity with applied voltage can readily be achieved.

The square-wave constant-amplitude output from the collectors is obviously useful, but the peculiar emitter excursions would appear to have little application. Between the two emitters of T_1T_2, i.e. across the capacitor, however, is a linear triangular waveform having

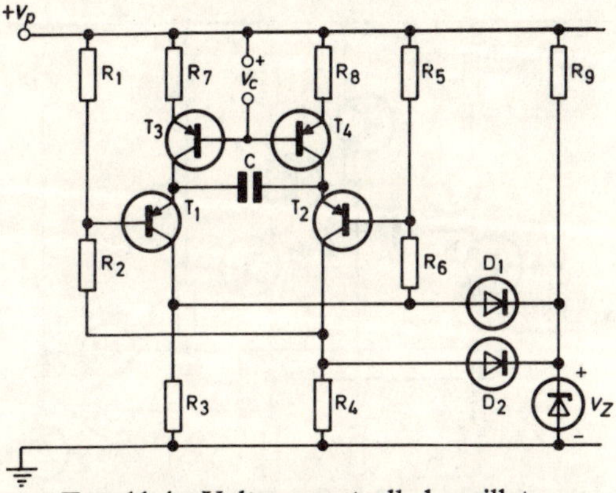

FIG. 11.4 Voltage-controlled oscillator

no noticeable spurious features. Since only a small amount of shaping is required to convert a triangle into a sine wave, the circuit can be used to produce a controlled sine-wave oscillator by adding a difference amplifier between the emitters of T_1T_2 and shaping the output. This is particularly useful at low frequencies such as 1 Hz and slower, since a normal 'tuned' oscillator cannot change its frequency in a time comparable with one cycle of oscillation, whereas the circuit used here can change its frequency at any time during a cycle (because changing I immediately changes the rate of rise of capacitor voltage).

Temperature effects in Fig. 11.4 which affect frequency most are V_{eb} variations in T_3T_4 which cause a linear increase in frequency corresponding to the same variation of V_C, i.e. 2 mV to 2·5 mV/degC.

Other effects are resistor drift and change in V_Z and in forward drop of the clipping diodes. Transistor collector leakage current in T_3T_4 and emitter leakage in T_1T_2 change the charging current of C, and β changes in T_1T_2 cause changes in their base potentials when conducting.

Most of these effects are small in a good design. V_{eb} changes in T_3T_4 are best balanced by another semiconductor junction; one method is shown in Fig. 11.5, the additional transistor having also the useful effect of reducing the loading on the control source V_c. Resistor drift need not exceed a few per cent and in a critical design precision types could be used giving 0·1 per cent. Drift in V_Z and the

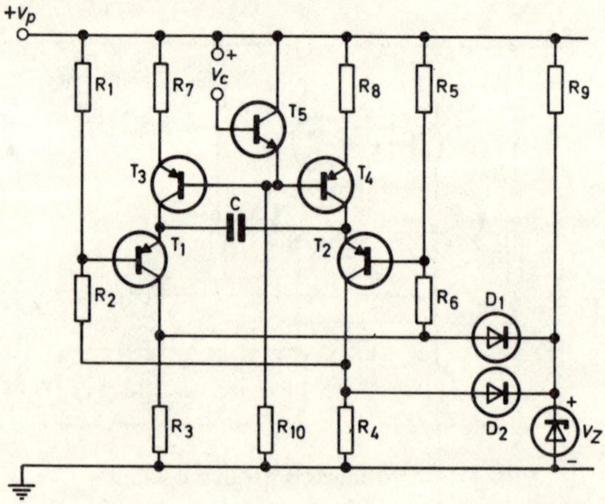

FIG. 11.5 Use of extra transistor for V_{EB} compensation

catching diodes more or less cancel each other if the Zener diode has a positive temperature coefficient of about 2 mV/degC, so its voltage should if possible be 6·8 or 8·2 nominally. Effects of transistor leakage depend on the values of I and resistors R_1, R_2, R_5, R_6; at maximum operating temperature the I_{cbo} of $T_{1,2}$ must be such that the voltage movement it produces on $T_{1,2}$ base, namely $I_{cbo}(R_5//R_6)$, is negligible and the I_{cbo} of $T_{3,4}$ must be much less than the minimum value of I; these requirements often dictate the use of silicon transistors. β should be so high that base currents in T_1, T_2 produce negligible voltage changes at T_1, T_2 base at the lowest temperature of operation.

Attention to the above points will normally reduce temperature effects to a frequency change of a few per cent over a 50°C temperature range.

Refinements. If the oscillator is required to operate at frequencies above 100 kHz it is advisable to add coupling capacitors in parallel with R_2 and R_6 to offset the input capacitance of T_1, T_2 and ensure sufficient loop gain during the transition. The time constant associated with these capacitors should be short compared with a half-cycle of oscillation or, alternatively, very large, so that the collector excursion is coupled in full to the opposite base. Another method is to replace R_2, R_6 by Zener diodes to ensure good coupling

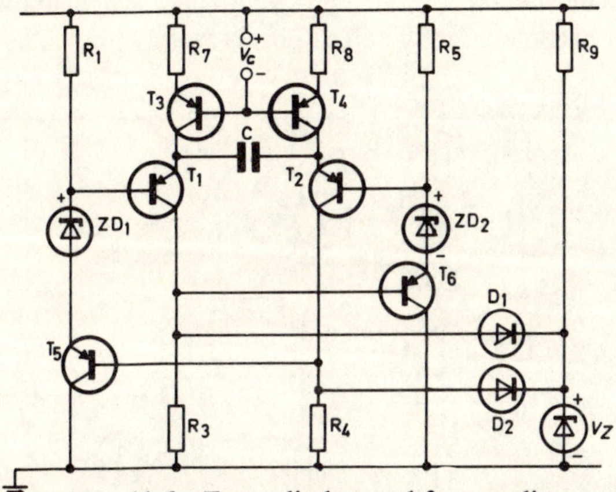

FIG. 11.6 Zener diodes used for coupling

at all frequencies, but since these need to operate at a few milliamps to be effective, emitter followers are then required as shown in Fig. 11.6.

It is desirable in some applications to have a means of setting up the clipping level of the collectors of T_1, T_2. The simple replacement of the Zener diode by an adjustable resistor chain is usually insufficient in practice, because the clipping level will rise as the value of I is increased; since the mean current into the clipping point is changing, decoupling the resistor chain does not effect a cure. In such a case the use of transistors instead of clipping diodes is necessary in order to avoid a very-low-resistance chain causing excessive drain from the supply (Fig. 11.7).

Transistor types. It is to be noted that T_1 and T_2 experience large reverse emitter–base potentials and should therefore be of alloy construction. Alternatively, protection diodes may be added in series with the emitters of T_1 and T_2. In other respects almost any transistor will be satisfactory, high-gain types being preferred for T_1, T_2 in order to have small base currents.

Frequency limitations. The lowest usable frequency is reached when (1) the leakage currents of the transistors begin to affect appreciably I_1, I_2, a practical lower limit for I_1, I_2 being 50 μA when operating up to 70°C; and (2) the value of C is so large that the required accuracy for the application cannot be obtained. Note that if an electrolytic capacitor is used, it must be of the reversible type.

The upper frequency limit is reached when the transition time becomes comparable to the half-cycle time, when amplitude and waveform become unpredictable. This is mainly limited by transistor

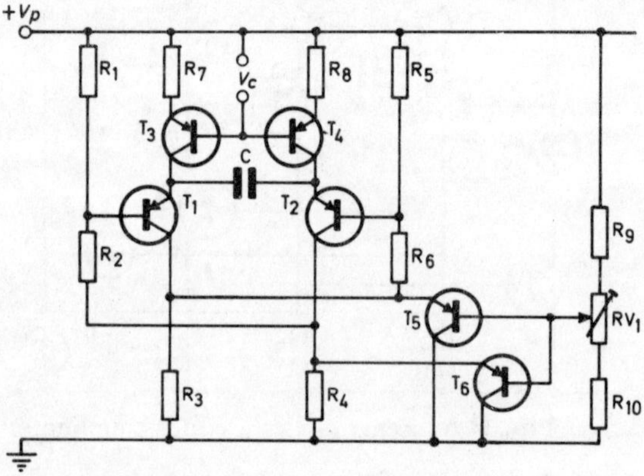

FIG. 11.7 Transistors used for clipping

f_T and collector base capacitance, both of which are partially offset by the use of coupling capacitors.

Triangular and sawtooth outputs. As pointed out at the beginning of the chapter, this circuit is useful in having a triangular waveform available across the timing capacitor C. This waveform is free from significant spurious content; in particular, no noticeable kink is present at the peak.

Because of the peculiar manner in which this triangle is generated, however, there are difficulties in converting this floating waveform into an unbalanced output, i.e. with one output connection to supply-common.

Figure 11.2 shows the two emitter waveforms, and it is clear that the difference between the two forms a triangle waveform with no vertical jumps at the transition, even in the asymmetrical case shown: the lack of symmetry merely causes the up and down slopes to be different. The problem arises in attempting to obtain the exact difference.

As was discussed in Chapter 4, a practical difference amplifier has certain deficiencies, the most important of which is the presence of 'push–push' as well as 'push–pull' gain. That is to say, the output will consist of the wanted signal, which is a multiple of the difference between the two input signals, and an unwanted signal which is a different, usually smaller, multiple of the sum of the two input signals.

For two sine wave input signals of the same frequency, the practical result is that two large signals give a larger output than two small input signals, even though the difference between the two inputs is the same in each case. Thus, the magnitude of the output is in error but no waveform distortion results.

The same result occurs, in fact, whenever the two inputs have the same shape at the same times, only the amplitudes being different: the output may be wrong in amplitude but is not distorted in shape.

In the case of the two emitter waveforms for T_1 and T_2, this condition does not hold, even in the perfectly symmetrical circuit, since when one emitter moves the other is stationary. The waveform likely to result from passing these signals through a difference amplifier of poor 'push–push' performance is shown in Fig. 11.8. Here the push–push and push–pull gains are assumed to be $A/10$ and A, respectively, and the output is seen to contain vertical steps at the triangle peaks, the magnitude of these steps being 1/10 of the peak-to-peak voltage of the triangle.

When examining this triangle waveform using a single input oscilloscope, these steps sometimes appear owing to the large capacitance which may exist between the oscilloscope 'common' and the transistor circuit power supply common through the mains wiring. This puzzling effect is avoided by operating the circuit from batteries or by the use of an oscilloscope with a balanced input, i.e. a differential amplifier.

The above special case in which the push–pull to push–push gain ratio is 10 to 1 may at first seem an improbable condition, since a typical rejection ratio for a straightforward emitter-coupled difference amplifier is 50 to 1, which would give only 2 per cent steps. In designing such an amplifier for extracting the triangular waveform, however, it soon becomes clear that unless special measures are taken the magnitude of the capacitor signal, usually several volts, will severely overload the emitter-coupled pair. Attenuating resistor networks cannot normally be used, because resistive loading on the capacitor will cause the charging current to vary, giving non-linear sides to the triangle.

The simplest solution is to add emitter resistance to the coupled pair (Fig. 11.9); correct values of all three emitter resistors then

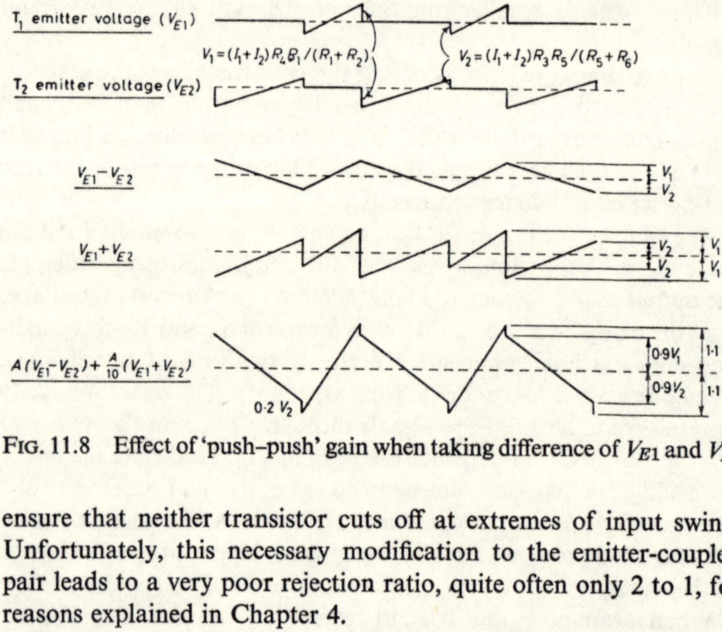

FIG. 11.8 Effect of 'push–push' gain when taking difference of V_{E1} and V_{E2}

ensure that neither transistor cuts off at extremes of input swing. Unfortunately, this necessary modification to the emitter-coupled pair leads to a very poor rejection ratio, quite often only 2 to 1, for reasons explained in Chapter 4.

Hence, it is further necessary to replace R in Fig. 11.9 by a constant-current transistor source which restores the rejection ratio to typically 500 to 1. The magnitude of the step is now likely to be only 1/500th of the total triangle. This differential amplifier is shown in Fig. 11.10.

Properties of the triangular waveform (given by the circuit of Fig. 11.10 combined with Figs. 11.4, 11.5, 11.6, or 11.7). (a) Each half-cycle is linear immediately after the transition, i.e. no rounding of corners. (b) Frequency is variable by change of C, variation of V_c,

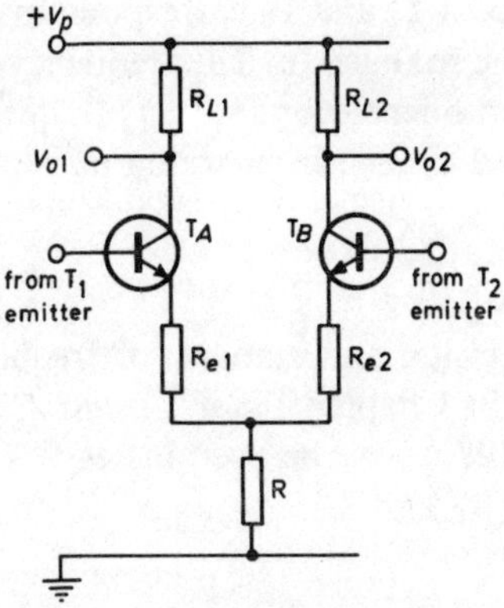

FIG. 11.9 Reduction of gain by extra emitter resistance, leading to poor rejection ratio

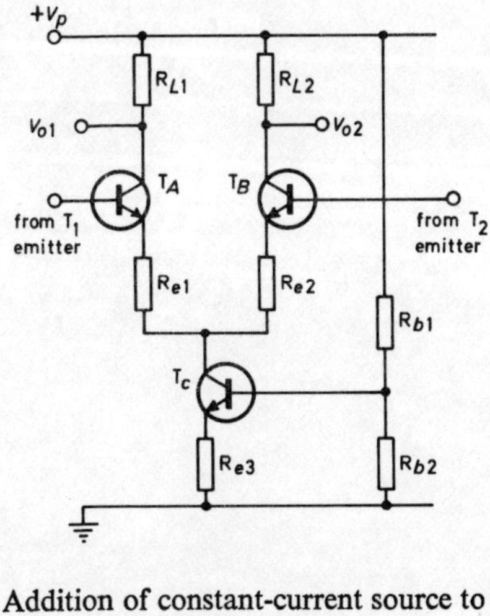

FIG. 11.10 Addition of constant-current source to restore good rejection ratio

or by change of R_7 and R_8. Amplitude remains constant unless V_c or R_7 or R_8 are so chosen to cause emitter current(s) to reduce below critical level, at which clipping of collector waveforms takes place. (c) The two slopes are respectively inversely proportional to R_7 and R_8, a ratio of 20:1 being readily obtained. (d) The rectangular collector waveforms of T_1 and T_2 correspond in their vertical edges to the two peaks of the triangle. (e) The frequency can be changed very rapidly (e.g. in a few microseconds at any frequency), since change in V_c produces a rapid change in charging currents as soon as T_3, T_4 respond.

Improvements

Where great precision and stability of frequency is required, the circuits described in Chapter 16 of *Circuit Consultant's Casebook* (Business Books, 1970) may be used instead of the simple constant current transistors T_3 and T_4.

Chapter 12

Operational amplifier applications

The advent of linear integrated circuits has enabled electronic system designers to regard complete amplifiers as single components. Some characteristics of operational amplifiers were discussed in Chapter 5. In the present chapter some of the common applications are described and simple procedures explained for predicting performance.

Choice of Amplifier

Most medium-performance operational amplifiers are available, with slightly differing type numbers, from most of the major integrated circuit manufacturers. Certain companies specialize in unusual or extra-high performance devices which are often worth considering when the standard product is inadequate and the alternative involves difficult discrete circuit design.

Some operational amplifiers such as the μA741 (*see* Chapter 5) are supplied already 'frequency-compensated'. This implies that a capacitor is incorporated within the device such that loop stability is guaranteed when all the output is negatively fed back to the input, i.e. circuit gain is unity, and the loop gain is the same as the open loop gain. This capacitor provides a dominant lag and so must have an associated time constant of at least A_0 times the next largest time constant in the amplifier (*see* Chapter 9), where A_0 is the d.c. open loop gain. Since A_0 is commonly 100,000 or more in a 'good' specimen, the time constant needs to be so large that it already produces a significant fall off in open loop frequency response even at 10 Hz, and at 1000 Hz the open loop gain has fallen to a mere 1000.

If such an amplifier is used when 100% feedback is *not* applied, the resulting frequency performance may limit its use to a few hertz only; moreover, the capacitor is then unnecessarily large for maintaining stability. For the type of application where the resulting bandwidth is inadequate, an uncompensated device such as the μA748 or LM101 may be used, the onus being upon the designer to specify the correct capacitor for the stability, i.e. freedom from oscillation, of his circuit. The manufacturer's data states the appropriate values according to the circuit gain of the amplifier stage, so that the maximum bandwidth compatible with certain stability is then achieved, within the limitations of compensation by a single capacitor.

This simple dominant-lag technique (*see* Chapter 9) does not result in optimum bandwidth and many of the high performance amplifiers provide means for more complex compensation.

Naturally, in choosing the right device, the simplest and lowest performance type adequate for the purpose should be selected.

Basic Amplifiers

The forms of amplifier shown in Figs 12.1(*a*) and (*b*) have already been described in Chapter 5, special attention being given to offset voltages and currents. Fig. 12.1(*c*) is a differential amplifier, giving an output dependent only upon the difference between v_1 and v_2. As before (Chapter 5), the simplest method to calculate v_o is as follows.

1 Check that the predominant feedback is negative. (Many an operational amplifier circuit has been connected with the input legs reversed.) Ignore input bias currents and input offset voltage (I_{B1}, I_{B2}, v_{os}) and assume infinite gain from the differential voltage between input legs and the output pin.
2 Calculate the applied voltage on one input leg. In many cases this is obvious by inspection to be zero or a simple fraction of the signal.
3 Assume the other input leg will be forced by feedback to take up the same potential.
4 Calculate v_o by Ohm's or Kirchoff's law.
5 Now consider the effects of I_{B1}, I_{B2}, v_{os} separately—by the principle of superposition of linear quantities they do *not* interact. Use values for I_{B1} etc. corresponding to worst case; ignore 'typical' values and thereby obtain the worst case offset.
6 Assume that in a good specimen $I_{B1} = I_{B2} = v_{os} = 0$, thus obtaining the best offset figure. The spread of performance between worst and best is now known. [This method is a simple way to deduce the properties of a feedback circuit when the loop gain is so high that the performance clearly intended by the

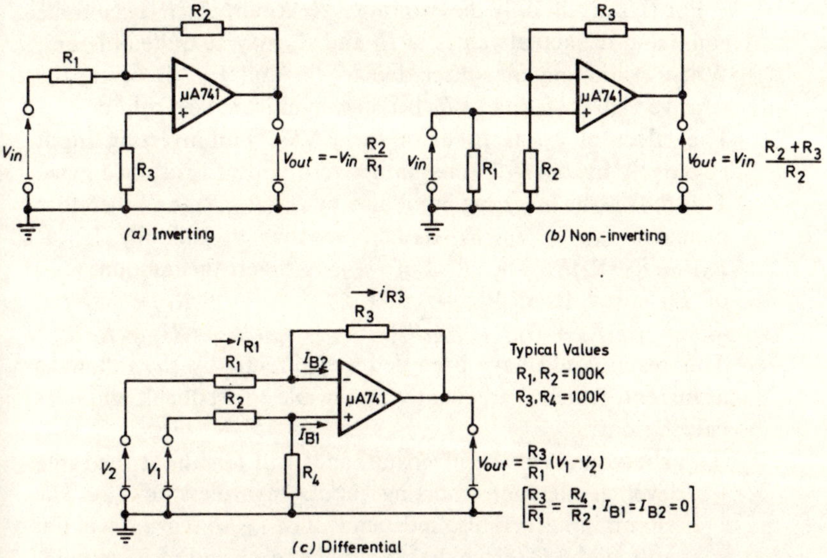

FIG. 12.1 Basic 741 amplifier forms

designer will be obtained. To find the exact behaviour, the form of calculation given in Appendix 4 should be carried out, with the open loop gain written as $A = A_o/(1 + jF/F_o)$.]

Taking the simple example of Fig. 12.1 (*c*):

1 Feedback is all negative via R_3.

2 The non-inverting input is at $v_1 R_4/(R_2 + R_4)$.

3 The inverting input must also be at $v_1 R_4/(R_2 + R_4)$.

4 i_{R1} is $[v_2 - v_1 R_4/(R_2 + R_4)]/R_1$, and this current must also flow in R_3 since the amplifier input current I_{B2} is assumed to be zero. Therefore

$$v_2 - v_o = (R_1 + R_3)i_{R1} = [v_2 - v_1 R_4/(R_2 + R_4)](R_1 + R_3)/R_1,$$

giving

$$v_o = v_1 \frac{R_4(R_1 + R_3)}{R_1(R_2 + R_4)} - v_2 \frac{R_3}{R_1} = v_1 \frac{1 + R_3/R_1}{1 + R_2/R_4} - v_2 \frac{R_3}{R_1}$$

$$= \frac{R_3}{R_1}(v_1 - v_2) \text{ if } \frac{R_3}{R_1} = \frac{R_4}{R_2}$$

This circuit therefore behaves as a differential amplifier and gives no output if identical inputs are applied.

For this result, only the two ratios R_3/R_1 and R_4/R_2 need to be equal and the actual values of R_1 and R_2 may be quite different. When examining the offset due to I_{B1} and I_{B2}, however, the relative values of R_1 and R_2 become significant (see below).

5 The effect of I_{B1} is to drive the μA741 non-inverting input negatively by $I_{B1}(R_2/R_4)$ assuming zero source resistance in v_1. This moves the inverting input also by $I_{B1}(R_2//R_4)$ and therefore increases i_{R1} by $I_{B1}(R_2//R_4)/R_1$, so that v_o falls by $(R_1 + R_3)I_{B1}(R_2//R_4)R_1$. The effect of I_{B2} is to divert this amount (I_{B2}) of i_{R1} away from R_3, so that $v_o \neq v_2 - i_{R1}(R_1 + R_3)$ but $= v_2 - [i_{R1}R_1 + (i_{R1} - I_{B2})R_3] = v_2 - i_{R1}(R_1 + R_3) + I_{B2}R_3$. This result could have been deduced directly by the following argument, which is frequently applicable to feedback amplifier calculations:

> The voltage at the inverting input will remain at the same level as the non-inverting input, regardless of I_{B2}. The circuit input v_2 is also independent of I_{B2} so that i_{R1} will not affected by I_{B2} since the potential at each end of R_1 remains constant. Therefore when I_{B2} flows it is the current in R_3 which must change by I_{B2}, causing v_o to rise by R_3I_{B2}. In other words since the inverting input voltage is fixed, any current flowing out of it (like I_{B2}) *appears to flow in* R_3.

The net effect of I_{B1} and I_{B2} is therefore to change v_o to

$$v_1\frac{1 + R_3/R_1}{1 + R_2/R_4} - v_2\frac{R_3}{R_1}$$

$$-(R_1 + R_3)I_{B1}(R_2//R_4)/R_1 + I_{B2}R_3$$

Now it is known that $I_{B1} \sim I_{B2}$, i.e. I_{os}, is noticeably less than I_{B1} or I_{B2}. When I_{os} is actually zero the minimum offset will therefore be obtained when

$$(R_1 + R_3)(R_2//R_4)/R_1 = R_3, \text{ i.e. } R_2//R_4 = R_1//R_3$$

Thus for minimum offset when $I_{os} \ll I_B$, the resistance 'seen' from input 1 $R_2//R_4$, will equal that seen from input 2, $R_1//R_3$, and the resulting output offset is $I_{os}R_3$. Combined with the results of (5), the conditino for a true differential amplifier with minimum offset is therefore $R_3 = R_4$, $R_1 = R_2$.

The effect of v_{os} is to make the voltages at the inverting and non-inverting inputs different by $\pm v_{os}$, so that the inverting input will be at $v_1R_4/(R_2 + R_4) \pm v_{os}$, adding to i_{R1} by $\pm v_{os}/R_1$, and therefore changing v_o by $\pm v_{os}(R_1 + R_3)/R_1$.

It is of course easy enough to derive these results in one algebraic calculation, but evaluating the individual effects of each contributor to offset makes it possible to assess instantly a practical circuit, with actual values attached. For example with the values specified in Fig. 12.1 (*c*), the nominal gain is $R_3/R_1 = 10$, for a differential input. The effect of I_{os} (worst case figure $\pm 0{\cdot}5\ \mu A$) is an output equal to $\pm I_{os}R_3 = \pm 0{\cdot}5 \times 0{\cdot}1\ V = \pm 50$ mV. The effect of $v_{os} = \pm 6$ mV is an output of ± 6 mV is an output of $\pm v_{os}(R_1 + R_3)R_1 = 66$ mV. Thus, the total offset referred to the input is 11·6 mV.

Other Useful Operational Amplifier Circuits

The following selection should indicate the versatility of the operational amplifier. At the same time some hints are given for optimizing performance by attention to design detail.

The integrator (Fig. 12.2)

Fig. 12.2 shows the 'standard' integrator in which a simple

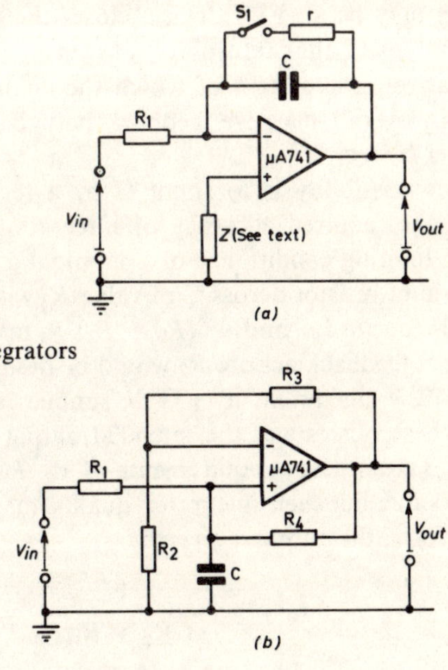

FIG. 12.2 741 integrators

amplifier configuration is used with a capacitor as the feedback element. In principle this circuit can be used with sine waves to obtain a 90° phase shift or, more commonly, to integrate a non-linear input function such as a square wave.

With the non-inverting input at zero, the current in R_1 is v_{in}/R_1 and this flows into C giving $dv_o/dt = -v_{in}/(CR_1)$, or $v_o = -(1/CR_1)\int v_{in}dt$; for sine waves $v_o = -v_{in}/(jwCR_1)$ [in operator form $v_o = -(1/pCR_1)v_{in}$].

It is particularly important to note that this relationship holds only provided feedback is maintained which is not the case if v_o reaches the amplifier limit. When this occurs the inverting input approaches v_{in} on a time constant R_1C and even after the reversal or zeroing of v_{in} takes several R_1C time constants to return to zero potential finally allowing the integrator to operate again. Since there is no d.c. feedback, the input needed to cause such limiting is very small and the amplifier's own offset voltage is sufficient although the limit may be reached only after several seconds when C has received the appropriate charge.

Therefore, a switch S1 (not forgetting the current limiter r to avoid damage to S1 by the capacitor discharge current) is often used when CR_1 is long enough to allow periodic resetting. Operation of S1, which may be an FET device, closes the feedback loop to d.c. and discharges C, thus returning v_o to zero.

Alternatively the system of which the circuit is part is made to provide an overall negative feedback path back to v_{in} to hold its mean level correct.

Another possibility is to shunt C by a resistor high enough to maintain the required accuracy of integration but low enough to avoid the limiting condition. For a nominal d.c. level of zero at the input, a shunt resistor across C of value R_2 will produce an offset of $\pm 0{\cdot}5R_2\,\mu$V due to I_{os}, and $\pm 6(R_2 + R_1)/R_1$ mV due to v_{os}. To maintain correct feedback action, R_2 would be designed to keep v_o below about ± 12 V maximum if ± 15 V supplies are used, taking into account these offsets and the intended output signal combined. To meet the d.c. accuracy requirements of v_o, R_2 may need to be less than this value but then integrator quality may be unacceptable. In operator form, the output is given by

$$v_o = -\frac{1}{pCR_1 + R_1/R_2}\,v_{in}$$

instead of

$$-\frac{1}{pCR_1}v_{in}$$

In general, therefore, the much quoted standard integrator is unusable in isolation and needs some means of holding its output level within suitable limits.

Effects of offsets on the integrator's behaviour at supply switch-on

The most frequent use of the integrator is in deriving a ramp from a step input, followed by the discharge of C either rapidly by S1 (an electronic switch) or at the ramp rate by input polarity reversal and continual repetition of the sequence. In such applications the events immediately following switch-on of power supplies are often important and the behaviour of Fig. 12.2(*a*) is in this respect dependent upon Z, the non-inverting external impedance.

Remembering that a capacitor is unable to change its charge in zero time enables the switch-on disturbance to be evaluated by inspection.

Assume that v_{in} is to be zero so that the v_o of an ideal integrator would remain at zero both immediately after switch-on and subsequently. If $Z = O$, i.e. the non-inverting input is directly grounded, then at switch-on the inverting input will leap to $\pm v_{os}$, and v_o will follow identically because C cannot change its charge. This jump will cause a current in R_1 of $\pm v_{os}/R_1$; also the inverting input will pass I_{B2} inwards. The sum of these currents begins to charge C so that v_o will, after the initial step, rise at a rate of $(I_{B2} \pm v_{os}/R_1)$ V/sec and this will continue until some limit is reached. This is shown in Fig. 12.3(*a*), where v_o lies somewhere within the shaded area depending on whether v_{os} has a positive or negative sign. The intended ramp will be added to v_o, when v_{in} is applied.

In an attempt to improve these unwanted effects, Z is often made equal to R_1 by analogy with the simple amplifier case in Fig. 12.1. The result [Fig. 12.3(*b*)] is to reduce to zero the slope at which v_o rises after any initial steps (if I_{os} is zero), but at the cost of an increased step at switch-on, now equal to $-I_{B1}R_1$ in addition to $\pm v_{os}$. If for example $R_1 = 100$ kΩ, the step would be $-0{\cdot}15$ V which may be more troublesome than was the spurious charging slope.

The correct approach is to follow even more closely the amplifier

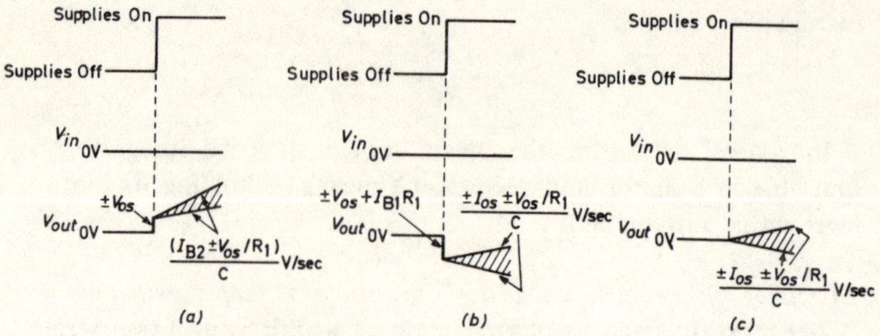

FIG. 12.3 Switch-on characteristics of Fig. 12.2(*a*) for (*a*) $Z = 0$, (*b*) $Z = R_1$, (*c*) $Z = R_1//C$

case and make $Z \equiv R_1//C$, which is known to be correct when 'C' is a resistor. Now we have only the $\pm v_{os}$ step at switch-on, since the non-inverting input no longer leaps to $-R_1I_{B1}$, and we have the final benefit of equalizing the input network resistance, giving no charging rate if I_{os} is zero.

Another snag with making $Z = R_1$ with no parallel C is that the non-inverting input is then a 'live' terminal, possibly of quite high input impedance (100 kΩ in the above example). The gain at moderate and high frequencies to the output v_o is unity, so that nearby circuits generating sharp pulses can cause v_o to produce similar waveforms in addition to the correct output. This form of interference is common when FET's are used as sample switches and driven with large fast gate pulses. The nearby integrator, supposedly capable of only slow output changes, leaps about in sympathy.

It is good practice, therefore, to short-circuit Z completely and accept the error in slope or use the correct $R_1//C$ combination. Only if R_1 is small and the switch-on step can be shown to be unimportant should the capacitor be omitted. One compromise is occasionally satisfactory—the use of a parallel C just for interference suppression, e.g. 0·01 μF, when the input step, which will not be eliminated by such a small value of capacitance, is considered unimportant.

Precautions

The integrator, in common with other circuits where large capacitors are attached to input legs, is a dangerous circuit in that it frequently causes operational amplifier destruction. The problem is most significant when C is large (> 1 μF) and when the power supplies are capable, when turned off, of falling rapidly to zero. This can happen simply because other heavy loads are present, causing fast discharge of residual voltages, or if the supply has 'crow-bar' protection in which a short-circuit is placed across the line in the event of a fault.

If the integrator of Fig. 12.2(*a*) has an output of, for example, +14 V at the moment when the supplies collapse suddenly, the output terminal falls to ground and the remote end of C falls rapidly towards −14 V. This causes the inverting input to be driven beyond the negative supply voltage (which at that moment is zero!) and destruction of the μA741 is likely. The same risk is present in the other direction if the output is sitting negatively at switch-off.

The elusive point is that an apparently comprehensive test may fail to reveal this weakness: the test supplies may collapse slowly and the collapse may not in any case coincide with a maximum v_o (with a large charge available on C), unless deliberately devised to do so. On the other hand a series of normal tests in production may be concluded satisfactorily and each production unit then destroyed (unnoticed!) as the test supplies are finally switched off, resulting in a complete batch of useless units.

It is difficult to devise in a simple way a theoretically correct solution to this problem. In practice a current-limit resistor R in the vulnerable inverting input lead (about 4·7 kΩ) or two shunt diodes from that input to ground are effective (Fig. 12.4). The theoretical

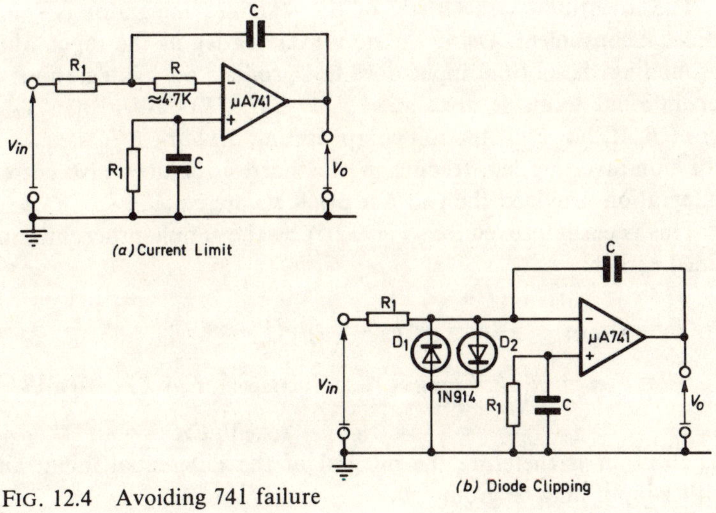

FIG. 12.4 Avoiding 741 failure

weaknesses are that no manufacturer gives a suitable figure for the safe current limit when inputs are driven in this way, and the input voltages are supposed not to exceed the supply voltage by any amount. Our current limiting of a few milliamps and the diode catch method (which allows the inputs to exceed the supply by a diode-drop) are based on favourable experience. Incidentally, if diodes are used and $Z = R_1//C$ care must be taken that neither diode can turn on due to the standing potential across it of $v_{os} - I_{B1}R_1$.

Stability against oscillation

The built-in capacitor in the μA741 should guarantee freedom from oscillation when used in an integrator circuit, but as with most amplifiers correct performance, including stability, depends on the provision of supply voltages of low impedance. When practical lead lengths are examined it is evident that the quality of the actual supply may be degraded en route to the amplifier.

It is essential therefore to decouple the supplies as close to the μA741 as possible, using 0·1-μF capacitors of good r.f. performance. Their omission may go unnoticed either because oscillation does not occur or because the visible effect of a VHF oscillator is an apparent d.c. drift whose real nature is unsuspected. Good practice dictates their inclusion.

Integration without inversion

The inverting characteristic of the standard integrator is sometimes inconvenient. Using the non-inverting leg as the input while grounding the normal input does not produce pure integration; in operational terms it gives $v_o = v_1(1 + 1/pCR)$ rather than $v_o = v_1/pCR$. If, however, the full compensating network $R//C$ is used in the non-inverting leg, feeding in via the resistor does give correct integration provided the two CR products are equal.

This is easily proved (see Fig. 12.5) by the simple procedure outlined earlier:

$$v_2 = v_1 = v_{in(1)}/(1 + pCR)$$

$$\therefore \quad i_R = (v_{in(2)} - v_2)/R = [v_{in(2)} - v_{in(1)}/(1 + pCR)]/R$$

$$\therefore \quad v_o = v_{in2} - i_R\left(R + \frac{1}{pC}\right) = v_{in(2)} - [v_{in(2)}(1 + pCR) - v_{in(1)}]/pCR$$

$$= [v_{in(1)} - v_{in(2)}]/pCR$$

The output is therefore the integral of the differential input, and either input may be grounded.

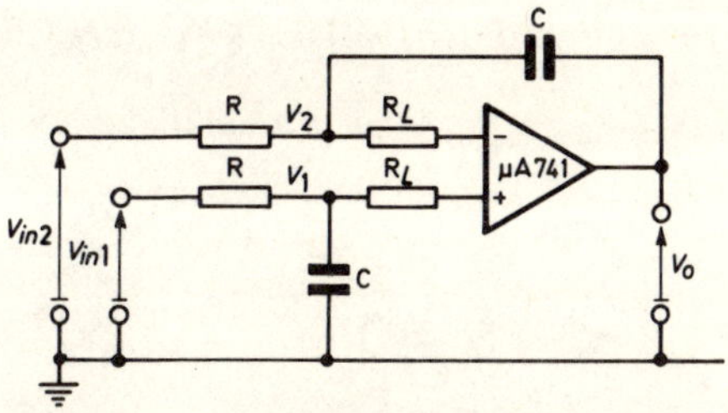

FIG. 12.5 Differential input 741 integrator

Inequality in the two *CR* networks gives a proportional departure from the intended law.

An alternative to the standard integrator

A problem in using the inverting standard integrator inside a feedback system, which is its main mode of application, is that as the input frequency increases the operational amplifier becomes ineffective as its gain decreases. This causes errors in the integration, and also allows the high frequency input signal to appear at the output by simply passing through R and C. Since the 'virtual earth' and low output impedance no longer exist due to absence of amplifier gain, this route provides a direct non-inverting path with little attenuation. Oscillation in the overall system is then likely and this is a much more serious result than the integrating error.

A more promising configuration would have the integrating capacitor grounded so that good h.f. attenuation is always present. Such a circuit, derived from the constant current circuits of Fig. 6.13 is illustrated in Fig. 12.6, where the load R_L of Fig. 6.13 has been replaced by capacitor *C*.

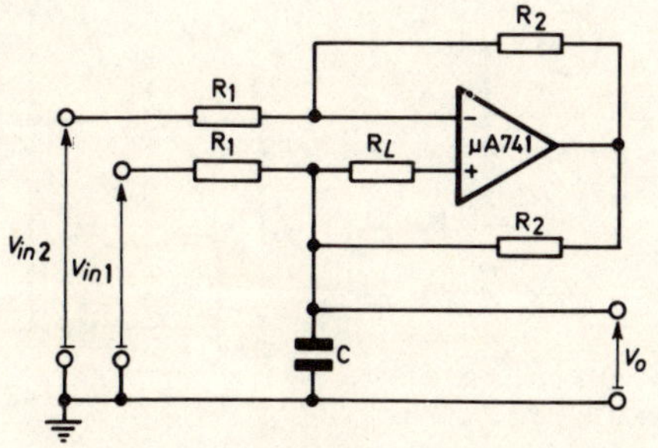

FIG. 12.6 Alternative 741 integrator

Since $$i_C = (v_{in(1)} - v_{in(2)})/R_1 \quad (\textit{see } \text{Chapter 6}),$$

$$v_o = \frac{1}{CR_1}(v_{in1} - v_{in2})\mathrm{d}t$$

or in operational terms

$$v_o = \left(\frac{1}{pCR_1}\right)(v_{in(1)} - v_{in(2)})$$

This circuit therefore has the advantage over the standard form that C continues to provide h.f. attenuation even when the amplifier has ceased to be effective in providing correct integration. By grounding either input terminal it may be inverting or non-inverting.

A further advantage over Fig. 12.5 is that correct integration depends on resistor rather than capacitor equality. Where long integration times are needed electrolytic capacitors may have to be used and their wide tolerance makes matching difficult, while resistor matching remains easy.

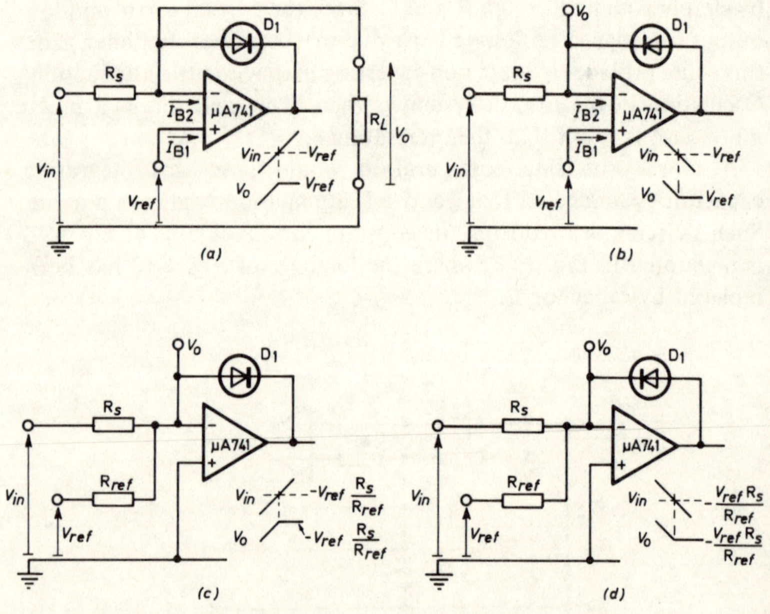

FIG. 12.7 Clipping circuits

Note the use of limiting resistors R_L in all of these circuits where capacitors are joined to input legs. As described in the section on the standard integrator, a charge, present on the capacitor when supplies collapse, can destroy the μA741 if no protection is provided.

Clipping Circuits

Diode and transistors used as rectifiers and for clipping (*see* Chapter 1) suffer from the disadvantage that a substantial voltage (0·1 V for germanium, 0·5 V for silicon) must be reached before the device operates; moreover this voltage has a variation of −2 to −2·5 mV/degC.

The clipping of a slowly moving voltage when it exceeds or falls below a reference level V_{ref} can be performed with improved accuracy with the help of an operational amplifier as in Fig. 12.7.

In Fig. 12.7(*a*) inputs more negative than V_{ref} cause the amplifier output to be highly positive, D1 being cut off and v_o being equal to v_{in}. When v_{in} just exceeds V_{ref}, the amplifier output swings negative but as soon as it reaches $(V_{ref} - V_f)$, V_f being the diode forward drop for conduction, D1 conducts and introduces negative feedback, holding v_o equal to V_{ref} so long as v_{in} exceeds V_{ref}. This gives the clipping action shown in the attached diagram. Naturally D1 may be reversed as in Fig. 12.7(*b*), when clipping occurs whenever v_{in} is more negative than V_{ref}.

This circuit is much more accurate and drift-free than the simple diode because the diode drop itself plays no part in determining clipping level (provided A_o is very high). A further advantage is that after the onset of clipping, v_o changes very little as v_{in} varies, because of the very heavy negative feedback (*see* Chapter 8). For the same reason v_o presents a very low resistance in feeding its load, but only after clipping level is reached.

The circuit does have certain defects, however. The first concerns the meaning of clipping level: it can be taken to mean the value of v_{in} at which v_o ceases to change; or it can be the value of v_o itself when it has ceased to change. Ideally both levels would have the value V_{ref}. In practice both will suffer a direct error of $\pm v_{os}$, but will suffer different effects due to I_{B1} and I_{B2}.

As drawn, Fig. 12.7(*a*) will produce a clipped output $v_o = V_{ref} \pm v_{os}$ but clipping will begin only after v_{in} has reached $V_{ref} \pm v_{os} + R_s I_{B2}$ if R_L is ignored, or, taking R_L into account, $[V_{ref} \pm v_o + I_{B2}R_s//R_L](R_s + R_L)/R_L$. In practical terms, v_{in} will itself pass the clipping level before clipping begins but the clipped output will have the correct value V_{ref} (except for $\pm v_{os}$).

To overcome the effect of I_{B2}, a resistor equal to R_s could be added in series with V_{ref} so that if $I_{B1} = I_{B2}$, i.e. $I_{os} = 0$, v_{in} will equal $V_{ref} \pm v_{os}$ at the onset of clipping. Unfortunately this results in v_o being equal to $V_{ref} - I_{B1}R_s \pm v_{os}$ so that one error has been changed for another; we now begin clipping at the right value of v_{in} but clip to an erroneous input level.

The choice must be made for the application in question and R_s made as small as practicable. The lower limit of R_s depends on how far beyond clipping level v_{in} will move, and what current can then be permitted to flow from v_{in} through R_s and D1 into the output terminal of the amplifier. As a guide, if R_s can be made <5 kΩ, the resulting error can probably be ignored.

Alternative circuits [Figs. 12.7(*c*) and (*a*)] can be used if v_{in} is to be clipped when it reaches a certain voltage of opposite polarity and v_o is then clipped to zero. In the circuits clipping begins at $v_{in} = -V_{ref}R_S/\text{Ref} - I_{B2}R_S//R_F \pm v_{os}$ and the clipped value of v_o is $\pm v_{os}$. These circuits differ also in that current always flows from v_{in} to V_{ref} before clipping so that both v_{in} and V_{ref} are more difficult to design accurately.

Rectifier Circuits

The rectifier circuits shown in Fig. 12.8 are related to the clipping circuits in using the high gain of the operational amplifier in a feedback connection to eliminate errors of rectification caused by diode forward drop. In Fig. 12.8(*a*) each positive input half cycle causes D1 to conduct and each negative half cycle causes D2 to conduct. Conduction begins accurately at the input waveform zero, with the usual errors due to I_{B2} and v_{os}. V_1 is therefore an accurate series of negative half cycles and V_2 positive half cycles, of peak magnitudes $\hat{V}_{in}R_2/R_1$ and $\hat{V}_{in}R_3/R_1$. One common application for this type of circuit is for an a.c. reading electronic voltmeter. For moving coil meters, high impedance or 'current driving' gives the fastest meter needle response. It is also more accurate since meters are calibrated in terms of current and have wide-tolerance coil resistances. An ideal drive circuit, again without diode errors, is shown in half-wave form in Fig. 12.8(*b*) and full wave in Fig. 12.8(*c*). In the half-wave circuit the average meter current is $\frac{1}{2}\overline{v_{in}}/R_1$, because each half cycle of a particular polarity [positive as shown in Fig. 12.8(*b*)] flows through

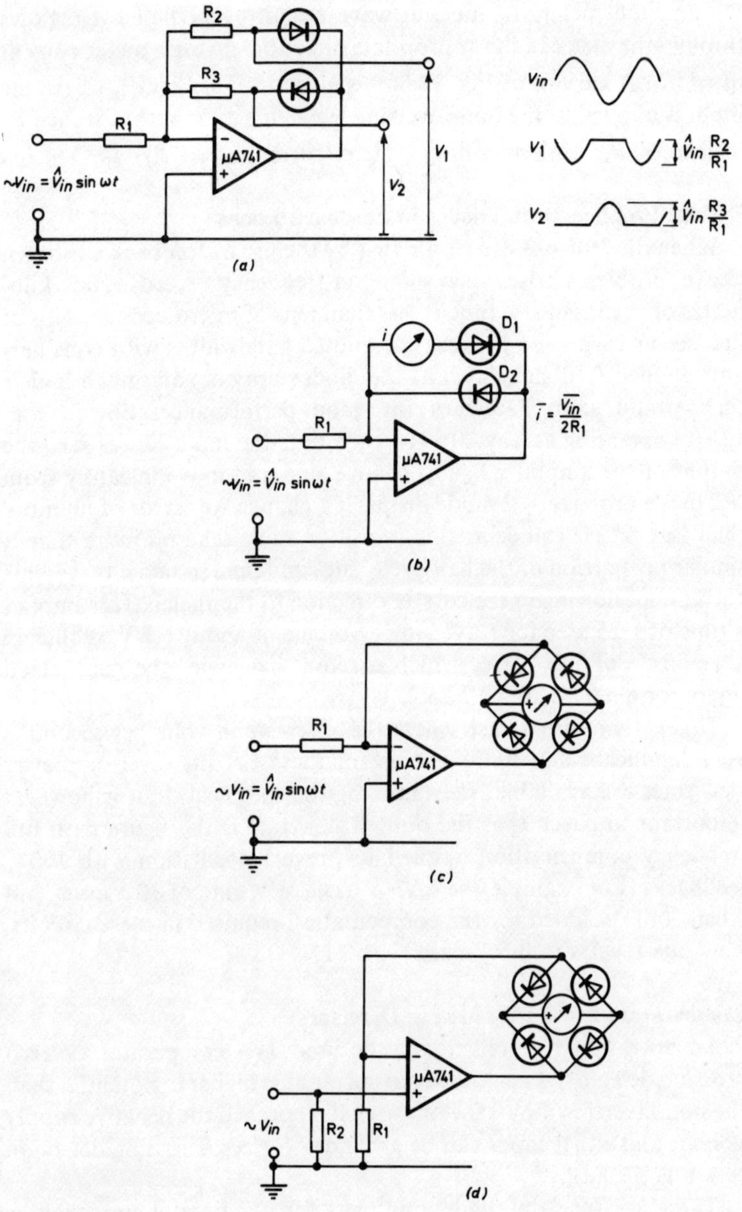

FIG. 12.8 741 rectifier circuits

D1 and the meter. In the full wave circuit every half cycle flows through the meter in the appropriate direction, giving a meter current of $\overline{v_{in}}/R_1$. A variant of Fig. 12.8(*c*) is given in Fig. 12.8(*d*), where the input is applied to the non-inverting leg giving an input resistance R_2 instead of R_1, while retaining a meter current of $\overline{v_{in}}/R_1$.

Special Problems with Diodes in Feedback Loops

When diode drops are eliminated by the use of feedback as shown above, problems arise when the input frequency exceeds a few kilohertzs or the input rise time is less than tens of microseconds. This at first seems surprising because the quoted bandwidths with considerable applied feedback (such as the diodes provide) are much higher than would seem necessary for good performance. But in Fig. 12.8(*c*) operating at, say 20 kHz, each time the input passes zero the output of the amplifier has to change sign and move instantly from +2 diode drops to −2 diode drops. If a changeover error of no more than, say 5% is to occur, this transition must take no more than a similar proportion of the half cycle since any time so taken represents current not flowing in the correct direction to the meter. This implies a time of 1·25 μsec for traversing a voltage of about 1·5 V, requiring a slew rate of 1·2 V/μsec which is more than twice the guaranteed figure for a μA741 (0·5 V/μsec).

This is therefore a case where the *slew rate* in volts per second is very significant and limits the use of these circuits severely. Better slew rates are available; in examining the specification it is however important to check that the quoted slew rate is the figure with full frequency compensation applied to prevent oscillation with 100% feedback. (For example the μA748 has a slew rate of 10 V/μsec, but when 30pF is added for the compensation required in these circuits, slew rate reverts to 0·5 V/μsec.)

Comparators and Zero-crossing Detectors

An operational amplifier may be used as a comparator or zero crossing detector: when the inverting input is higher in potential than the non-inverting input, the output will approach the negative supply voltage and either input can be grounded for zero-crossing detection [see Fig. 12.9(*a*)].

There are specially made amplifiers for comparator use, such as the LM111 or 311. These have faster slew rates than the usual operational amplifier and in fact provide a final output transistor which

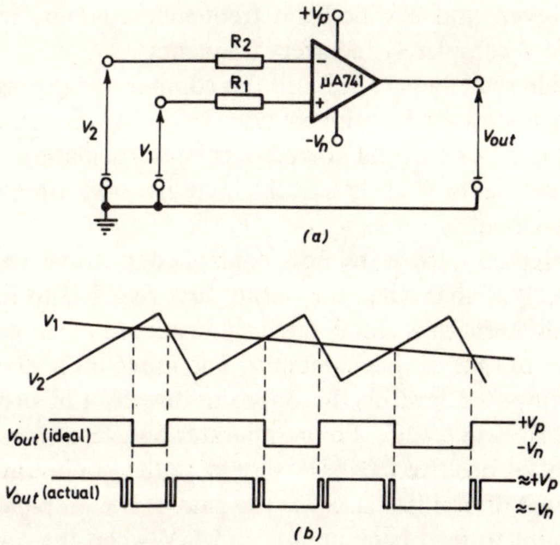

FIG. 12.9 Comparator performance with hysteresis omitted

can be connected either as an emitter follower or as a grounded emitter amplifier. They are not generally usable as feedback amplifiers, being difficult to stabilize against feedback oscillation.

Whichever device is used the effects of source resistances R_1 and R_2 are to cause errors of $I_{B1}R_1$ and $I_{B2}R_2$ in the voltages reaching the input pins. Equality of R_1 and R_2 reduces the effect of these errors to zero if $I_{os} = I_{B1} \sim I_{B2} = 0$. The offset error $\pm v_{os}$ is always present.

These circuits are very simple but are unsatisfactory when the crossing rate of one input signal relative to the other is slower than, say, 1 V/10 μsec. In that case, which represents the majority of usage of the comparator, the output oscillates during each changeover with an amplitude equal to the total swing [Fig. 12.9(*b*)]. This occurs because during the time that $v_2 - v_1$ is passing through the near-zero region the amplifier is in a non-limiting condition. Its own noise, possibly combined with external interference, is effectively in series with $v_2 - v_1$ and v_{os}, and it causes the output to switch positive and negative until $v_2 - v_1$ is large enough to hold the output one way or the other. When the amplifier is an LM311 these transitions occur in a few tens of nanoseconds and at slow input rates such as 1 kHz are likely to pass unnoticed when examining the square wave output on an oscilloscope. The circuit following the comparator might well react, however, and a wide-band frequency counter, for example, would read a completely incorrect frequency.

A possible remedy is to smooth the comparator output by means of a *CR* network or by adding 'compensation' capacitance inside the amplifier as one would correct feedback oscillation. This is not generally satisfactory as it usually increases rise times to an unreasonable extent.

The preferred cure is to add controlled positive feedback [*see* Fig. 12.10(*a*)] so that when the output first switches to its new level, the input differential is simultaneously increased in the correct sense to hold the output in that direction. The input must now overcome this superimposed level in the opposite direction in order to cause the output to switch back. For comparator and zero crossing circuits, the amount of positive feedback should be the minimum compatible with removal of oscillation. Since the gain of the amplifier is >1000, it is sufficient to feed back about ± 15 mV when the output swings ± 15 V giving $R = 1000r$.

The circuit now possesses 'hysteresis' or 'backlash' in that a rising $v_2 - v_1$ causes switch over when $v_2 - v_1 = +15$ mV, but a falling $v_2 - v_1$ switches back when $v_2 - v_1 = -15$ mV. This represents an error in comparison between v_2 and v_1 but this is inevitable in such circuits. Without the introduction of the positive feedback which causes this backlash the circuit is basically unsound.

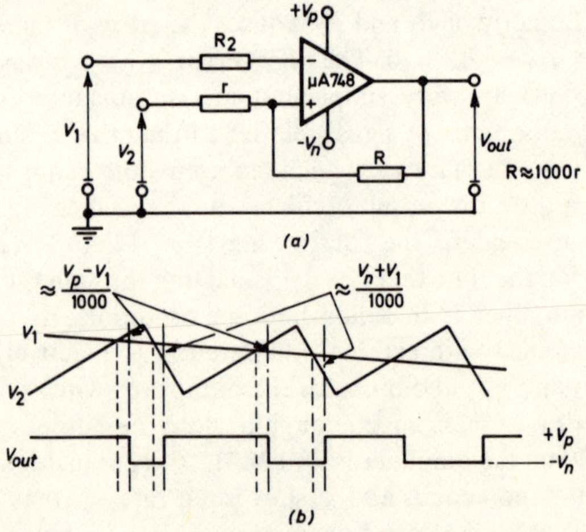

FIG. 12.10 Comparator performance with hysteresis

Schmitt Triggers

A 'Schmitt trigger' was originally used to describe a transistor circuit with snap action similar to the integrated version in Fig. 12.10. Some integrated circuits such as the μA710 are so described and are usually intended for driving the 'clock' inputs of logic circuits (which require fast rise time) from slow analogue signals.

In signal processing circuitry there is frequently a need for a circuit of the type shown in Fig. 12.10 but with considerable hysteresis of several volts. At the same time both triggering levels may need to be accurate to a few per cent. In principle this can be provided by grounding v_1 and designing the ratio R/r to give the appropriate levels. However, the swing of v_o is badly defined and the desired triggering levels may be unequal and of the same sign.

A useful circuit for accurate trigger levels is shown in Fig. 12.11, where the very low saturation potentials of the transistors give only small errors. R_B is designed to give optimum base drive to T1 and T2 for minimum $V_{ce(sat)}$ (*see* Chapter 8 on choppers). If V_{Z1} and V_{Z2} differ greatly, giving unequal drives to T1 and T2 base, R_B may be shunted by a series diode (connected in the appropriate sense) and resistor as shown to equalize the base currents. This need be only an approximate equality so that each base current lies in the 1/3 to 1 mA region, provided resistors $R + r$ pass less than about 200 μA (*see* Chapter 8). In this mode of connection T1 and T2 saturate to within a few millivolts, not the normal $V_{ce(sat)}$ of 0·25 V. The use of R and r is optional but is convenient if the desired trigger levels are small, enabling realistic values of V_Z to be used.

If performance immediately following transitions is unimportant, T1 and T2 may be replaced by the clipper circuits of Fig. 12.7(*a*) and (*b*), as shown in Fig. 12.12. It is not necessary either in Fig. 12.11 or

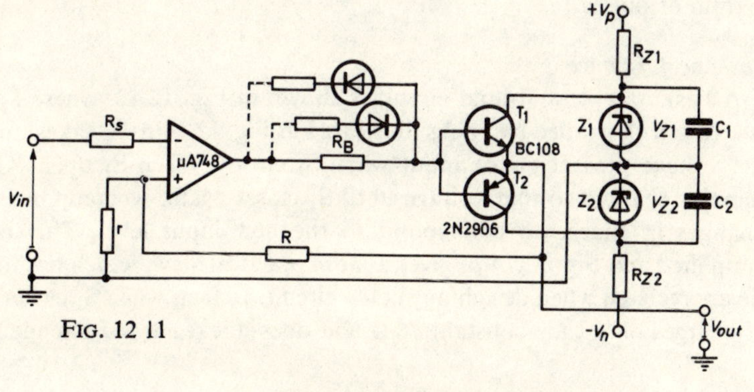

FIG. 12 11

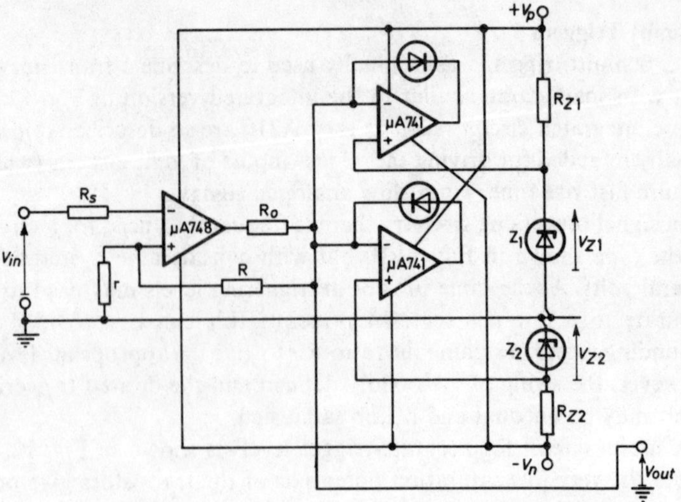

FIG. 12.12 Accurate Schmitt trigger for LF operation

12.12 that V_{Z1} and V_{Z2} be opposite in polarity provided the μA748 swings above V_{Z1} and below V_{Z2}.

Sample-and-hold Circuits

These circuits are used when the level of an input signal at a particular moment is to be measured, remembered or used for some other part of a signal processing system, even though the signal is changing.

It follows that there are two inputs to a sample-and-hold circuit, namely the signal to be measured and the pulse which decides when the measurement is to be made. The latter may be a single pulse or a train of pulses.

Basic performance

A basic sample-and-hold circuit is shown in Fig. 12.13, where S_1 is a switch controlled by v_2. As illustrated in Fig. 12.14(*a*), C takes up the voltage present at the input when S_1 closes. When S_1 opens C remains charged to that voltage until S_1 closes again, whereupon C changes in charge to correspond to the new input level. This is simplified and Fig. 12.14(*b*) gives a more practical view, which has to be appreciated when designing such a circuit. In fact, when S_1 closes, C charges on a time constant R_SC and does not reach the intended

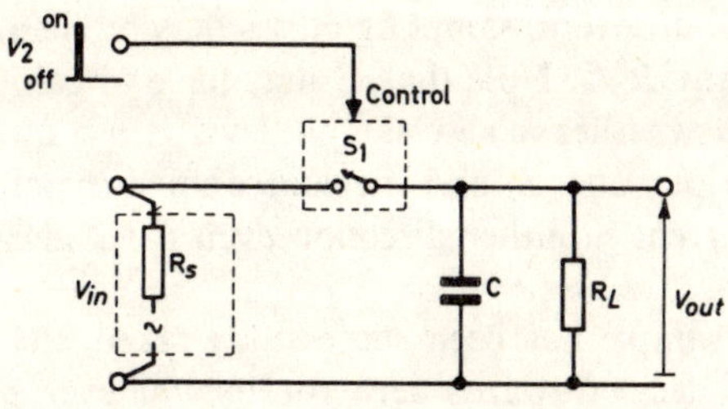

FIG. 12.13 Basic sample and hold circuit

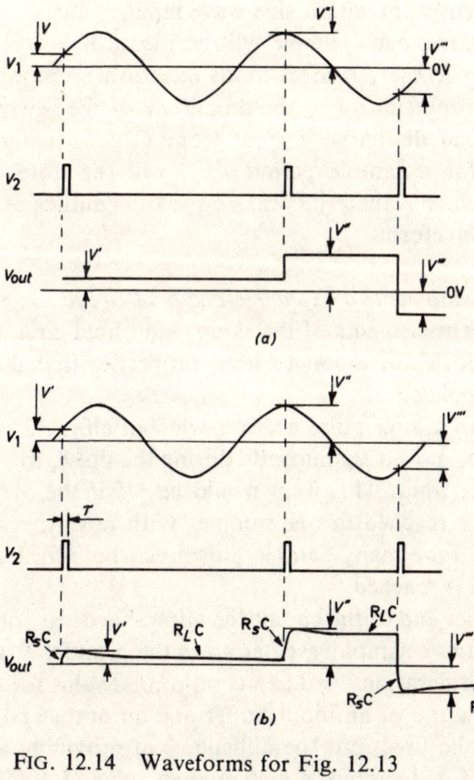

FIG. 12.14 Waveforms for Fig. 12.13

value until about $5R_SC$ has elapsed. If this value is to be reached in just one sample then τ, the closure time of S_1 must exceed $5R_SC$. It will be seen that if several sample pulses occur either while v_1 remains constant or at such times, e.g. multiples of the period of v_1, that each input sample is identical, then the correct level will eventually be reached on C but several sampling pulses may by then have occurred if τ is less than $5R_SC$. Note that C may have to gain or lose charge depending on whether a new sample level is more or less positive than the previous one. S_1 and the source must therefore be capable of passing current in either direction even if v_{in} always remains of the same sign.

When the sample has been successfully taken, the load R_L on C causes v_o to decay towards zero (or to whatever potential R_L is returned) on the tim econstant R_LC. Thus v_o becomes less accurate as the interval between samples increases. If sampling occurs regularly in synchronism with a sine wave input v_1 and the input amplitude is constant, each sample will be identical but because of the decay due to R_LC, v_o is a sawtooth waveform. Assuming the decay is a small proportion of v_o the amplitude of the sawtooth is easily predicted. The discharge current from C is virtually constant at v_o/R_L and for a sample period T, C will therefore discharge by $v_oT/(CR_L)$ which is then the peak-to-peak magnitude of the sawtooth or 'ripple' waveform.

Practical requirements of a sample-and-hold circuit

The different elements of the sample-and-hold circuit can be seen from the above discussion to need properties that depend on the particular application.

For a given sample pulse width τ, which itself is chosen so that the input has not moved significantly during the pulse, the product R_SC has an upper limit. This limit would be $\tau/5$ if the correct sampled level must be reached in one sample. With repetitive sampling one must decide how many sample pulses can be afforded, before the correct level is reached.

At the other end of the circuit the allowable decay, or ripple in the case of repetitive sampling, determines the product R_LC.

These considerations lead to a compromise value for C and almost always to the use of an input buffer and an output buffer. Discrete buffers may be used, but the difficulties of providing a low enough value of R_S for the input, a high enough value of R_L for the output and low offset for both usually dictate the use of integrated circuit operational amplifiers.

One requirement sometimes overlooked is the charging current needed for C when the input has changed from its maximum to its minimum level (or vice versa) between samples. The voltage across

R_S at the second of these samples is the difference between these levels, and the resulting charging current may be beyond the capability of the input source.

Note particularly that even if all sampled levels are, say, positive, the charging current may be in either direction depending only on whether a higher or lower level is sampled first. Therefore the input buffer must be capable of supplying rapidly a charging current of V_{in}/R_S in either direction where V_{in} is the largest difference between input voltages of successive samples. If this condition is not observed, the buffer cuts off and R_S becomes much larger.

To control the behaviour of the circuit it is advisable to add a resistor in series with S_1 to prevent unnecessarily high charging currents. With very low R_S and a high current capability from the buffer, the peak charging current, which must flow through the capacitor leads, can cause interference due to the resulting electromagnetic field.

The switch S_1 may also have appreciable resistance which adds to R_S. If a bipolar transistor is used the current capability will be different according to current polarity and the device may leave saturation in one direction. The FET, either MOS or junction type, is preferred.

The output buffer has as its main requirement a very low input current I_B. The lower I_B, the smaller can be C for a given droop or ripple specification. This in turn allows higher R_S for sufficietnly rapid sampling.

Typical design (*see* Fig. 12.15)

Supposing an input voltage varying between +5 V and −5 V is to be sampled every 10 msec for 0·1 msec with a resultant output droop of <20 mV between samples. If the output buffer is to be a μA741, I_B is 1·5 μA max (at −55°C) giving a droop between samples of $TI_B/C = 1{\cdot}5 \times 10^{-8}/C$. If this is to be less than 20 mV, C must exceed 0·075 μF. In turn this requires effective source resistance R_S during sampling given by

$$5R_SC \leqslant \tau, \text{ i.e. } R_S \leqslant 266\ \Omega.$$

This is within the capability of the input μA741 buffer and, using the 2N4859A FET with its 'on' resistance of 25 Ω, allows a current limiter τ of at least 100 Ω.

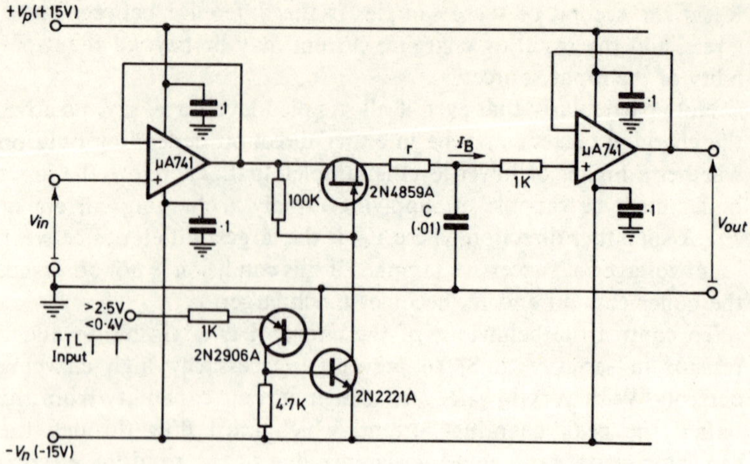

FIG. 12.15 Typical sample and hold circuit

Critical design features

A more stringent specification in almost any respect, such as shorter sample pulse, longer interval between samples, or lower permitted droop, would have required either a faster input buffer or a better output buffer.

The use of an FET input device for the output buffer, e.g. μA740, allows much smaller C values to be used, though below 0·1 μ care must be taken by screening to avoid pick-up of interference such as mains hum on the capacitor. The smaller C makes possible operation with smaller sampling pulse widths since $R_S C$ is reduced. Where still smaller pulse widths in the microsecond region are to be used the input buffer must be changed to a faster type, e.g. μA715, because the output resistance of the μA741 is poor at high speeds.

Manufacturers now produce complete sample-and-hold circuits in one package where very tight specifications are to be met, e.g. HA2425. The basic principles of operation are unchanged but the amplifiers used have much better high-frequency performance and overall feedback is applied to reduce offset errors.

Nulling Circuits

Reference has frequently been made to the errors caused by v_{os} and I_{os}. For any particular set of conditions it is possible to reduce this offset to zero. This can be done by adding a suitable signal to an input terminal or by making use of the 'offset null' terminals. Either way involves a potentiometer which is set manually to give zero output for zero signal input.

Unfortunately such nulling adjustments will be incorrect when the ambient temperature changes and unless provision for resetting is made offset errors will appear.

In military and unattended industrial applications the use of nulling potentiometers is not normally of sufficient benefit to justify the extra cost of the component (as expensive as the operational amplifier!) or the cost of setting it.

For laboratory usage, temperature variations are small enough to enable nulling circuits to be effective and the manufacturer's recommendations should be followed.

When good nulling is vital and must be held over large temperature changes, nulling loops can be used. In effect these carry out automatically, by switching circuits, the manual procedure. The input signal is periodically earthed; the output is compared with zero and an input applied until the output settles at zero. The input which had to be applied is stored, either digitally or on a capacitor, and the signal input reconnected leaving the correction signal still applied. The interval between nulling switching is frequent enough to correct for offset changes which are due to aging and temperature.

Usually, only military systems, with the wide temperature range of operation and the need to operate unattended, are able to justify the expense of nulling loops.

Vulnerability of the Operational Amplifier

Reference was made earlier, in connection with the integrator, of the danger of destruction of the operational amplifier caused by the charged integrator capacitor when switching off.

This damaging mechanism is always a possibility when a large capacitor is connected to an input terminal of an operational amplifier. It may need imagination to determine just what combination of signal levels, supply switch-on or off can cause damage. It can be hundreds of switching operations later before the damaging condition occurs. The damage may even then not be obvious until a more critical input has to be used.

The only safe practice is to protect against inputs exceeding supply magnitudes by including current limits or volage clipping (*see* Fig. 12.4). Similarly damaging conditions can occur without the presence of a capacitor but the danger is then generally obvious. It is however good practice at the conclusion of a design of any circuit to examine it for vulnerability of this kind.

Non-standard Connections

To understand operational amplifier behaviour it is instructive to study the internal circuit diagrams supplied with the data sheet. If carried too far this can, however, be misleading since not only is the circuit sometimes incomplete (spurious elements being present in the actual device) but also completely different circuits may be used by other manufacturers or in a subsequent model by the same manufacturer.

Therefore, terminals designed for the connection of compensation capacitors or offset nulling potentiometers should not be used for other than the intended purpose.

Summary

Operational amplifiers are inexpensive, versatile, easy to use and they enable larger and more complex systems to be built than would be practical with all-discrete designs.

Some care in their use still has to be exercised particularly when large capacitors are connected to input terminals. Failure to add protection can result in mysterious destruction of devices when in service.

Chapter 13

The transistor pump

The title refers to a circuit described by J. Willis and P. L. Burton* (p. 293) as an improved version of the well-known diode pump, which is often used as a 'staircase' generator and as a counting-type frequency discriminator.

In order to appreciate fully the transistor circuit it is essential to understand the operation of the apparently simple diode pump.

THE DIODE PUMP AS A STAIRCASE WAVEFORM GENERATOR

When used as a staircase generator the diode pump circuit is as shown in Fig. 13.1.

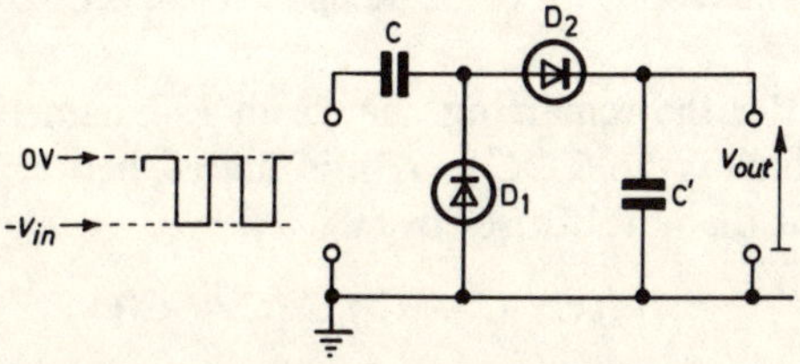

FIG. 13.1 Diode pump staircase generator

Since the output waveform, neglecting initial transients, will be independent of the d.c. component of the input waveform, it will be assumed for convenience that the input square wave starts at zero volts and falls periodically by V_{in} volts (Fig. 13.2), and that initially V_{out} is zero.

When the input falls to $-V_{in}$, D_1 conducts and C charges by V_{in} volts, provided the forward voltage drop of D_1 is negligible compared with V_{in}. Referring again to Fig. 13.1 the left-hand connection

* *Wireless World*, March 1958, 'Some Unusual Transistor Circuits'.

of C is now at $-V_{in}$ and its right-hand connection approximately at zero potential; C has therefore acquired a charge of CV coulombs.

When the input returns to zero the right-hand connection of C rises and D_2 conducts, connecting C to C′. Again ignoring forward diode voltage drops this connection of C to C′ occurs immediately the input rises, so that the rise of V_{in} in volts is shared between C and C′ giving a rise of $V_{in}C/(C + C')$ at the output.

On the second cycle of operation, conditions are different. The input falls to $-V_{in}$ and C recharges as before, but when the input rises towards zero, D_2 does not conduct until the input has moved by $V_{in}C/(C + C')$, bringing the left-hand connection of D_2 equal to

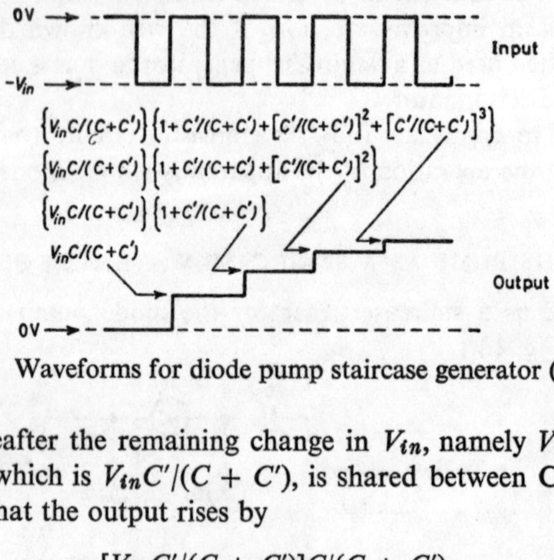

FIG. 13.2 Waveforms for diode pump staircase generator (Fig. 13.1)

V_{out}. Thereafter the remaining change in V_{in}, namely $V_{in} - V_{in}C/(C + C')$, which is $V_{in}C'/(C + C')$, is shared between C and C′ as before so that the output rises by

$$[V_{in}C'/(C + C')]C/(C + C')$$

to a new voltage of

$$V_{in}C/(C + C') + [V_{in}C'/(C + C')]C/(C + C'),$$

that is

$$[V_{in}C/(C + C')][1 + C'/(C + C')]$$

Succeeding cycles continue to increase V_{out} in increments to give the output waveform shown in Fig. 13.2. It is shown in Appendix 5 that after n input cycles the output voltage is

$$V_{in}[1 - (C'/(C + C'))^n]$$

which for small values of n or small ratios of C/C' approximates to

$V_{in}nC/(C + C')$. Under these conditions, therefore, the output rises by $V_{in}C/(C + C')$ for every input cycle.

When the above conditions for n and C/C' are not obeyed, examination of the expression for V_{out} shows that the steps of output voltage become successively smaller and that the maximum output after an infinite number of input cycles is V_{in}.

Applications

Within the limitations just mentioned this circuit is useful in the generation of a staircase waveform; these limitations mean, however, that if all steps of the staircase are to be substantially equal, either the number of such steps will be small, or capacitor ratio C/C' will be small, implying that the final value of V_{out} will be much less than V_{in}.

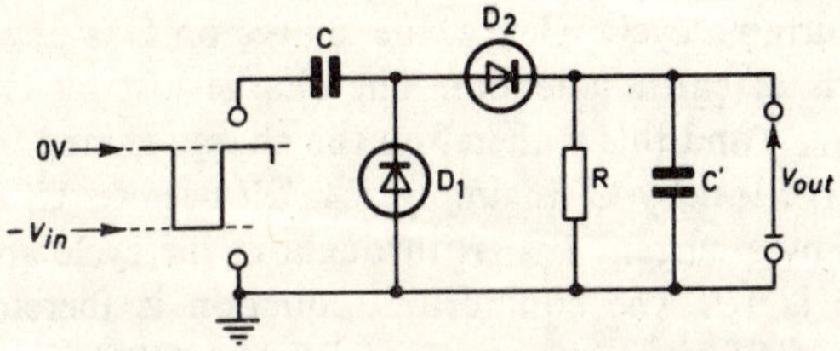

FIG. 13.3 Diode pump frequency discriminator

This circuit can be converted into a frequency divider by adding a trigger circuit at the output, which operates an electronic switch arranged to discharge C' when V_{out} reaches a certain voltage level. The accuracy of the triggering level and of the capacitor ratio, and the magnitude of V_{in} determine the maximum frequency division ratio at which operation will be reliable. Because of the successive reduction in each step discussed above, the limitations are severe and a frequency division of 10 to 1 is near the reliable limit for normal input voltages.

A more commonly used form of the diode pump is shown in Fig. 13.3, where a resistive load R is added in parallel with C′. In this circuit C' is made very large in comparison with C, so large that virtually no voltage steps are present across C' compared with the magnitude of V_{in}.

In this circuit the capacitor C charges to V_{in} each time the input goes negative and discharges into C′ as before, so that C′ must acquire charge each positive input half-cycle. Since the values of C'

and R are so large that no significant waveform is visible across C′, the implication is that C′ must lose this acquired charge at some time during the cycle. The only path for discharge of C′ is through R, and for such a discharge to occur C′ must carry a steady potential, since it is known to carry no alternating waveform.

For an input of frequency f and magnitude V_{in} there is therefore a d.c. output V_{out} the magnitude of which is calculable by equating the charge gained and lost by C′ during a complete cycle.

The charge gained by C′ can be calculated as follows. When the input goes negative by V_{in}, C acquires a charge CV_{in} since, as in the staircase circuit, D_1 conducts. When the input now rises by V_{in}, D_2 conducts and charge flows from C into C′. At the end of this half-cycle, when the input is about to fall, the right-hand connection of C is at a potential V_{out}, since it has been assumed that V_{out} is unchanged during a cycle. Hence, the charge on C is now CV_{out} since the input is at earth potential. The charge lost by C is therefore $C(V_{in} - V_{out})$ and this is therefore the charge gained by C′.

The charge lost by C′ is simply V_{out}/Rf because C′ continuously discharges by a current V_{out}/R throughout the cycle and the period of a cycle is $1/f$. The equilibrium equation is therefore $C(V_{in} - V_{out}) = V_{out}/Rf$, giving $V_{out} = V_{in}CRf/(1 + CRf)$. This implies that for values of CRf much less than unity, $V_{out} = CRfV_{in}$, i.e., the circuit of Fig. 13.3 produces a d.c. output proportional to input frequency.

The above condition, that $CRf \ll 1$, is equivalent to $V_{out} \ll V_{in}$. If the condition is not satisfied, then although the output voltage still rises with input frequency, the relationship is non-linear, the ratio V_{out}/f becomes less and eventually V_{out} approaches the value V_{in}.

This circuit is often used as a direct reading-frequency meter by replacing R by a milliammeter. Provided V_{in} is more than a few volts in magnitude, the non-linearity is no more than a few per cent with a typical meter voltage drop of 100 mV.

SUMMARY

The diode pump can be used as a staircase generator, as a frequency divider and as a frequency discriminator.

For most applications where linear characteristics are desirable it is necessary to accept an output voltage which is always much less than the input level, whichever form of the circuit is being used.

This forms a severe limitation at high frequencies, where large inputs may be difficult to provide and at any frequency if exceptional linearity is desired.

TRANSISTOR PUMP

In looking for a method of improving the diode pump, it is helpful to recall that the non-linear characteristic and low maximum output level are both caused by the influence of the output level on the charge and discharge of C.

The clearest approach to improve this situation consists in isolation of the charge and discharge circuits for C from the output voltage.

First Version

One method is shown in Fig. 13.4, where D_2 becomes the emitter-base diode of a transistor and C′ is placed in the collector circuit.

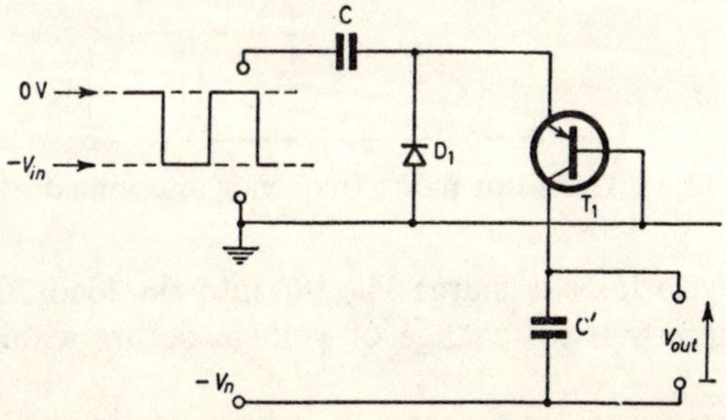

FIG. 13.4 Transistor pump staircase generator (1)

When the input goes negative, C charges to V_{in} and when the input returns to zero, C discharges completely into the emitter of T_1. This current flows out of the collector and charges C′ to a voltage of $V_{out} = CV_{in}/C'$ ($C'V_{out}$ must equal CV_{in} since all the charge from C transfers to C′).

This action is repeated on every cycle, since the change taking place at T_1 collector has virtually no influence on the charge/discharge process.

Hence, a staircase output waveform results in which each step is equal until the transistor bottoms, after which no further change occurs. The maximum output is therefore limited not by the magnitude of V_{in}, but by the supply voltage, which in turn is limited by the maximum permissible V_{cb} for the transistor and the available supply voltage.

The staircase is therefore equally stepped, of magnitude limited only by transistor ratings, and can therefore be used as a frequency divider for reliable division even at high ratios, 50 to 1 being quite practicable.

Consider now the addition of a resistor R in parallel with C′, $C'R$ being very large compared with an input period (*see* Fig. 13.5). For an input frequency f, C charges to V_{in} on every negative input swing and discharges its charge of CV_{in} into T_1 emitter on every positive swing. Consequently, C′ receives a charge CV_{in} on every cycle and

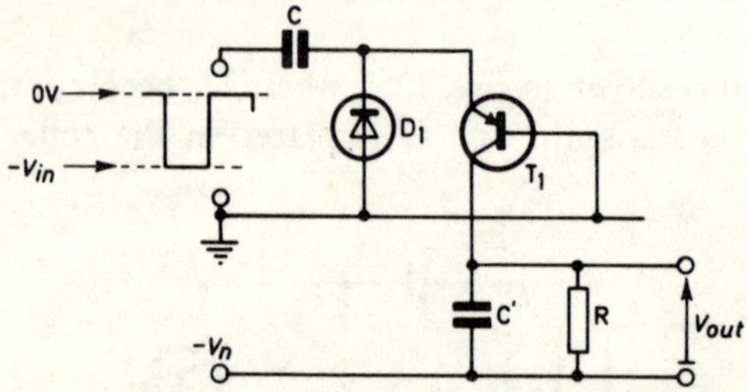

FIG. 13.5 Transistor pump frequency discriminator (1)

during the cycle loses a charge V_{out}/Rf into the load R, assuming $C'R$ is so large that no change of voltage occurs within any one cycle.

Hence, $CV_{in} = V_{out}/Rf$, or $V_{out} = CRfV_{in}$, so that V_{out} is directly proportional to input frequency until, as for the staircase, V_{out} is so large that T_1 bottoms.

The transistor circuit therefore forms a linear discriminator with an output voltage which can be considerably larger than the input swing and is limited only by transistor rating and available supply voltage.

The circuits of Fig. 13.4 and 13.5 therefore correct the main deficiencies of the simple diode pump by the principle of preventing the output voltage from influencing the charging of C.

Second Version

The circuit of Fig. 13.6 shows the form of transistor pump usually quoted. In this circuit the first negative swing charges C, through the base–emitter diode of T_1 to V_{in} and the succeeding positive swing to zero causes C to pass charge to C′, so that $V_{out} = V_{in}C/(C + C')$.

The second negative swing charges C to $(V_{out} + V_{in})$ volts, since the right-hand connection of C is caught at V_{out} by T_1. When the

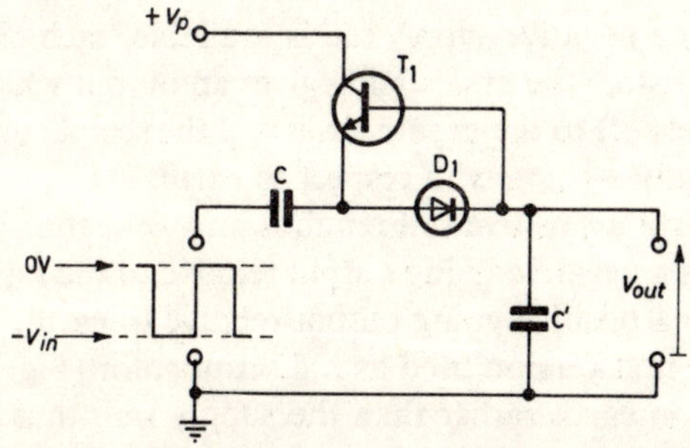

FIG. 13.6 Transistor pump staircase generator (2)

input voltage starts to return to earth for the second time D_1 conducts immediately (unlike the diode pump where the input had to rise by $V_{in}C/(C + C')$ before D_1 would conduct). The full rise of

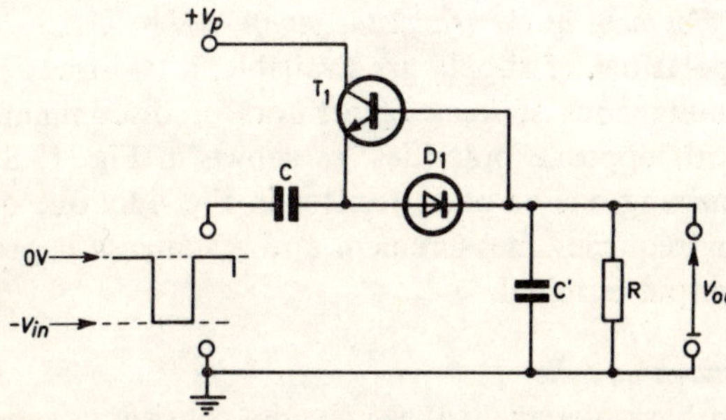

FIG. 13.7 Transistor pump frequency discriminator (2)

V_{in} is therefore shared by C and C′ as on the first positive swing and the output therefore rises, as before, by $V_{in}C/(C + C')$.

This process is repeated so that each step is $V_{in}C/(C + C')$ and the staircase has a linear characteristic as in the first version of the transistor pump.

The addition of a load resistor R (Fig. 13.7) with large C' still results in a performance equivalent to that of the first transistor pump, since, as before, equal increments of charge are transferred on every cycle.

Comparison Between the Two Circuits (Fig. 13.4, 13.6)

The main practical differences between the two circuits are as follows.

(1) If only a negative supply rail is available, each circuit requires a *p–n–p* transistor; the first version gives an output which is positive-going with respect to the negative rail and the second gives an output which is negative-going with respect to earth.

(2) Similarly a positive rail requires an *n–p–n* transistor; the first version gives a negative-going output relative to the rail, whereas the second gives a positive-going output relative to earth.

(3) In the first version used as a discriminator (Fig. 13.5) C′ may be returned to earth rather than the supply rail, thus confining the input signal current paths away from the supply rails. This is advantageous when used at frequencies above 1 Mc/sec., and cannot be achieved in such a simple manner in the second circuit (Fig. 13.7).

(4) In Fig. 13.4, step magnitude is $V_{in}C/C'$; in Fig. 13.6, $V_{in}C/(C + C')$.

Further development of the transistor pump

If both polarities of supply are available, a two-transistor pump enables simultaneous staircase generators or discriminators to be obtained with opposite polarities, as shown in Fig. 13.8 and Fig. 13.9. By omitting one or other resistor in Fig. 13.9 one output can be used for frequency measurement and a staircase is produced at the other output terminal.

Design of pump circuits

Although basic design of these circuits consists in satisfying the simple formulae for V_{out}, there are a few further points which should be numbered.

First, it has been assumed throughout that C charges and discharges to a steady state between each input swing and that the input is a square wave. The essential point is that C shall always complete its charge and discharge before the input reverses. In a practical circuit there will exist some source resistance R_s, and D_1 and D_2 (or T_1 in the transistor pump) will have forward resistance R_{D1} and R_{D2}. Provided the total series R is such that $(R_s + R_{D1})C$ is less than one-fifth of the duration of the negative input step and $(R_s + R_{D2})C$ is less than one-fifth of the duration of the positive step, the errors will be less than one per cent. A further factor of 2 in these time constants will make errors from this cause negligible.

It is not necessary that the input should be a symmetrical square

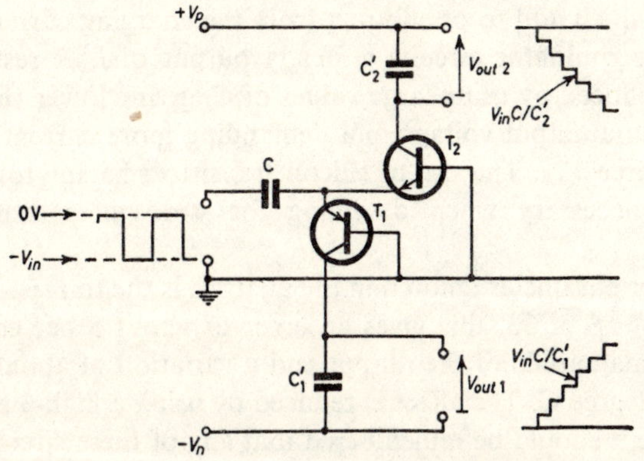

FIG. 13.8 Double-output staircase generator

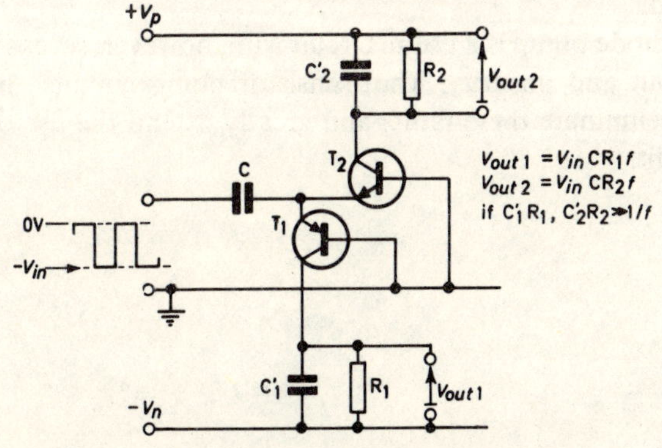

FIG. 13.9 Double-output frequency discriminator

wave; so long as the input remains in each state long enough for C to charge to the correct level, this is sufficient.

The turn-on voltages required by the diodes and transistor have been ignored in the analysis. In practice these voltage drops cannot be ignored and their effect is equivalent to a drop in the magnitude of V_{in}. This equivalent drop varies with temperature at a rate of

-2 to $-2{\cdot}5$ mV/degC and can thus cause considerable error in an equipment operating over a wide temperature range. The effect can be minimized by using a large input voltage, by arranging a controlled input variation with temperature or, for extreme stability, by oven-controlling the temperature of the diodes.

Temperature errors are also caused by diode and transistor leakage currents which add to or subtract from the charging currents to C′. In the discriminator circuits a steady output of $I_{co}R$ results. This error is reduced by using large values of C, giving lower values of R for the same output voltage, but demanding more current from the signal source V_{in}. The use of silicon transistors having low leakage is often necessary when designing for wide temperature range variation.

Another parameter ignored in the analysis is the transistor current gain. For a β of 33, this gives an error of about 3 per cent in the absolute magnitude of the output and a variation of about 0·03 per cent per degree C. The effect is reduced by using a high-β transistor.

Finally, it should be remembered that any of these circuits can be changed by reversing each diode and type of transistor from *p–n–p* to *n–p–n* or vice versa, provided the power supply polarity is also changed.

The diode pump is a useful circuit with, however, severe limitations in output and linearity. The transistor pump circuits almost completely eliminate these faults and greatly extend the use of this type of circuit.

Chapter 14

The transistor cascode

The circuit to be described in this chapter is a transistor version of the well-known valve cascode arrangement. Although not all cascode characteristics are common to both the valve and transistor circuits, it will be useful to recall the problems of low-noise high-frequency amplification which led to the development of the valve cascode.

PROBLEMS IN HIGH-FREQUENCY VALVE AMPLIFICATION

At high frequencies it becomes difficult to prevent oscillation in a

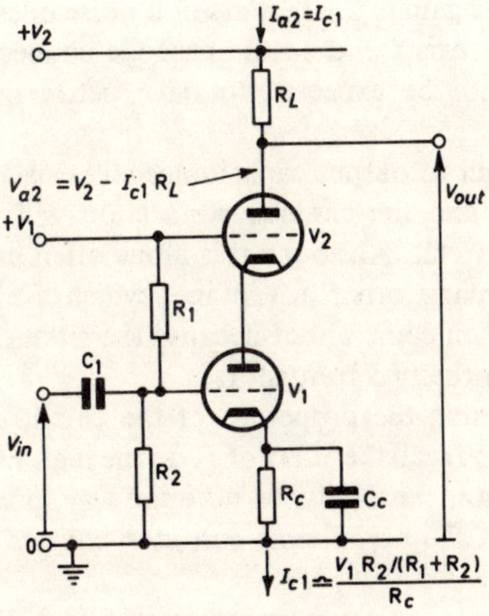

FIG. 14.1 Valve cascode

triode-tuned amplifier because of relatively large grid-anode capacitance giving unwanted coupling. A pentode stage overcomes the difficulty since direct anode-grid coupling is very small; moreover, higher stage gains are obtained. However, a pentode, because of its multielectrode structure, produces a high degree of partition noise.

There appears in such stages therefore to be a choice between a triode with poor gain and stability or a pentode with bad noise performance.

The cascode, which uses two triodes in a special configuration (Fig. 14.1), provides, at the cost of an extra valve, the stability and gain of a pentode and the low noise of a triode.

Examination of the circuit shows that the output electrode is well-screened from the input, and since the anode load of V_1 is low compared with the r_a of V_1 ($1/g_m \ll r_a$), $i_{a1} = g_{m1}v_{g1}$ provided $X_{Cc} \ll 1/g_m$. Also, $i_{a2} = i_{a1}$, so that $V_{out} = g_{m1}v_{g1}R_L$, which is the same result as if V_1/V_2 were replaced by a single pentode valve with a g_m equal to g_{m1}.

TRANSISTOR CASCODE

At first sight the transistor version of Fig. 14.1 appears to offer little advantage over a single-transistor amplifier stage, since in many respects a transistor has pentode characteristics, and, as for the pentode, its stage gain is g_mR_L. Partition noise does not occur, as no fourth electrode exists and so the cascode connection of two such devices could not be expected to have better noise performance than one.

However, input to output capacitance still exists in a simple transistor amplifier and the cascode connection will provide isolation as in the valve circuit. Although this alone often justifies its use, the cascode offers many other advantages which are not immediately obvious and which come about because the circuit requirements are divided between the two transistors.

Before examining the properties of the circuit of Fig. 14.2, it is desirable, first, to recall the effect of f_α on the high-frequency response of an amplifier and, secondly, to note the way in which base circuit conditions affect the permissible collector–emitter voltage rating of a transistor.

First, the base-to-collector current gain β is 3 dB down at a frequency of f_α/β and the emitter-to-collector current gain α is 3 dB down at a frequency f_α. Hence, an earthed emitter amplifier is 3 dB down at $1/\beta$ of the frequency at which an earthed base amplifier is 3dB down provided the source resistance is high (*see* Chapter 4).

Secondly, the collector–emitter voltage rating of a transistor usually depends on the base conditions; the permissible voltage is higher if the base resistance is low and is often halved if the base resistance becomes as high as 1 kΩ.

Returning now to Fig. 14.2, it is clear that if the transistors are similar, the $f_{\alpha 2}$ of the output transistor T_2 has little influence on the high-frequency performance of the circuit and can have a value as

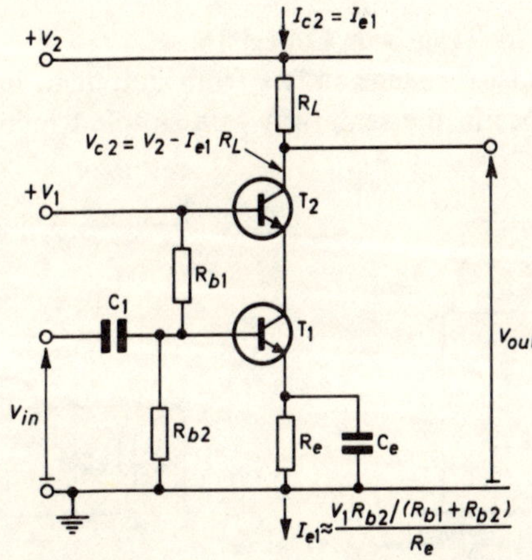

FIG. 14.2 Transistor cascode

low as $f_{\alpha 1}/\beta$ before beginning to contribute to the frequency fall-off.

Hence, for good high-frequency performance T_1 must be chosen as if the circuit were a normal earthed emitter amplifier stage, and T_2 can have an f_α which is roughly β times worse.

With regard to voltage ratings, T_1 must have a V_{ce} rating of at least V_1 under its particular base resistance conditions, and T_2 must have a rating of $(V_2 - V_1)$ under zero base resistance conditions (i.e. the favourable case).

Since V_1 can be small in view of the small signal amplitudes on T_1, and $(V_2 - V_1)$ may have to be large where large output voltage swing is required, V_{ce} rating can be low for T_1 but must be high for T_2. The same applies to the power ratings since currents are equal.

These results are useful because the division of duties between T_1 and T_2 proves to be ideal: T_1 must have good f_α and T_2 must have high voltage and high power rating. For a given application this implies that two inexpensive transistors can replace one expensive type which in a simple amplifier would require not only high voltage and power ratings but also good high frequency performance.

Although the above suggests that the cascode is primarily of interest as a high frequency amplifier it must be remembered that the term 'high frequency' applies whenever the frequency of operation approaches the f_α/β for the transistors used.

Effects Caused by Transistor Capacitance

A simple transistor stage suffers from output to input (collector–base) capacitance in the same way as a simple triode amplifier. For

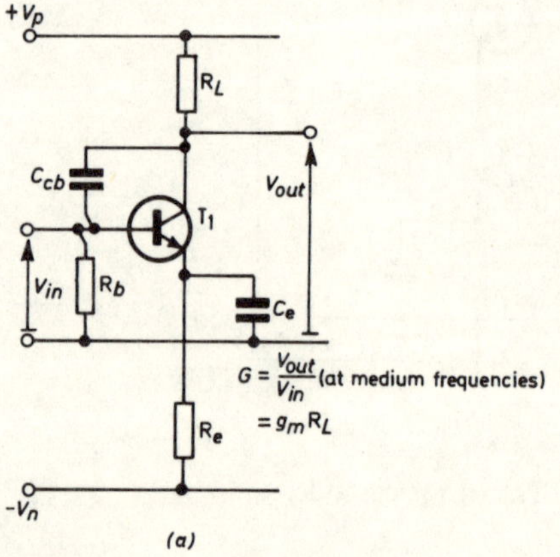

FIG. 14.3 Miller effect: (*a*) connection of C_{cb}, (*b*) equivalent effect of C_{cb}

a stage gain $-G$ the effect of collector–base capacitance C_{cb} is equivalent to the addition of a capacitor of value $(1 + G)C_{cb}$ in shunt with the base-emitter path (*see* Fig. 14.3). This is known as Miller effect, and the above result is easily calculated.

This is often the major cause of falling high-frequency response in a transistor amplifier and conversion to a cascode connection in such a case will bring about a dramatic improvement.

Another high-frequency problem is that of designing an isolating stage to prevent, for instance, an oscillator being detuned by changes in the load circuit. At low frequencies a natural choice for the isolating stage would be an emitter follower, since any change of load current in the emitter would cause a much smaller change in the base current. At high frequencies, however, transistor base–emitter capacitance forms a low impedance connection from source to load and isolation is poor.

Again, the cascode is useful as no direct back-coupling exists and buffering is effective to several hundred megahertz.

Variations of the Cascode

The addition to the basic cascode of a resistor R′, as shown in Fig. 14.4, enables the operating current of the two transistors to be

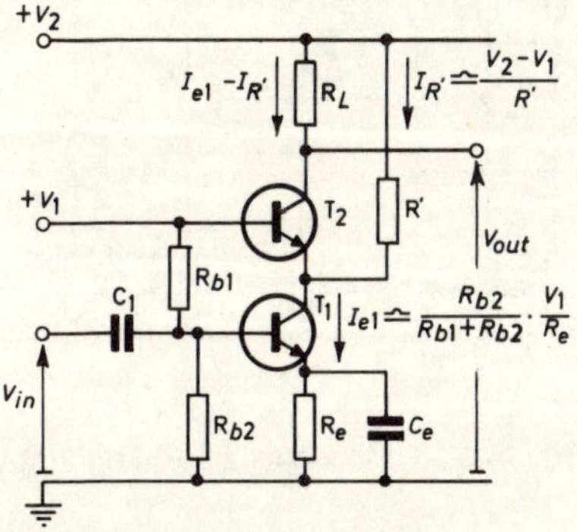

FIG. 14.4 Cascode with different emitter currents for T_1 and T_2

different. Thus T_1 can be run at high current, making its g_m high, and T_2 can pass a small current, equal to $I_{T1} - (V_2 - V_1)/R'$, so that R_L can be large. In this way high gain is obtained at the cost of less accurate determination of T_2 collector voltage. This occurs because T_2 collector current, which determines this voltage, is the difference between T_1 current and the current in R′, two independent variables. Changes in either of these currents therefore produces a larger percentage change in T_2 collector current. In practice T_2 current can rarely be made less than one-third of T_1 current without introducing too large an uncertainty in T_2 current.

Fig. 14.5 illustrates the use of a multiple cascode circuit as an adding amplifier. The standing currents in T_1, T_3, T_4, etc. all add in T_2 emitter so that in a k-input symmetrical design each g_m will be $1/k$ of the value which would be obtained in a single-input stage (since g_m is proportional to current). Hence, the gain from any input is $1/k$ of that of a single-input stage running at the same T_2 collector current. This circuit is particularly useful in having very good isolation between the several input terminals.

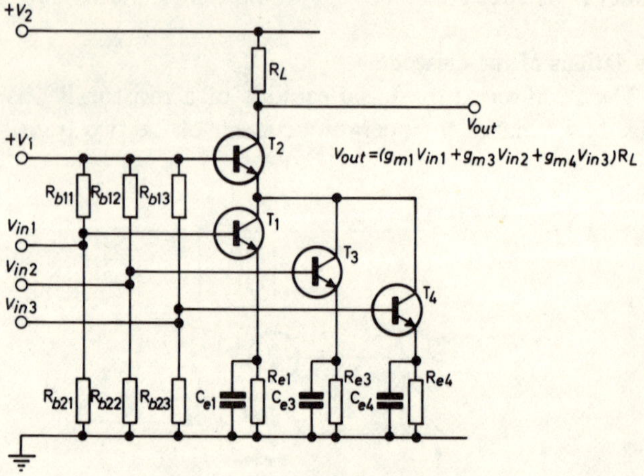

FIG. 14.5 Cascode as an adding amplifier

Fig. 14.6 shows how another form of multiple cascode enables large voltage output swings to be obtained from transistors of much lower rating. The network of resistors and capacitors maintains sharing of output voltage between the transistors; capacitors are necessary for transient operation and must be much larger than the input capacitances of the associated transistors. Since saturation of any transistor would upset this capacitor–resistor potential divider, precautions, such as prelimiting, must be taken to prevent overload. Switch-on and switch-off transients which often cause difficulty in the operation of voltage-sharing circuits are in this case harmless since the capacitors ensure slow rise and decay of the applied voltages.

A complementary version of the cascode is shown in Fig. 14.7.

The major disadvantage of this version is of course the need for resistor R_x to supply the total current of T_1 and T_2.

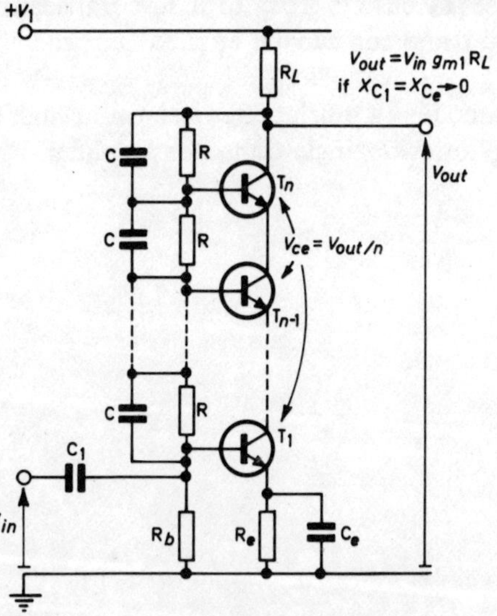

FIG. 14.6 High-voltage cascode

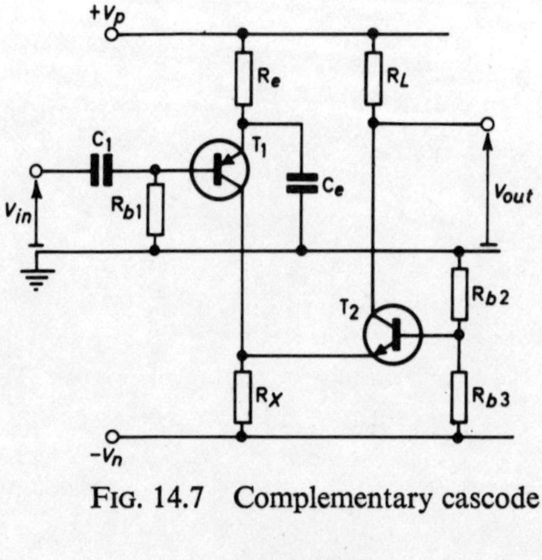

FIG. 14.7 Complementary cascode

SUMMARY

(1) In high-frequency high-output applications the transistor requirements are split; the input transistor must be high frequency but can be low voltage, and the output transistor must be high voltage but can be β times worse in frequency response.

(2) Miller effect, often the major cause of falling high frequency response, is virtually eliminated.

(3) Because T_2 base is fixed to a low impedance point the V_{cb} rating of the transistor can be applied (not merely the lower V_{ce} rating).

(4) The cascode is a much better isolator at high frequencies than an emitter follower or single-transistor amplifier.

Part Three

USEFUL TECHNIQUES

Chapter 15

Bootstrapping

The technique known as 'bootstrapping' involves a positive feedback loop which causes a point in the circuit to be 'pulled up as if by its own bootstraps'. This principle is often used in linear sweep generation (*see* Chapter 5) and also in high input impedance amplifiers, in driver circuits for power stages, and in reducing the effects of stray capacitance.

In this chapter are discussed the last three applications; design procedure and also practical difficulties are given.

HIGH INPUT IMPEDANCE BY BOOTSTRAPPING

When an amplifier is required to have a high input impedance of, for example, 1 MΩ, the obvious circuit configuration to consider is the emitter follower either in its simplest form or compounded. In defining the transistor operating conditions, however, there is a maximum value which the input base resistor may be given without the risk of excessive temperature drift due to base current variations. In practice this value can rarely exceed a few kilohms for germanium, or some tens of kilohms for silicon transistors, so that even if the circuit is inherently capable of providing high input impedance, the external base resistor spoils its performance.

In some applications the resistance of the source may be sufficiently low to act as the base resistor, so avoiding the need for the shunting effect of a separate resistor, but in many cases the source must be capacity coupled. A very common use for a high input impedance amplifier is in conjunction with a crystal transducer, such as an accelerometer, a record pick-up, or microphone; such a transducer is roughly equivalent to a source of e.m.f. in series with a capacitor of several hundred picofarads, shunted by hundreds of megohms. In this and many other applications there is no escape from the use of base resistors.

In such circumstances the bootstrap technique can be used to make the low-valued base resistor appear to have a much higher value from the point of view of the input signal.

To understand the principle involved here it is helpful to consider the true meaning of high input impedance. It means that when the input signal changes by a small amount the change of current taken by the circuit from the source is small. In the circuit of Fig. 15.1, assume for the moment that connection AB is not made. The input impedance at T_1 base is $(R_{b1} + R_{b2})$ in parallel with the impedance of the transistor circuit, and the current taken by $(R_{b1} + R_{b2})$ is clearly $\delta V_b/(R_{b1} + R_{b2})$ when T_1 base voltage V_b is changed by δV_b.

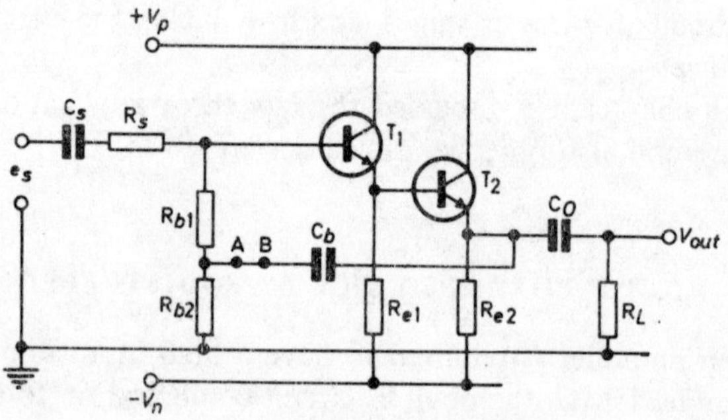

FIG. 15.1 Simple bootstrap feedback

Now consider the effect of connecting link AB, assuming that C_b is very large and that R_{b1}, R_{b2} cause no appreciable loading on T_2. It is easy to write a set of equations involving V_b, V_{out}, etc. and this is essential in discovering some of the side effects, but the best way to see the object of the feedback is to assume that V_b is moved by δV_b and deduce roughly what change in input current takes place.

If the gain V_{out}/V_b is exactly unity and if C_b is so large that $V_A = V_{out}$, then V_A will change in the same way as V_b, i.e. $\delta V_A = \delta V_b$. The change in current in R_{b1} is therefore zero, since there is no nett change across it; here R_{b1} behaves as if it were of infinite size in terms of its loading effect on the source.

This is the principle of bootstrap arrangements designed to increase the apparent impedance of a component: a method is found which makes that connection of the component which is remote from the source vary as nearly as possible in sympathy with the source. The apparent impedance is raised since the current taken from the source is reduced.

In an actual design of the circuit of Fig. 15.1, the gain V_A/V_b will in general be slightly less than unity and can be represented by $(1 - \lambda)$, where λ is a small fraction. The result now of an input change δV_b is to produce a change $\delta_{V_A} \Rightarrow (1 - \lambda)\delta_{V_b}$ so that the current taken by R_{b1} is $[\delta_{V_b} - (1 - \lambda)\delta_{V_b}]/R_{b1} = \lambda\delta_{V_b}/R_{b1}$. The apparent resistance of R_{b1} is therefore R_{b1}/λ, which, as was shown previously, tends to infinity as λ tends to zero.

The apparent magnitude of an impedance is therefore multiplied, by applying bootstrap feedback, by a factor which is the inverse of the fraction by which bootstrap gain falls short of unity.

If the emitter followers of Fig. 15.1 are replaced by a circuit having a gain of more than $+1$, e.g. $(1 + \lambda)$, R_{b1} now appears to be $-R_b/\lambda$. For gains slightly in excess of unity, the bootstrap therefore converts the bootstrapped component into a higher impedance of opposite sign. A capacitor would under these conditions appear to be inductive and vice versa, and a resistance becomes a negative resistance. Note that the presence of 'negative resistance' does not necessarily imply that the circuit will oscillate, since other parallel resistances (possibly the source) may produce overall positive resistance.

D.C.-COUPLED BOOTSTRAP

Where high input resistance in a circuit such as that of Fig. 15.1 is required down to zero frequency, C_b can be replaced by a direct coupling network. The point to remember is that R_{b1} appears to be large not only to the signal source but also to the transistor base current; if no source is connected, the temperature drift and quiescent behaviour of the circuit will be just the same as in a simple (i.e. non-bootstrap) circuit, but with R_{b1} replaced by its large bootstrapped apparent value.

For a particular transistor type, run under stated conditions, the drift/input resistance ratio at zero frequency is unaltered by bootstrapping; in this context 'zero frequency' implies the very low frequencies at which temperature changes occur.

Special Design Points for the Circuit of Fig. 15.1

The design for this circuit is affected by several considerations.

(1) As shown, only R_{b1} is bootstrapped, so that other components of the input impedance, not normally important, may cause appreciable lowering of the predicted value for Z_{in}.

(2) For d.c. bias purposes the base resistor is $(R_{b1} + R_{b2})$, but only the R_{b1} component becomes bootstrapped. Hence, an amplifier gain of $\left(1 - \frac{1}{30}\right)$, giving an input impedance due to R_{b1} of $30R_{b1}$, leads to a raising of Z_{in} of only 15 times if $R_{b1} = R_{b2}$.

(3) The amplifier has to supply its normal load and R_{b2} in parallel. Strictly, the bootstrapped value of R_{b1} should also be included, but this will be of no significance in most circuits.

(4) When fed from a capacitive source, as shown, the bootstrap loop has a gain V_{out}/e_s which always exceeds unity at a frequency

$$f = \frac{1}{2\pi\sqrt{(C_s C_b R_{b1} R_{b2})}}$$

(neglecting R_s). The gain at this frequency is given by

$$V_{out}/e_s = \sqrt{1 + \frac{C_b}{C_s}\frac{R_{b1}R_{b2}}{(R_{b1} + R_{b2})^2}}$$

which can be very large (e.g. 20 times!). The presence of resistance between V_b and earth, and of R_s, makes little difference to this result; nor does the fact that V_{out}/V_b is less than the unity value used in the calculation. These results are derived in Appendix 6.

Unfortunately, when this circuit is used for its most usual purpose, namely audio amplification from a crystal transducer, the peak usually occurs within the audio band, and its magnitude is often 2 or 3 times the mid-band gain of unity. A second unlucky point is that a designer who is being particularly careful will naturally give C_b an overlarge value to ensure that the bootstrapping will be effective at low frequencies, where, indeed, the high input resistance is especially needed when using a capacitance source. As the equations show, this leads to an even higher peak than normal, and at a lower frequency.

There are four basic approaches to make this circuit usable in spite of this problem. One is to make C_b less than normal design would suggest; this reduces the peak and tends not to upset low-frequency gain because of the presence of this remaining peak. Low-frequency phase response is, however, greatly affected. Another is to add a network which damps the effect: the frequency equation shows that the whole amplifier, following C_s, if replaced by an inductor of magnitude $R_{b1}R_{b2}C_b$, would produce the same equation. This suggests that a parallel resistance from T_1 base to earth would damp the peak—this works but degrades the input impedance. An alternative damping idea is to add a capacitor and parallel resistor in series with the bootstrap feedback (e.g. between points A and B), so that at very low frequencies feedback is reduced. The third method is to make C_b so large that the peaking frequency is below the low frequency cut-off required by the specification. This is the usual remedy but can be very dangerous. If C_s cannot be similarly increased (because it is outside the designer's control), the peak can be extremely large. Even if peak frequency signals are not passed to the output, various troubles now arise. A change of input source by switching, or the turn-on of amplifier supplies, can result in violent saturation or cut-off in the preamplifiers for a quarter-cycle at the peak frequency, i.e. several seconds. If R_{b1} and R_{b2} are returned to a supply line (rather than supply common), the peculiar gain characteristic can lead to very-low-frequency oscillation through a loop which includes the supply lines. Often this effect leads to a damped oscillation, or 'ring', and is accepted as being the time required by the circuit to settle after switch-on; it is, however, likely to recur even after a momentary disturbance. The fourth idea is to make R_{b1} as small as possible consistent with a high enough Z_{in}.

The point is that the peak gain depends on

$$(C_b/C_s)[R_{b1}R_{b2}/(R_{b1} + R_{b2})^2]$$

$(R_{b1} + R_{b2})$ is constant at a value determined by thermal drift. C_bR_{b2} is constant for a given low-frequency performance, so that peak gain depends on R_{b1}, which should therefore be small. The snag this time is that lowering R_{b1} requires more gain (λ smaller) for a specified Z_{in}, and the constancy of Z_{in} becomes worse, since, the nearer the gain is to unity, the greater is the effect of a slight change of gain.

Design Procedure

Except for the calculation of the response peak, which should be done as a matter of course with this capacity-fed circuit, design is simple. T_2 operating conditions are chosen according to the load and output voltage swing, giving R_{E2} ; T_1 current must exceed I_{co} for T_2 and be sufficiently large to ensure that β_1 is adequate. $(R_{b1} + R_{b2})$ can be determined as usual, although some advantage can be taken of the fact that T_1 and T_2 can be allowed to drift more than normal in this circuit before waveforms are affected.

The only 'bootstrap' design points are the correct splitting of R_{b1} and R_{b2}, and the choice of C_b. If R_{b1} is made much smaller than R_{b2}, then the bootstrapped value of R_{b1} may be insufficiently large; on the other hand, if R_{b2} is very small it adds considerably to the amplifier loading, so that its reflected value at T_1 base ($\approx R_{b2}/\beta_1\beta_2$) may seriously reduce the input impedance; the unwanted peak gain will also be higher than for low R_{b1}. A small value of R_{b2} may also cause T_2 to cut-off on positive signal swings if the current in R_{E2} is too small to supply the extra load of R_{b2}; then R_{E2} has to be reduced and this again reduces the input impedance. The practical solution is a compromise and the optimum ratio R_{b1}/R_{b2} usually lies in the region 1/4 to 3/4. The optimum as far as input impedance is concerned occurs when the bootstrapped value of R_{b1} and the reflected value of R_{b2} are equal as seen by the signal source, but the graph relating Z_{in} to R_{b1}/R_{b2}; has only a flat maximum.

Having decided R_{b1} and R_{b2}, C_b is now designed to give adequate bootstrap action at the lowest frequency of interest, but is made no larger than necessary, i.e. $\omega C_b R_{b2} \approx 1$. The peak frequency and gain V_{out}/V_b are now calculated, and, knowing that the gain at zero frequency is zero and the gain at medium frequencies is almost unity, the approximate response curve can be predicted. If unsatisfactory, C_b or R_{b1}/R_{b2} may be changed until a suitable value is found giving the smallest peak while maintaining adequate low-frequency performance.

Typical Design

To illustrate the procedure, assume the circuit is to be driven from a crystal pick-up having 500 pF capacitance and into a load of 10 kΩ, the required low-frequency cut-off being 1 kHz. Supply rails of ±10 V are available and maximum signal output voltage is 1 V peak.

The current in T_2 must be at least $\hat{V}_{out}/R_L$, i.e. 100 μA. In addition T_2 has to supply R_{b2}, which the designer can only estimate at this stage as being no less than perhaps 5 kΩ, i.e. another 200 μA.

Finally, T_2 must supply its own supply resistor R_{E2}, which for 300 μA would be 30 kΩ, requiring another 33 μA. To allow wide tolerance margins, R_{E2} may be 18 kΩ, giving a supply current to T_2 of roughly 10/18 = 0·55 mA; ignoring V_{be1}, V_{be2} and the drop in R_{b1} and R_{b2}, the current required is still 100 μA for R_L, 200 μA for R_{b2} and now 1/18 = 55 μA for R_{E2}, totalling 10·35 mA. Hence, 18 kΩ is satisfactory for R_{E2}. Now R_{E1} current must exceed I_{cbo} for T_2 (which could be 10 μA at 50°C) and should be sufficient to give reasonably high β in T_1. Here R_{E1} is 39 kΩ.

($R_{b1} + R_{b2}$) must be low enough to cause negligible effects due to drift. The base current of T_1 is highest at low temperatures in the inward direction and is roughly

$$[(V_1/R_{E2})/\beta_2 + V_1/R_{E1}]/\beta_1,$$

i.e. $(0{\cdot}55/\beta_2 + 0{\cdot}25)/\beta_1 \approx 10$ μA if $\beta_{1(min.)}$ is 25. The I_{cbo} component of base current flows inwards and can be 10 μA, so that the drift of V_b is from $+10(R_{b1} + R_{b2})$ μV at low temperature (0°C) to -10 $(R_{b1} + R_{b2})$ μV at high temperature (50°C). The tolerable drift in this circuit is determined mainly by the danger of T_2 cutting-off if the drift is negative; in the positive direction a drift of even 5 V would be harmless. ($R_{b1} + R_{b2}$) can therefore be 5/100 MΩ, i.e. 50 kΩ, giving a drift from $-0{\cdot}5$ to $+0{\cdot}5$ V. V_{be} drifts aggravate this by another -5 mV/degC, giving a total drift at T_2 emitter of $-0{\cdot}5$ to $+0{\cdot}75$ V from 0 to 50°C.

The ratio R_{b1}/R_{b2} can be taken as approximately unity at this stage, giving $R_{b1} = R_{b2} = 22$ kΩ. Note that this value for R_{b2} is 4 times higher than the value used in calculating R_{E2}, thus improving the safety margin considerably.

Z_{in} must now be calculated assuming no loss in C_b, the first step being to find V_{out}/V_b. From the transistor equivalent circuit given in Part 1, T_2 emitter behaves like a source of e.m.f. V_b with series resistance $[(1/g_{m2}) + (1/\beta_2 g_{m1})]$ when fed from a zero impedance source V_b (which is the required condition to calculate V_{out}/V_b). g_{m2} is approximately 50 Ω at 0·55 mA and g_{m1} is 100 Ω at 0·25 mA, giving a total of about 54 Ω if β_2 is 25. This effective source resistance from T_2 emitter is loaded by $R_{E2}//R_L//R_{b2}$, i.e. $18//10//22 \approx 5$ kΩ. Hence, $V_{out}/V_b = 5/5{\cdot}05$, giving $\lambda = 0{\cdot}01$, or $1/\lambda = 100$. The input impedance is therefore 100 $R_{b1} = 2{\cdot}2$ MΩ, in parallel with the r_c of $T_1(\approx 0{\cdot}5$ MΩ) and the reflected values of all loads, i.e. $(5\beta_1\beta_2)$ kΩ = 3 MΩ ($R_{E2}//R_L//R_{b2}$, $\beta_1 = \beta_2 = 25$) and $\beta_1 R_{E1} = 1$ MΩ. Total input impedance is 2·2//0·5//3//1 MΩ = 260 kΩ, and since C_s = 500 pF, the response is 3 dB down at $\omega 500 \times 10^{-12} \times 260 \times 10^3 = 1$, i.e. f = 1225 Hz.

C_b would normally be calculated so as not to affect this figure, but, as shown previously, this leads to excessive peaks in the response. Hence, C_b should be chosen so that the input resistance is beginning to fall at 1225 Hz; if the coupling from C_b to R_{b2} is allowed to fall 3 dB (which is reasonable because the peak in the response will keep the gain higher) then C_b is given by $\omega C_b R_{b2} = 1$, giving

$$C_b = \frac{1}{2\pi 1225 \times 22 \times 10^3} = 0{\cdot}006\ \mu\text{F}$$

The peak must now be calculated, and, if unsatisfactory, the value of C_b, and possibly R_{b1}/R_{b2}, may require to be changed.

Peak frequency

$$f_p = \frac{1}{2\pi\sqrt{(R_{b1}R_{b2}C_sC_b)}} = \frac{1}{2\pi 22 \times 10^3\sqrt{(500 \times 10^{-12} \times 0{\cdot}006 \times 10^{-6})}}$$

$$= 4{\cdot}18\ \text{kHz.}$$

Peak gain

$$G_p = \sqrt{\left[1 + \frac{C_b}{C_s}\frac{R_{b1}R_{b2}}{(R_{b1} + R_{b2})^2}\right]} = \sqrt{(1 + 12/4)} = 2$$

This being unsatisfactory, a ratio for R_{b1}/R_{b2} of 1/4 can be tried, giving R_{b1} = 10 kΩ, R_{b2} = 39 kΩ.

Input impedance is now 1//0·5//3·4//1 = 233 kΩ, a 10 per cent drop, which is tolerable. Low-frequency 3 dB point is given by $\omega C_s\, 233 \times 10^3 = 1$, i.e. f = 1350 Hz.

C_b is given by

$$\frac{1}{2\pi 1350 \times 39 \times 10^3} = 0{\cdot}003\ \mu\text{F}$$

$$f_p = \frac{1}{2\pi\sqrt{(10 \times 10^8 \times 39 \times 10^3 \times 500 \times 10^{-12} \times 0{\cdot}003 \times 10^{-6})}}$$

$$= 6{\cdot}58\ \text{kHz.}$$

$$G_p = \sqrt{\left(1 + 6\,\frac{10 \times 39}{49 \times 49}\right)} = \sqrt{2} = 1{\cdot}4$$

These figures show what the designer is up against—the lowering of peak gain is relatively slight, and the peak frequency is much higher than the lowest signal frequency.

An alternative, as indicated earlier, is to make C_b very large, so that the peak frequency is below the signal band in the hope that the high gain is unimportant.

If $C_b = 1\ \mu F$, and the original choice for R_{b1}/R_{b2} is taken, then

$$f_p = \frac{1}{2\pi 22 \times 10^3 \sqrt{(500 \times 10^{-12} \times 10^{-6})}}$$

$$= 323\ \text{Hz}.$$

and $$G_p = \sqrt{\left(1 + \frac{10^{-6}}{500 \times 10^{-12}} \frac{1}{4}\right)} = 22{\cdot}35$$

This may well be the best solution, but the designer must be certain that any following circuits have large attenuation at 323 Hz and that the bootstrap stage cannot receive a step input (or an input at 323 Hz) more than one twenty-secondth of its normal (mid-band) overload level; in the present design this would restrict step inputs to a few hundred millivolts.

Refinements to Fig. 15.1

As mentioned earlier, the circuit of Fig. 15.1 increases only that part of the input resistance due to R_{b1}. Evidently there are other contributors to input resistance which now become significant. Referring to Fig. 15.1, the base–collector resistance of T_1 appears directly in shunt with the input signal and therefore adds a parallel component r_{c1} to the input resistance. The resistance due to T_1 base circuit is also important and is given by $\beta_1/g_{m1} + \beta_1 R_E^*$, where R_E^* includes all T_1 emitter loading. This in turn is given by

$$R_{E1}//[\beta_2/g_{m2} + \beta_2(R_{E2}//R_{b2}//R_L)]$$

Unless further stages are added, nothing can be done to reduce the effect of R_L or R_{E2}, since in the limit the input has to supply $1/\beta_1\beta_2$ of the load current and R_{E2} and R_L are equally to be considered as loads.

The influence of R_{E1} is basically due to the change of current which takes place in it when the input varies; the input source then has to supply $1/\beta_1$ of this current. If the voltage across R_{E1} is maintained constant, this part of the input loading will therefore vanish.

Similarly, if T_1 collector is made to follow T_1 base, no current will flow in the collector–base path and its effect will also vanish. As in all bootstrapping systems, imperfection means that the improvement produced is finite; typical factors of improvement are 20–100.

Figure 15.2 shows how R_{E1} and T_1 base–collector can be bootstrapped to raise the input impedance.

As before, the bootstrapped elements are split to enable the feedback to be applied, so that the maximum benefit is somewhat lower than might be anticipated. In design R_L and R_{E11} are made large enough not to overload the output, but small enough to maintain normal bias conditions on T_1.

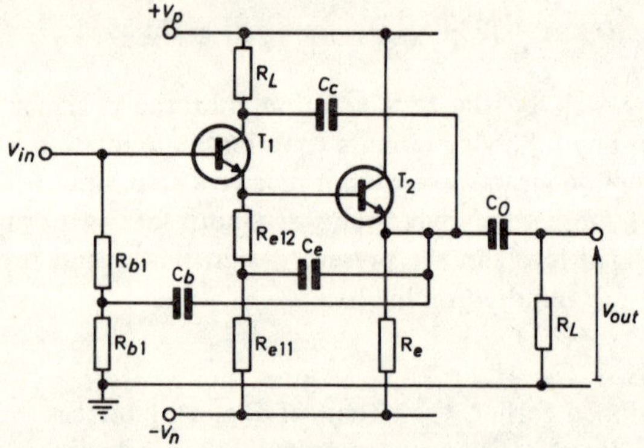

FIG. 15.2 Multiple bootstrap

Note that the apparent collector–base capacitance in T_1 is reduced for exactly the same reason that r_c is increased, namely that the current flowing into these elements due to input signal has been reduced by the bootstrap feedback.

An even simpler way to avoid the loading of R_{E1} is to omit this resistor; this can be done provided that the base current for T_2 is always outwards, even at high temperature, when I_{cbo} could exceed I_{E2}/β_2. If this is not so, then T_1 will cut-off. Even when this condition is met, T_1 emitter current may be so low that β_1 is poor. In practice R_{E1} can usually be omitted if T_2 is silicon (low I_{cbo}) and T_1 is planar (high β at low current).

Further improvements can be made to Z_{in} if R_L is increased, thus allowing R_{E2} to be higher, raising not only the reflected values of these resistors but also the bootstrap gain. Although R_L can normally not be changed, a further buffer stage (emitter follower) interposed between T_2 and R_L will produce the same effect. Using this arrangement, Z_{in} can be made as high as 1000 MΩ.

Bootstrapping in Power-drive Stages

The principle described in this section is worth bearing in mind whenever swings of voltage have to be obtained which are comparable with the supply rail voltage.

The designer's problem in such cases is shown in Fig. 15.3, where T_1 collector rests at about +11 V in the quiescent state and is driven between bottoming and cut-off by V_{in}. With only light loading on

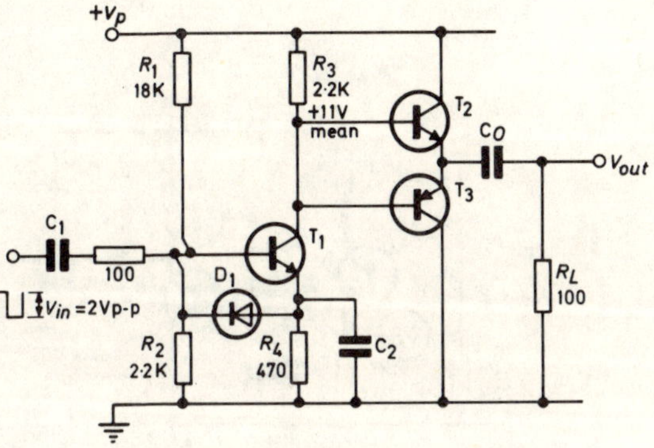

FIG. 15.3 Power output circuit

T_2T_3, V_{out} is therefore a square wave of maximum level about +20 V, and minimum level about +2 V.

When R_L is 100 Ω, the +2 V level remains unchanged, but when T_1 cuts off and T_1 collector rises, T_2 emitter takes current and T_2 base current increases, causing a voltage drop in R_3. V_{out} does not reach +20 V but has a maximum value governed by the equation

$$20 - \frac{R_3(V_{out(max.)})}{\beta R_L} = (V_{out(max.)})$$

i.e.

$$V_{out(max.)} = \frac{20\beta R_L}{R_3 + \beta R_L}$$

To put this into words, the load requires a peak current of about 90 mA, implying a T_2 base current of about 3 mA. This would cause a drop in R_3 of 6·6 V, so that the required positive swing will not be attained. This difficulty often occurs in the design of audio power output stages; in other circumstances the simplest solution is to connect R_3 to a more positive rail, but in audio design there is often no such rail available. It is here that bootstrapping becomes useful.

In Fig. 15.4, R_3 has been split and the junction bootstrapped by the output voltage through C_3. If R_{3A} and R_{3B} are each equal to 1 kΩ, then the voltage at their junction normally rests at about +15·5 V and T_1 collector at +11 V, the current in R_{3B} being about 4·5 mA. When T_1 cuts off, T_1 collector voltage rises, but because of the bootstrap connection and the near-unity gain of T_2, the voltage

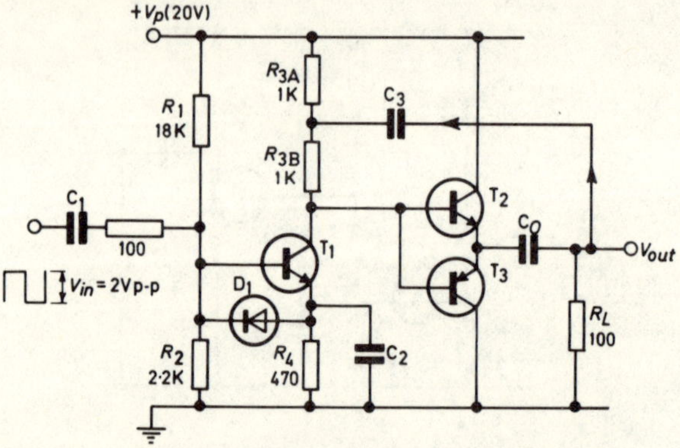

FIG. 15.4 Power output circuit with bootstrap

across R_{3B} remains at 4·5. This means that R_{3B} still carries 4·5 mA, and since T_1 is cut-off, this current must flow into T_1 base, Assuming β is 30, as before, T_2 emitter current can reach 120 mA. This cannot occur unless V_{out} rises to +23 V (from its initial +11 V), and so V_{out} will reach +20 and limit.

This circuit configuration is similar to the bootstrap sweep generator referred to previously. Because this circuit is to carry signals of each polarity from the quiescent state, R_{3A} cannot here be replaced by a diode since this would load V_{out} on negative swings. However, a diode can with advantage be put in series with R_{3A}, thus reducing the loading on V_{out} on positive swings.

Although square wave input was considered above, this was only for simplicity and the principle applies equally well for sine waves. The only design points are: (1) that the quiescent voltage across R_{3B}

must be such that the current in R_{3B}, when multiplied by $(\beta_2 - 1)$* is greater than the maximum load current; (2) C_3 must be such that the voltage drop across it while bootstrapping is negligible compared with the quiescent voltage drop across it.

* $(\beta_2 - 1)$ rather than β_2, because R_{3B} is itself part of the load.

Chapter 16

Prototype testing

Even the experienced designer usually finds it necessary to construct and test his designs before pronouncing them fit for production.

There are many reasons why this is necessary: first, he may wish to evaluate the performance of the circuit in certain respects which were not directly controlled in the design (e.g. a particular d.c. level which has no specific limits put on it but of which it is useful to have some knowledge); secondly, in an involved circuit it is easy to make mistakes of arithmetic or to forget the influence of a certain transistor parameter; thirdly, since the circuit as drawn cannot be achieved in practice because of the unwanted resistance, inductance, and capacitance of connections and components, it is often necessary to specify a particular layout after optimizing this by practical experiment.

The following notes are intended to help the designer who lacks practical experience to avoid the most common errors in construction and testing, and also to suggest test procedures for the various types of circuit considered in the previous sections.

GENERAL PRINCIPLES

Although failure to meet a specification (or complete failure of a circuit to operate at all) is sometimes caused by bad layout, especially at high frequencies, experience has shown that the usual cause of failure is simply that the constructed circuit does not agree with the circuit design.

It is strange that an engineer, after spending some hours in designing a circuit with great care, will often make an elementary wiring blunder, so that the circuit fails to operate. He then immediately assumes the design was faulty and wastes much time in re-checking the arithmetic (usually finding an error!); finally the wiring mistake is discovered, usually by another person who does not even understand the circuit.

This sequence is, unfortunately, quite common among all engineers and is analogous to the inability of many mathematicians to make arithmetical calculations. The simple operations of wiring and arithmetic are subconsciously regarded as being too trivial to justify much expenditure of time. This attitude, and eagerness to test the new creation, combine to produce mistakes.

In another class altogether are those wiring errors where the connections are correct but where the moving of a wire by an inch along the same conductor means the difference between success and failure. This problem is by no means limited to high-frequency circuits and is often the reason for excessive hum level or loop oscillation in stabilizer circuits and audio amplifiers. The principles described in the following sections will be found helpful in curing these troubles, but in high-frequency circuits considerable practical experience is often necessary to optimize the layout.

It would be useful if in the initial stages of testing several sets of components could be used to ensure that the design is not marginal owing to miscalculation or to component spreads outside expected tolerances. This is rarely feasible because of the time required for such tests, and the designer must usually await the first production run. However, two tests which to a large extent give the same information are the effects of temperature and supply variation.

A good general rule in testing any circuit is, then, to subject it to changes of about 10 per cent in each individual supply rail (or more if required by the specification) and to ambient temperature changes of at least 10 degC. If the circuit is designed for smaller changes in these conditions, the tests may put the unit outside its normal performance specification. This is to be expected; on the other hand, complete failure to operate under these slightly changed conditions usually indicates unsound design. A circuit so marginal in operation that a 10 degC rise or 10 per cent supply change causes failure is equally likely to fail when a different set of components is used.

Even if measurements indicate that a circuit is operating correctly, it is always advisable to check transistor operating voltages and waveform amplitudes throughout the circuit. This does not normally require accurate measurements, and a direct-coupled oscilloscope is extremely useful in checking how near to bottoming or cut-off a circuit is operating by examining d.c. + a.c. levels.

It is always important to examine signal waveforms even in a circuit which is intended to contain none, since it often happens that a d.c. stabilizer circuit will oscillate without disturbing the mean d.c. levels. Thus, normal d.c. tests would indicate correct operation and the next unit constructed could well oscillate much more violently, causing failure or excessive ripple.

When designing a circuit it is usually assumed that the d.c. power supply, often a transistor-stabilized rectified supply, will behave like a perfect battery. That is, it is assumed to have negligible internal resistance or inductance and to be purely d.c. In practice the power source may be a dry battery having appreciable internal resistance, and even if a highly stabilized transistor supply is used, the long connecting leads can have enough inductance or resistance to cause trouble; moreover, unwanted ripple may be present.

The precautions to be taken here are, first, to design the circuit so that supply resistance and ripple have little effect, usually achieved by decoupling-circuits; and secondly, to take care that prototype testing is done with various possible supply lead lengths and ripple content to confirm that these are not critical. If these factors are critical, then the circuit must be modified to accept any desired length of supply lead and ripple content.

Failure at Switch-on or Switch-off

Another, often infuriating, problem is transistor failure at the moment of switch-on or switch-off. The difficulty here is that no practical evidence can be obtained about the cause, except that surge voltages or currents must be responsible. Even more annoying is the circuit which sometimes survives switch-on and switch-off transients but once in a while fails; this is often missed in prototype testing but can be relied upon to show up in the entire production batch which follows.

There are two main causes of this type of transient failure. The switching on of supplies can initially cause excessive voltages or currents to flow before the intended operating conditions are reached, often dependent upon the order in which the various rails are connected. On the other hand, although the supplies themselves may cause no harm, the initial action of the circuit itself can be the cause of excessive transients.

Concerning the first of these possibilities, it is often recommended that for test purposes the supplies should be either switched simultaneously by a single on–off switch, turned on in a particular sequence known to be safe or turned on at low level and brought up smoothly. The on–off switch method is best avoided (as a cure for transistor failure), since inevitably some time interval exists between the closure of various poles of the switch and this time changes with usage. Sooner or later the interval can be so long that simultaneity is lost, and failure will recur if one is unlucky in the resulting sequence of connection.

The other methods are usually successful and can be justified when measurements have to be made to confirm quickly that the circuit is basically correct. Matters must not, however, be allowed to rest there, and since circuit modifications which will follow are likely to affect normal performance, it is unwise to take detailed accurate measurements until they are incorporated.

These switch-on–off problems are not confined to transistor failure; in many circuits capacitors receive surges of several times their running voltage at these instants; in others, switching surges cause no permanent damage but cause faulty circuit operation, especially in trigger circuits.

Curing the surge troubles is not always easy and sometimes requires a completely different approach to part of the circuit. It is naturally important to understand how the damage can be caused, and this can be done easily provided the designer understands the basic principles of *CR* charging circuits. (Occasionally inductance is involved but, in general, capacitors which have to change charge to reach operating level are the cause of destructive transients.)

To analyse switch-on behaviour it is assumed that each capacitor in the circuit is fully discharged, having zero volts between its plates. It is then assumed that the supply line is switched in zero time to its operating value and the transistor conditions are examined just after this value is reached. For switch-off analysis, each capacitance is assumed to be at its normal circuit operating level. The supply voltage is then removed in zero time and immediately afterwards transistor conditions are again examined. In some subtle cases switch-on can be damaging only if the capacitors are partly discharged so that there is a critical interval between switch-off and switch-on which leads to failure. Since each capacitor is likely to have its own discharge rate, the calculation required here to be confident of safety can be tedious.

One example is given here to show how even the simplest circuit can be difficult to assess. Figure 16.1 illustrates a simple pulse-operated light relay. The transistor OCP71 is a 'photo-transistor', which is basically a normal *p–n–p* alloy germanium with a clear envelope which enables light to reach the collector–base junction. Light here produces the same effect as heat in that it increases transistor I_{cbo}, and since with the base open circuit the collector current is βI_{cbo}, the collector voltage varies considerably according to light falling on the junction.

In Fig. 16.1 the light source is intended to be present whenever the unit is operating so that the collector voltage of the OCP71 is about zero (i.e. there is enough light to cause T_1 to saturate). T_2 emitter and base potentials are therefore also zero, and so is T_3 base. C therefore has zero charge and relay A/2 is not energized, implying that its contact A_1 is in the position shown.

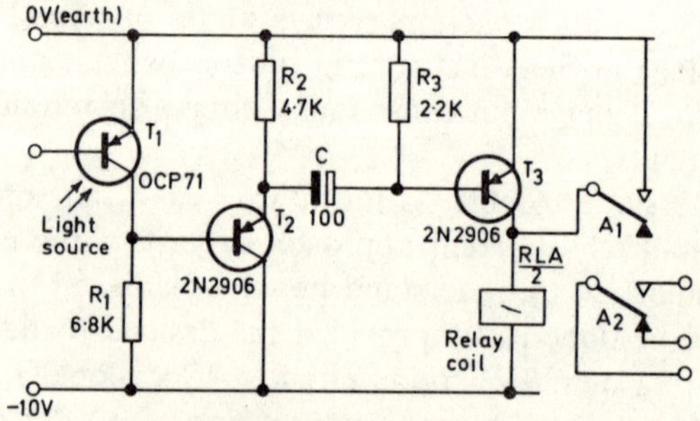

FIG. 16.1 Pulse-operated light relay

If the light beam is interrupted, the current in T_1 falls below saturation level, so that T_1 collector potential falls and T_2 base and emitter fall. Since C cannot immediately change its charge, T_3 base also falls, turning on T_3 and energizing relay A/2. This causes A_1 to change over, holding the relay in.

Return of the light beam now results in T_3 cutting off but relay A/2 remains energized because of the connection of A_1. Other contacts of A/2 can therefore be used to give a permanent alarm signal indicating that the beam was broken. This circuit can be used for intruder detection, critical level detection, and similar applications.

From this rather detailed account it is clear that when the supply is switched on it is essential that T_3 should not conduct even momentarily, since contact A_1 would then turn on the relay for all time.

In a particular application the lamp and circuit were energized from the same supply and sometimes the relay would lock over at switch-on; at the other times correct action would be obtained, depending upon the distance between lamp and photo-transistor.

This apparently mysterious and random effect was readily explained by circuit examination on the lines suggested above. Assume C is not charged; now apply −10 V to the circuit instantaneously, and also energize the lamp. To assess the voltage build-up on C, the conditions on T_1 and then T_2 must be found. There are two possibilities here: (1) the lamp output causes T_1 to saturate before the −10 V succeeds in making T_1 collector negative: (2) the lamp is slow to give enough output to saturate T_1 immediately, so that T_1 collector initially falls towards −10 V.

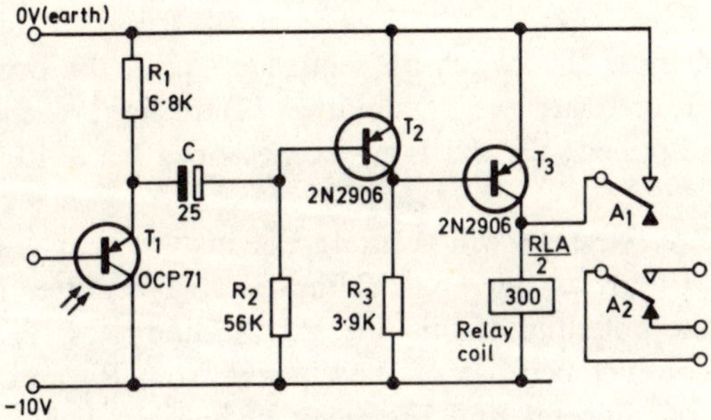

FIG. 16.2 Improved light relay

It is now clear that in the event of (2) the voltage on T_2 base and emitter falls and this is communicated through C to T_3, turning on the relay. In the case of (1), however, T_2 emitter remains at earth potential and no trouble arises.

The problem therefore resolves itself into whether the lamp or the −10 V supply arrives first at T_1. Clearly, a supply line which is fast to rise can give trouble, especially if the lamp is distant, and unless this is recognized in the design of the supply the relay is likely to operate.

Although this is not a case of destruction caused by switching, the difficulties which arise are typical of switch-on transients. The cure in this case can be achieved either by adding a time constant to the supply line to limit its rate of increase or by a re-design of the circuit which inverts the effect of the supply and the lamp (Fig. 16.2).

The operation of this circuit is as follows. In the quiescent condition with the lamp lit and the -10 V supply connected T_1 is saturated, so that T_1 emitter is at almost -10 V. The base current for T_2, which is approximately $10/R_2$, is designed to be sufficient to make T_2 bottom, thus cutting off T_3 (provided $V_{EC(sat.)}$ for T_2 is less than the voltage needed to turn on T_3). Hence, the relay is not energized and contact A_1 is in the position shown.

If the light beam is interrupted, T_1 emitter rises, causing T_2 base to rise above T_2 emitter. T_2 therefore cuts off and, provided R_3 is correctly designed, T_3 now bottoms, turning on the relay. Contact A_1 closes and holds in the relay. When the beam is restored, T_1 saturates, bringing T_2 back into conduction and T_3 to cut-off. However, A_1 keeps the relay held in.

Thus, in normal operation, this circuit behaves in the same way as the first.

Consider now the switch-on sequence. As in the original design of Fig. 16.1, there are two possibilities: (1) the light reaches T_1 before -10 V is applied; (2) the lamp comes on after -10 V has been applied.

In the first case T_1 will saturate the instant that any voltage is applied, so that, as the -10 V line rises, T_1 emitter immediately takes up the potential of this line. As the line rises, T_2 is therefore made to conduct because of the current from R_2 and from C. T_3 therefore remains off and the relay de-energized, so that correct conditions are established.

In the second case T_1 remains at first cut-off or slightly conducting. Application of -10 V then causes T_2 to saturate immediately because of the current in R_2. When the lamp lights, T_1 saturates, bringing T_2 even harder into saturation until C becomes charged (eventually supporting about 10 V potential difference). Again, correct starting conditions have been obtained.

The reason for the success of this circuit change is that initial transient currents have been made to assist, rather than oppose, the desired starting conditions. In particular, the charging current required by C is in such a direction as to turn on T_2 even harder whether lamp or supply line appears first.

This second circuit is fundamentally better than the modified form of the first which can be obtained by adding a time constant in the supply circuit (Fig. 16.3), since no real tolerance can be put on the effective rise time of the light source. This time depends on the

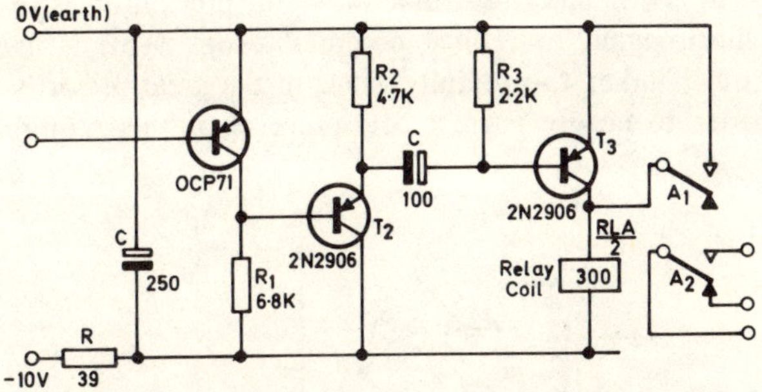

FIG. 16.3 Alternative method to improve light relay

applied lamp voltage, the particular lamp used, the distance between lamp and photo-transistor and the sensitivity of the photo-transistor. All the designer could do in these circumstances would be to use a value for R' and C' (Fig. 16.3) so large as to be adequate under any conditions. This does lead to a safe design, but the waiting time before correct quiescent conditions are reached is then excessive.

It is hoped that the above example illustrates the approach required when a circuit has switch-on difficulties and shows how a re-design can sometimes be the only satisfactory solution. There is therefore good reason to assess such transient behaviour as soon as the geometry of a new design is envisaged, and before working out component values.

In addition to the above general points regarding circuit testing, a number of difficulties arise which apply to particular types of circuit, and these are dealt with below.

Power Circuits and Stabilizers

Special care must be taken in the construction of circuits involving high currents, because even short lengths of interconnecting wire can cause potential differences of many millivolts. If a high-gain amplifier is used in the circuit, these potential differences may be amplified, thus giving large unwanted outputs.

The main precaution to be taken is correct routing of the wiring so that although the unwanted potential differences may occur, they do not become coupled to sensitive parts of the system. A second and more obvious precaution is to choose wire of adequate gauge to reduce the magnitudes of these potentials. This second method should not be carried too far, however, since in production it may be undesirable to use really heavy wire; in fact, printed circuit track may have to be used. The third, much quoted, precaution is to keep all leads short, again to reduce potential drops. This is even more dangerous if taken too literally, since in the great majority of cases it is better to ensure correct routing, even if this requires longer wires.

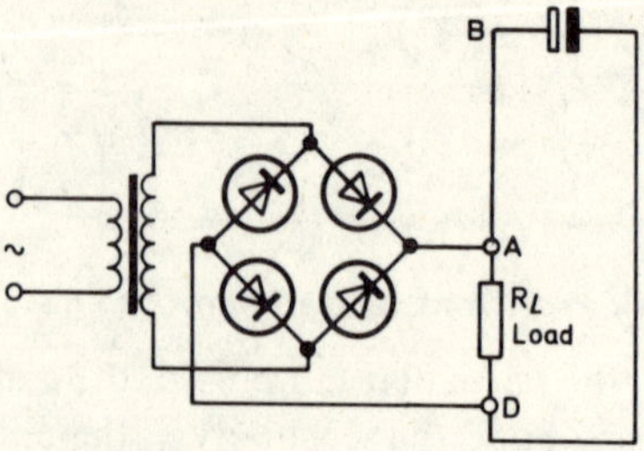

FIG. 16.4 Power supply with load remote from reservoir capacitor

In Fig. 16.4, which represents a simple unstabilized d.c. supply, a quite common method for capacitor connection is shown. The arrangement is not really satisfactory, however, because of the resistance of the capacitor connecting leads. The load voltage waveform is not then the normal sawtooth charging waveform but has added to this the peaky capacitor current waveform, as shown in Fig. 16.5.

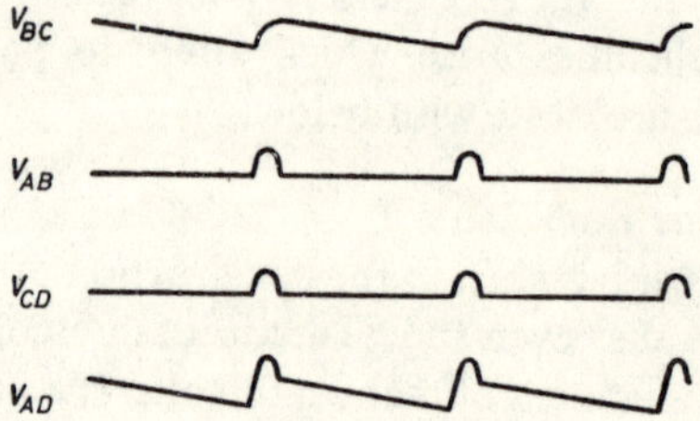

FIG. 16.5 Ripple waveforms for Fig. 16.4

Apart from the mental worry of the designer in observing this waveform instead of the expected sawtooth, the consequences of this effect are slight, because with practical values of ripple peak current and connection resistance, the extra voltage peaks are only of the same order as the sawtooth. For example, a 250 mA supply using a 5000 μF capacitor with 50 Hz bridge rectification will have a peak-to-peak ripple given approximately by $v = it/C$, where t is $\frac{1}{2}(1/50)$ sec, i.e. 10 msec. Ripple voltage is therefore $\frac{1}{4}(10 \times 10^{-3})/5 \times 10^{3} \times 10^{-6}$, that is 0·5 V. If the resistance of AB and CD totals 100 mΩ (representing 10 in. of typical printed circuit track), the additional ripple is 0·1 I_{peak} where I_{peak} is the peak charging current for C, typically 10 times the mean load current, or in this case 2·5 A. Thus, the spurious ripple is (0·1) (2·5), or 0·25 V, raising the total ripple to 0·75 V peak-to-peak instead of 0·5 V.

Real trouble, however, can occur when several circuits which are powered by such a supply are attached at various points on AB or CD. This is a very common problem with stabilizer circuits and can result in the stabilized output having a higher hum content and worse stability against load changes than the original supply!

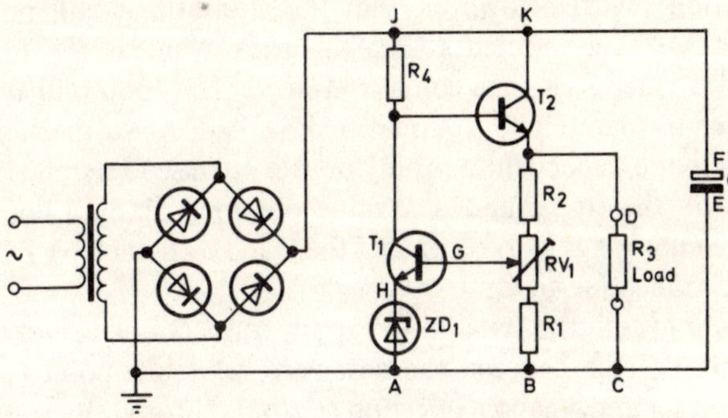

FIG. 16.6 Simple stabilized supply

In Fig. 16.6 the actual connecting points are to be taken as if the lines represent actual wires and the stabilizer connections are deliberately placed very badly.

In the normal operation of this circuit the load voltage V_{CD} is stabilized by the following action. A fraction of V_{CD}, namely V_{BG}, is applied between T_1 base and earth, and a stable Zener voltage V_{AH} is applied between T_1 emitter and earth. Should any change in load or supply line cause V_{AH} to differ from V_{BG} (ignoring V_{be} for T_1), then T_1 collector current will change, modifying V_{CD} in the right direction to restore V_{AH} to equal V_{BG}.

For correct action therefore it is essential that V_{BG} is a simple fraction of V_{CD}, where V_{CD} is the actual voltage to be maintained constant. In the drawing of Fig. 16.6 it is evident that any potential drop along BC will change this situation. Since the wire BC carries the ripple current of C, V_{BC} could well have a peak-to-peak ripple magnitude of several millivolts and a d.c. value also of several millivolts, since all the load current passes along BC. Thus—and this is the all-important point in the understanding of these problems—if the loop performs correctly, it will cause V_{BD} (*not* V_{CD}) to be held constant, so that V_{CD} will contain all the unwanted d.c. and a.c. variations of V_{BC}.

To remedy this fault, the temptation is to shorten BC or to use thicker wire, but although this will improve the performance, the real cure is to remove the connection between R_1 and B and connect this wire direct to C.

The situation now is that V_{CG} is a true fraction of V_{CD}, and so if the loop keeps V_{CG} constant the output V_{CD} is also constant. Further examination reveals, however, that the situation is still not ideal, because ZD_1 is connected to 'earth' at A whereas the load and potential divider R_1R_2 are connected at C. The loop will therefore endeavour to maintain V_{AH} equal to $V_{AC} + V_{CD}$, so that again the output is made dependent not only on the intended reference voltage but also on the ripple and d.c. value obtained along a connecting wire. The cure is again to reconnect the leads by detaching ZD_1 from A and reconnecting to C.

Examining now the remaining supply connections it is clear that any ripple voltage between the new zero reference point (or earth connection C) and J causes injection of ripple through R_4 to the base of T_2 and, hence, to the output. If no loop were present, the magnitude of this output ripple would be roughly equal to that existing at J, i.e. the full ripple on reservoir C_1. Although this effect will be reduced by the gain in the loop T_1T_2, it can still not be ignored, so that the ripple at J should at least not be allowed to be any larger than that of C_1 at F (V_{EF}). Resistor R_4 should therefore be removed from J to F, and CE should be minimized in length.

Ripple injected at K will have even less effect, but again its effect can be reduced by reconnecting to F.

The new circuit layout is shown in Fig. 16.7. It will be realized that, as suggested by the drawing, many of the wires will be greatly increased in length by these changes. This is quite permissible provided either that the currents in the long wires are negligible (RC, SC, TD, UF) or that the voltage drops have no significant effect (PC, QF).

This example has been given in considerable detail to show how simple is the reasoning behind correct layout in such cases. The sceptic finds it difficult to believe that these effects are even measurable, let alone of great significance, but a practical test on a stabilizer designed for, e.g. 30 V, 1 A with 1 mV ripple, soon shows up all these dangers, spurious drifts and ripple levels of many tens of millivolts being quite usual with wrong layout.

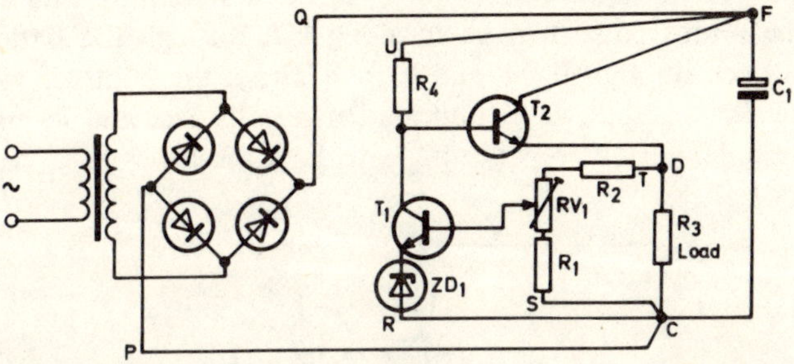

FIG. 16.7 Correct routing of connections for circuit of stabilized supply

Amplifiers

Similar problems arise in all types of amplifier but predominate where high currents are involved, as in power audio amplifiers and high-frequency amplifiers.

The same principles with regard to correct routing still apply, but in the case of high-frequency amplifiers at frequencies in the tens of megacycles there is an additional difficulty: routing has to be correct but wires must also be short. The reasons for shortness are twofold: first, stray capacitance between wires, or to ground, can cause feedback or signal loss to occur because a few picofarads is a very low impedance at these frequencies; secondly, the inductance of even a centimetre of wire has considerable reactance at high frequencies, causing loss or, worse, resonant effects. This is the basic reason why high-frequency practice still has to be regarded as an art. Only intuition and long experience will lead to the optimum arrangement of wiring, although the general principles of dealing with unwanted signal paths are still applicable.

Since amplifiers require similar treatment to the stabilizer previously discussed, only one example will be given (Fig. 16.8).

This is a simple amplifier with emitter follower output, and at first sight there is no difficulty in layout; indeed, in most applications of this circuit little thought need be given to cable routing. If, however, this circuit is used at high frequencies or for driving heavy loads, the situation is quite different. In either case the effect of load currents is likely to increase in significance because at high frequencies the inductance of wires causes larger voltage drops, and with heavy currents the resistance causes similar large voltage drops.

In this circuit load current flows through T_2 emitter and collector, the return path then being along HJ, through the battery to A and back through B, D, and E to F. From the point of view of T_1, the signal is that which appears between its base and its emitter, i.e. the potential difference from C to E, if the losses in C_1 and C_2

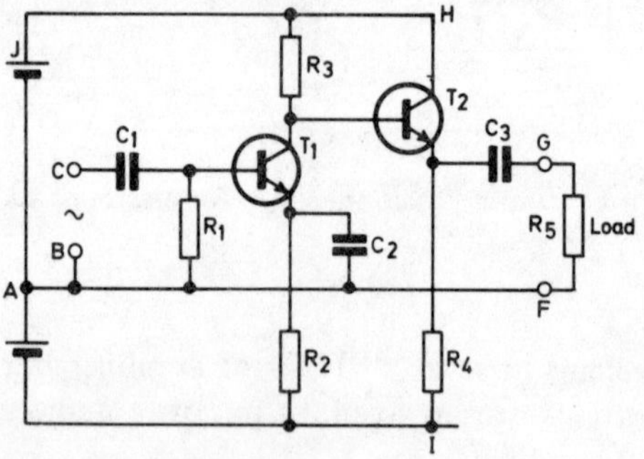

FIG. 16.8 Simple amplifier circuit

are ignored. The input is, however, intended to be V_{CB}, so that T_1 has a spurious input of V_{BE}. Since the load current flows along BE, this spurious input can be large and may add to or subtract from the true input according to the frequency concerned. (At medium frequencies the feedback would be positive, giving increased gain and probable oscillation.)

There are several remedies. Perhaps the simplest is to remove the link AB and connect A direct to F so that no load current flows along BE. Another is to reconnect capacitor C_2 to B instead of E (and preferably take R_1 to B also). It may happen that because of some existing printed board the above changes cannot be made. A possible solution then is to add R_5C_4, as shown in Fig. 16.9, in T_2 collector circuit with C_4 returned direct to F. This last modification has the further advantage of avoiding load currents in the positive line and is therefore useful in protecting any other circuits, which may have to share the line, from stray effects.

The best arrangement, to be used where possible, would be a combination of all three methods, as shown in Fig. 16.9.

In more complicated amplifiers employing several stages the difficulties considered above become multiplied in number and magnitude: in multistage high-gain circuits the voltage drop caused by output load current flowing in a few inches of printed circuit track can be as large as the intended input signal to the amplifier. Since it is not always possible to avoid such a voltage drop, it is essential to use methods similar to those just described in order to isolate the critical stages.

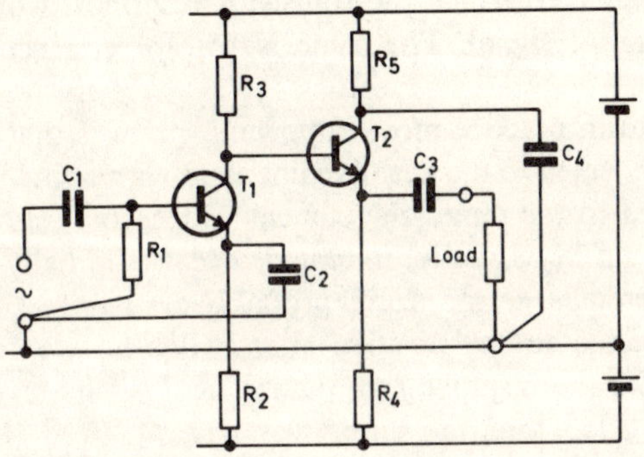

FIG. 16.9 Improved layout for simple amplifier circuit (Fig. 16.8)

It would clearly be impossible to illustrate remedies to suit all amplifier circuits, but the following hints should prove helpful.

The problem usually has to be approached from two directions. First, the output end of the circuit must be so arranged that the heavy currents flowing into the load and supply lines are localized and do not cause unwanted voltage drops in the more sensitive stages. Secondly, the sensitive stages should be sensitive only to the intended signal input and not to spurious voltage drops in connecting leads or track. This is easier said than done, but much can be achieved by simply ensuring that the emitter, base, and signal return paths of any one stage are each separately returned to one point in the wiring; if they have to be routed to different points along the same wire, then no other currents, especially large-signal currents, should be allowed to flow along the same path.

This latter point brings up a question which is really a circuit design rather than a layout problem, but since its understanding requires a similar approach it seems appropriate for discussion here. Referring to Fig. 16.9, the question which could have arisen in initial design is whether C_2 should be returned to earth as shown, or to the negative line, or the positive line. After considering the above rules, it is obvious that 'earth' is here the correct choice, since only in this way can the emitter, base, and signal returns be made to the same point. Connection to the negative line would mean that any signal content on that line (caused for instance by the change of current which occurs in R_4 when the output appears) would directly add to or subtract from the input, i.e. the true base–emitter signal voltage, which is the signal to which the transistor responds, would not be the intended input signal. The same is true of connection to the negative line.

Another, and in practice more important, point is that at switch-on with C_2 connected to the negative line, and with a low-impedance source attached to the input, the current which could flow through C_2 and T_1 would be virtually unlimited, because initially C_2 would remain with zero charge and would probably destroy T_1. Conversely, with C_2 connected to the positive line, switch-on would drag T_1 emitter down to the negative rail, damaging T_1 if its reverse base–emitter rating is less than the supply voltage.

Chapter 17

Consequential damage charts

When electronic equipment is found to be defective during production testing or subsequently in normal use, the faulty component, joint, or track is located and corrected. Provided the equipment then meets its routine room-temperature tests it is usually assumed fit for service.

Often the application of power to the equipment with the fault present will have caused stress to other components. This could result in changes in their electrical characteristics so that the equipment fails its routine test after clearing the original fault. In some cases the replacement for the faulty component is itself damaged because of these changes to other components. When these effects occur, intelligent diagnosis by the test engineer or the designer will usually yield a solution; if this proves too time-consuming then complete circuit boards can be scrapped. In either event, the final equipment can be assumed to be reliable, since all faulty components will have been replaced.

A much more difficult problem arises when the damage to other components is insufficient to cause failure on a routine room-temperature test. Consequential damage would not be investigated, and the equipment could prove unreliable in service owing to premature failure of components or parametric changes at temperature extremes due to their earlier abuse from the original fault.

To avoid this unreliability it is desirable that all components overstressed due to the original fault should be replaced. The extra cost involved is small compared with the cost of clearing a subsequent fault. If the train of consequential damage is very long, it may be deemed advisable to scrap the board because of the mechanical consequential damage likely to be caused during repair. In either case the building into a final equipment of suspect components will be avoided.

Consequential Damage Charts

It is becoming standard practice in many companies to prepare a Consequential Damage Chart for each circuit board used in production equipment. This chart indicates the additional components to be replaced when power has been applied to the board and a faulty component found during routine functinoal board tests.

An understanding of the circuit operation is essential in chart preparation. If, during chart preparation, long trains of damage are found, a weakness in the design may be inferred and a change of circuit configuration or possibly the addition of protective components considered. From both these aspects it is therefore desirable that the designer should himself prepare the chart and this is increasingly being regarded as part of his normal work.

Without some simplifying rules, chart preparation could be very time-consuming; by adopting the suggestions in the next section this preparation can be carried out reasonably rapidly. It is often assumed that computer analysis could give a much quicker result, but attempts to produce a simple computer routine have so far been unsuccessful, while the manual method is easy and quick.

Chart Preparation

The following procedures apply to individual printed boards and similar circuit units in which components may be faulty, of the wrong type or value, or connected in the wrong manner but to the correct track pads. The possibilities of wrong interconnection patterns, solder bridges between tracks and incorrect external connections are not considered here. In order to make chart preparation practical, the following principles are recommended.

1 Begin by listing all the components in the circuit, and suggest replacement of all other components damaged consequentially as a fault in the listed component or a wrongly fitted value.

2 Assume the extreme possibilities implied by the original fault. If a resistor is high, assume it is open-circuit (o/c). If a resistor is low, assume a short-circuit (s/c). Assume similar extremes for other components. For a semiconductor assume that any combination of short- and open-circuits between electrodes may have occurred before reaching the known faulty condition.

3 For deriving trains of consequential damage, again assume the extremes of possible damage for any component overstressed.

4 Assume that supplies may have been switched on or off as step functions in any order before or after the known fault had occurred. The slow rise and fall of practical supplies cannot readily be taken into account in preventing damage where instant rise and fall would be definitely damaging, and in practical testing momentary board disconnection and reconnection is a common occurrence.

5 Assume that any over-rating of a component causes damage. This includes, for instance, reversal of polarity of electrolytic capacitors or the subjecting of semiconductor junctions to voltages beyond the stated rating even if current limiting may appear to prevent damage.

If the above procedures result in unreasonably large numbers of consequentially damaged components, such that complete board replacement may be more economic, then the design of the board is suspect in this context. One extra resistor or a slight change of circuit configuration often provides a simple solution. A typical analogue circuit yields a maximum of about four damaged components as a result of one fault.

Validity of Recommended Procedure

The above method appears unduly pessimistic, implying that any faulty component has been more faulty than is at all likely, and that even slight over-ratings cause similar destruction to other components.

The reason for this policy is simplicity: it enables charts to be prepared manually and rarely occupies more than a few man-hours for a 12-transistor circuit.

Example (*see* Fig. 17.1)

Starting at R_1, assume R_1 is found low or high due either to damage, or to the fitting of a wrong value. Consider only the two extremes, i.e. s/c or o/c. If s/c, T_1 base current and D_1 current may be exceeded, since source power is unknown. If o/c, no damage will occur. If T_1 and D_1 are both damaged no further damage occurs. If only D_1 were damaged initially and D_1 became o/c, then T_1 would

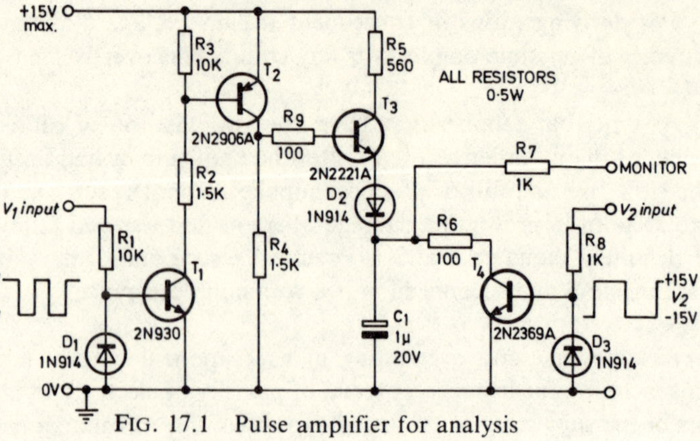

FIG. 17.1 Pulse amplifier for analysis

then be damaged by excessive emitter base reverse voltage on negative inputs.

The first chart entry is therefore:

FAULTY	REPLACE ALSO
R_1	D_1, T_1

The next devices are D_1 and T_1, already dealt with, giving the next entries:

FAULTY	REPLACE ALSO
D_1	T_1
T_1	—

Next is R_2, its function being to limit I_{C1} and I_{B2}. If faulty and o/c, no damage. If faulty and s/c, T_1 and T_2 are damaged by excess current (NB: no attempt is made to assess whether the drive to T_1 is adequate to turn on a damaging collector current, since T_1 may be out of specification on h_{FE} (max), normally unimportant with R_2 present.)

Faulty T_1 has no further effect except to maintain, if s/c e–c, the damaging base drive to T_2. Faulty T_2 has no effect since the most it can achieve is a 0 or 15 V drive to T_3, a normal condition. R_3 causes no damage whether s/c or o/c.

Next chart entries are therefore:

FAULTY	REPLACE ALSO
R_2	T_1, T_2
T_2	—
R_3	—

R_4 s/c will damage T_2 (again, no dissipation calculation for T_2 with 10 mA drive should be made for assessing risk as h_{FE} (max) may be out of specification). R_4 o/c has no damaging effect.

Entry:

FAULTY	REPLACE ALSO
R_4	T_2

Next T_3. If o/c no damage results. If s/c, and T_4 is on (for how long is unknown) and therefore a permanent state is assumed, then R_5 and R_6 are driven by 15 V less V_{D2} and $V_{CE(sat)4}$. Ignoring the diode and transistor drops, R_5 dissipates nominally 225/560, i.e. 0·4 W, a safe level.

Entry:

FAULTY	REPLACE ALSO
T_3	—

If R_5 is itself s/c as fitted then R_6 will be damaged when T_3 and T_4 turn on and if both R_5 and R_6 are finally s/c, T_3, T_4 and D2 will be damaged.

If R_6 is s/c as fitted then T_4 will be damaged due to the undefined collector current from C_1 as T_4 turns on.

FAULTY	REPLACE ALSO
R_5	R_6, T_3, T_4, D2
R_6	T_4

Now D_2; if o/c no damage. If s/c T_3 base emitter will reverse by 15 V if T_2 is first on, then off, as C_1 remains charged (T_4 off throughout). T_3 is assumed thereby damaged, even though R_4 limits the maximum dissipation in T_3 to about 40 mW.

Entry:

FAULTY	REPLACE ALSO
D_2	T_3

C_1 whether s/c, o/c or reversed causes no damage. Entry:

FAULTY	REPLACE ALSO
C_1	—

R_7 is for monitor point protection and, if s/c as fitted, allows s/c monitor equipment to cause malfunction but not damage.

FAULTY	REPLACE ALSO
R_7	—

T_4 o/c is not damaging. T_4 s/c is a normal circuit condition ($V_{CE4} \simeq 0$).

D_3 s/c causes no damage, but o/c allows excessive reverse V_{be4}, damaging T_4. R_8, s/c as fitted, damages D_3 and T_4 either directly or consequentially.

With R_8 s/c, D_3 can become o/c, resulting in T_4 B–C s/c, driving R_6 to -15 V. This reverse-biases and damages C_1, and overdissipates R_6 with their previously determined consequences. Note that this sequence is only true if R_8 is s/c.

R_9 is a current limit for T_2 as C_1 charges and if s/c results in damage to T_2 and T_3.

Entries:

FAULTY	REPLACE ALSO
D_3	T_4
R_8	D_3, T_4, C_1, R_6
R_9	T_2, T_3

Part Four

APPENDICES

Appendix 1

Half-wave diode rectification

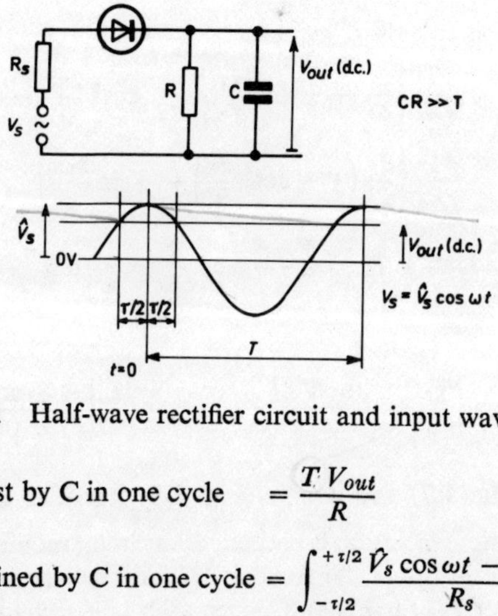

FIG. A1.1 Half-wave rectifier circuit and input waveform

$$\text{Charge lost by C in one cycle} = \frac{T\,V_{out}}{R} \qquad \text{...(A.1)}$$

$$\text{Charge gained by C in one cycle} = \int_{-\tau/2}^{+\tau/2} \frac{\hat{V}_s \cos \omega t - V_{out}}{R_s}\, dt$$

$$= \frac{1}{R_s}\left[\frac{\hat{V}_s}{\omega} \sin \omega t - t V_{out}\right]_{-\tau/2}^{+\tau/2}$$

$\left(\text{Note: } T = \dfrac{2\pi}{\omega}\right)$

$$= \frac{1}{R_s}\frac{\hat{V}_s T}{\pi} \sin \frac{\pi\tau}{T} - \frac{\tau}{R_s} V_{out} \qquad \text{...(A.2)}$$

Equating expressions (A.1) and (A.2)

$$V_{out} = \frac{\hat{V}_s \sin(\pi\tau/T)}{\pi[R_s/R + \tau/T]} \qquad \text{...(A.3)}$$

Also $$V_{out} = \hat{V}_s \cos(\pi\tau/T) \text{ directly,} \qquad \text{...(A.4)}$$

therefore $$\frac{R_s}{R} = \frac{1}{\pi}\left[\tan\frac{\pi\tau}{T} - \frac{\pi\tau}{T}\right] \qquad \text{...(A.5)}$$

From equations (A.3) and (A.5), $(V_{out}/\hat{V}_s)$ may be plotted against R_s/R and against τ/T (*see* Fig. 1.9).

$$\text{Power in } R_s \text{ in 1 cycle, } P_{Rs} = \frac{1}{R_s T}\int_{-\tau/2}^{+\tau/2}(V_s - V_{out})^2\,dt$$

$$= \frac{\hat{V}_s^2}{R_s T}\int_{-\tau/2}^{\tau/2}\left(\cos\pi t - \cos\frac{\pi\tau}{T}\right)^2 dt$$

$$= \frac{V_s^2}{R_s T}\left(\frac{t}{2} + \frac{1}{2}\frac{T}{4\pi}\sin\frac{4\pi t}{T} + t\cos^2\frac{\pi\tau}{T} - \frac{2T}{2\pi}\cos\frac{\pi\tau}{T}\sin\frac{2\pi\tau}{T}\right)_{-\tau/2}^{+\tau/2}$$

Therefore

$$P_{Rs} = \frac{V_s^2}{\pi R_s}\left[\frac{\pi\tau}{2T} + \frac{1}{4}\sin\frac{2\pi\tau}{T} + \frac{\pi\tau}{T}\cos^2\frac{2\pi\tau}{T} - \sin\frac{2\pi\tau}{T}\right]$$

$$= \frac{V_s^2}{\pi R_s}\left[\frac{\pi\tau}{T}\left(\frac{1}{2} + \cos^2\frac{\pi\tau}{T}\right) - \frac{3}{4}\sin\frac{2\pi\tau}{T}\right]$$

$$\text{Now, power in load} = P_L = \frac{V_{out}^2}{R}, \text{ and } \frac{R_s}{R} = \frac{1}{\pi}\left[\tan\frac{\pi\tau}{T} - \frac{\pi\tau}{T}\right].$$

Therefore

$$\frac{P_{Rs}}{P_L} = \frac{P_{Rs}R}{V_o^2} = \frac{(\pi\tau/T)[\frac{1}{2} + \cos^2(\pi\tau/T)] - \frac{3}{4}\sin(2\pi\tau/T)}{\cos^2(\pi\tau/T)(\tan(\pi\tau/T) - (\pi\tau/T))}$$

Full-wave (Fig. 1.7) and Bridge (Fig. 1.8) Rectifier Circuits

Assuming input isolation is needed, each circuit requires an input transformer. The transformer T_{r1} for Fig. 1.7 must deliver twice the voltage of the transformer T_{r2} for Fig. 1.8. Since each half-secondary of T_{r1} passes current only on alternate half-cycles, identical cores and bobbins may be used for T_{r1} and T_{r2}, the latter using secondary wire of twice the area but half the number of turns.

For economy, Fig. 1.7 is thus preferred, since diode cost is halved.

However, in Fig. 1.8 R_s is effectively half that of Fig. 1.7 if transformer secondary resistance predominates over other contributors to source resistance.

Technically, Fig. 1.8 is then superior, having in the limit a regulation performance twice as good as that of Fig. 1.7.

Appendix 2

Analysis of bias circuit (Fig. 2.5)

Assuming an emitter current I_e, collector current is $(\alpha I_e + I_{cbo})$, giving a base current (Kirchhoff's first law) of $(1 - \alpha)I_e - I_{cbo}$ in the direction shown.

Therefore

$$V_b = R_b[(1 - \alpha)I_e - I_{cbo}]$$

Now,

$$I_e = \frac{V_e - V_{be} - V_b}{R_e}$$

Therefore

$$I_e = \frac{V_e - V_{be} - R_b[(1 - \alpha)I_e - I_{cbo}]}{R_e}$$

i.e.

$$I_e = \frac{V_e - V_{be} + R_b I_{cbo}}{R_e + R_b(1 - \alpha)} \qquad \text{...(A.6)}$$

Now,

$$I_c = \alpha I_e + I_{cbo}$$

Therefore

$$I_c = \frac{\alpha(V_e - V_{be}) + I_{cbo}(R_e + R_b)}{R_e + R_b(1 - \alpha)} \qquad \text{...(A.7)}$$

and

$$V_L = I_c R_L$$

i.e.

$$V_L = \frac{R_L}{R_e + R_b(1 - \alpha)}[\alpha(V_e - V_{be}) + I_{cbo}(R_e + R_b)] \qquad \text{...(A.8)}$$

Special Cases

(*a*) *When*

$$R_b = 0$$

$$I_e = \frac{V_e - V_{be}}{R_e} \qquad \text{...(A.9)}$$

$$I_c = \alpha\frac{V_e - V_{be}}{R_e} + I_{cbo} \qquad \text{...(A.10)}$$

and $$V_L = R_L \left[\alpha \frac{V_e - V_{be}}{R_e} + I_{cbo}\right] \quad \text{...(A.11)}$$

(*b*) *When*

$$R_b \to \infty$$

$$(1 - \alpha)I_e - I_{cbo} = 0$$

Therefore

$$I_e = \frac{I_{cbo}}{1 - \alpha} \quad \text{...(A.12)}$$

$$I_c = I_e = \alpha I_e + I_{cbo}$$

$$I_c = \frac{I_{cbo}}{1 - \alpha} \quad \text{...(A.13)}$$

and $$V_L = R_L \frac{I_{cbo}}{1 - \alpha} \quad \text{...(A.14)}$$

(*c*) *When*

$$R_e = 0$$

$$R_b[(1-\alpha)I_e - I_{cbo}] + V_{be} = V_e$$

Therefore

$$I_e = \frac{V_e - V_{be} + R_b I_{cbo}}{R_b(1 - \alpha)} \quad \text{...(A.15)}$$

Also, $$I_c = \alpha I_e + I_{cbo}$$

Therefore

$$I_c = \frac{\alpha(V_e - V_{be}) + R_b I_{cbo}}{R_b(1 - \alpha)} \quad \text{...(A.16)}$$

and $$V_L = R_L \left[\frac{\alpha(V_e - V_{be}) + R_b I_{cbo}}{R_b(1 - \alpha)}\right] \quad \text{...(A.17)}$$

Notes

Special case (*a*), $R_b = 0$. I_e is independent of α, I_{cbo} and V_{be} provided that $V_e \gg V_{be}$.

I_c somewhat dependent on I_{cbo} and α: only slight dependence if $I_c \gg I_{cbo}$ and $\alpha \approx 1$.

Special case (*b*) $R_b \to \infty$. I_e and I_c critically dependent on α and directly proportional to I_{cbo}.

Special case (*c*) $R_e = 0$. I_e and I_c critically dependent on α. Dependence on I_{cbo} is great if $R_b I_{cbo} \gg V_e$, negligible if $R_b I_{cbo} \ll V_e$.

Appendix 3

Analysis of emitter follower and earthed emitter amplifier

Fig. A.3.1 is the simplified equivalent circuit (as Fig. 4.10) which enables the impedance and gain figures for both the emitter follower and the earthed emitter amplifier to be calculated.

$$i_1 = \frac{V_1 - V_e}{Z_s + \beta/g_m} = \frac{V_e}{Z_e} + i_o$$

$$\therefore \qquad V_1 = V_e \frac{Z_e + Z_s + \beta/g_m}{Z_e} + i_o(Z_s + \beta/g_m) \tag{1}$$

$$V_e - V_L = (i_o + g_m V_{be})/h_{oe}$$

But $$V_{be} = \frac{V_1 - V_e}{Z_s + \beta/g_m} \beta/g_m$$

$$\therefore \; (V_e - V_L)h_{oe} = i_o + \beta \frac{V_1 - V_e}{Z_s + \beta/g_m} \tag{2}$$

From (1) and (2),

$$V_L h_{oe} + V_1 \frac{\beta - h_{oe}Z_e}{Z_e + Z_s + \beta/g_m} + i_o \left[1 + \frac{Z_e(\beta + h_{oe}\{Z_s + \beta/g_m\})}{Z_e + Z_s + \beta/g_m}\right] = 0 \tag{3}$$

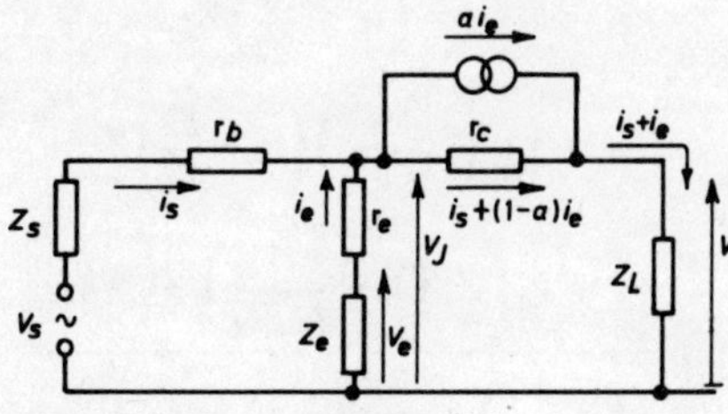

FIG. A3.1

Grounded Emitter

Now for grounded emitter stage,

$$\textit{Output impedance} = Z_o = -\frac{\partial V_L}{\partial i_o} = \frac{1}{h_{oe}}\left[1 + \frac{Z_e(\beta + h_{oe}\{Z_s + \beta/g_m\})}{Z_e + Z_s + \beta/g_m}\right] \quad (4)$$

$\therefore$ if $\quad Z_e \to \infty,\ Z_o \to \frac{1}{h_{oe}}(1+\beta) \simeq \beta/h_{oe}$

and if $\quad Z_e \to 0,\ Z_o \to \frac{1}{h_{oe}}$

Gain $\quad V_L = i_o Z_L \quad (5)$

$\therefore$ from (3) and (5),

$$\text{Gain} = \frac{V_L}{V_1} = -\frac{(\beta - h_{oe}Z_e)/(Z_e + Z_s + \beta/g_m)}{h_{oe} + \dfrac{1}{Z_L}\left[1 + \dfrac{Z_e(\beta + h_{oe}\{Z_s + \beta/g_m\})}{Z_e + Z_s + \beta/g_m}\right]}$$

$$= -\frac{(\beta - h_{oe}Z_e)Z_L}{(1 + h_{oe}Z_L)(Z_e + Z_s + \beta/g_m) + Z_e(\beta + h_{oe}\{Z_s + \beta/g_m\})}$$

If $\quad \frac{1}{h_{oe}} \gg Z_L$ and $\frac{1}{h_{oe}} \gg Z_e/\beta, \quad (6)$

then gain $\quad = \frac{V_L}{V_1} = -\frac{Z_L}{Z_e + 1/g_m + Z_s/\beta}$

Input Impedance

From Fig. A3.1

$$V_e - V_L = (i_o + g_m V_{be})/h_{oe}$$

But $i_o = V_L/Z_L$ and $V_{be} = i_1\beta/g_m$

$$\therefore \quad V_e = V_L\left(1 + \frac{1}{h_{oe}Z_L}\right) + \frac{1}{h_{oe}} i_1\beta$$

Since $\quad V_L = i_o Z_L = \left(i_1 - \frac{V_e}{Z_e}\right) Z_L,$

$$V_e = \left(1 + \frac{1}{h_{oe}Z_L}\right)\left(i_1 - \frac{V_e}{Z_e}\right)Z_L + \frac{1}{h_{oe}} i_1\beta$$

$$\therefore \quad V_e = \frac{i_1\left[Z_L\left(1 + \dfrac{1}{h_{oe}Z_L}\right) + \beta/h_{oe}\right]}{1 + \dfrac{Z_L}{Z_e}\left(1 + \dfrac{1}{h_{oe}Z_L}\right)} \quad (7)$$

But $$V_1 = V_e + i_1(Z_s + \beta/g_m)$$

$$\therefore \text{ Input impedance } Z_{in} = \frac{V_1}{i_1} - Z_s = \beta/g_m + \frac{Z_L\left(1 + \frac{1}{h_{oe}Z_L}\right) + \beta/h_{oe}}{1 + \frac{Z_L}{Z_e}\left(1 + \frac{1}{h_{oe}Z_L}\right)}$$

$$= \beta/g_m + Z_e \frac{Z_Lh_{oe} + 1 + \beta}{1 + (Z_e + Z_L)h_{oe}}$$

If $(Z_e + Z_L)h_{oe} \ll 1$, this reduces to

$$Z_{in} = \beta(Z_e + 1/g_m) \tag{8}$$

If $Z_L \to \infty$,

$$Z_{in} = \beta/g_m + Z_e$$

Emitter Follower

The basic equations are identical to those derived above for the earthed emitter amplifier, but Z_L is generally zero and the output terminal is the emitter.

$\therefore$ $Z_{in} = \beta(Z_e + 1/g_m)$ as for earthed emitter.

$$\text{Gain} = \frac{V_e}{V_1}$$

$$V_1 - V_e = i_1(Z_s + \beta/g_m)$$

From (7),

$$V_e = \frac{Z_e(1 + \beta + Z_Lh_{oe})}{1 + h_{oe}(Z_e + Z_L)} \cdot \frac{V_1 - V_e}{Z_s + \beta/g_m}$$

$$\therefore \quad \frac{V_e}{V_1} = \frac{\dfrac{Z_e(1 + \beta + Z_Lh_{oe})}{\{1 + h_{oe}(Z_e + Z_L)\}\{Z_s + \beta/g_m\}}}{1 + \left\{\dfrac{1}{Z_s + \beta/g_m}\right\}\dfrac{Z_e(1 + \beta + Z_Lh_{oe})}{\{1 + h_{oe}(Z_e + Z_L)\}}}$$

$$= \frac{Z_e(1 + \beta + Z_Lh_{oe})}{Z_e(1 + \beta + Z_Lh_{oe}) + \{Z_s + \beta/g_m\}\{1 + h_{oe}(Z_e + Z_L)\}}$$

$$\therefore \quad \text{Gain} = \frac{Z_e}{Z_e + 1/g_m + Z_s/\beta} \tag{9}$$

if $h_{oe}(Z_e + Z_L) \ll 1$ and $Z_Lh_{oe} \ll \beta$, $\beta \gg 1$.

Output Impedance

Let load current in Z_e be i_e

Then $$i_e = i_1 - i_o$$

Now $$i_1 = \frac{V_1 - V_e}{Z_s + \beta/g_m}$$

and $$V_e - V_L = (i_o + g_m V_{be})\frac{1}{h_{oe}} = [i_o + i_1\beta]\frac{1}{h_{oe}}.$$

$\therefore$ $$i_o = h_{oe}(V_e - V_L) - \beta i_1$$

$\therefore$ $$i_e = i_1 - i_o = \frac{V_1 - V_e}{Z_s + \beta/g_m}[1 + \beta] - h_{oe}(V_e - V_L)$$

$$\therefore \quad V_e\left[h_{oe} + \frac{1+\beta}{Z_s + \beta/g_m}\right] = V_L h_{oe} + V_1\frac{(1+\beta)}{Z_s + \beta/g_m} - i_e$$

$$\therefore \quad Z_{out} = -\left.\frac{\partial V_e}{\partial i_e}\right|_{V_L,\, V_{1\,const}} = \frac{1}{h_{oe} + \dfrac{1+\beta}{Z_s + \beta/g_m}} \simeq Z_s/\beta + \frac{1}{g_m} \qquad (10)$$

Appendix 4

Analysis of feedback amplifier

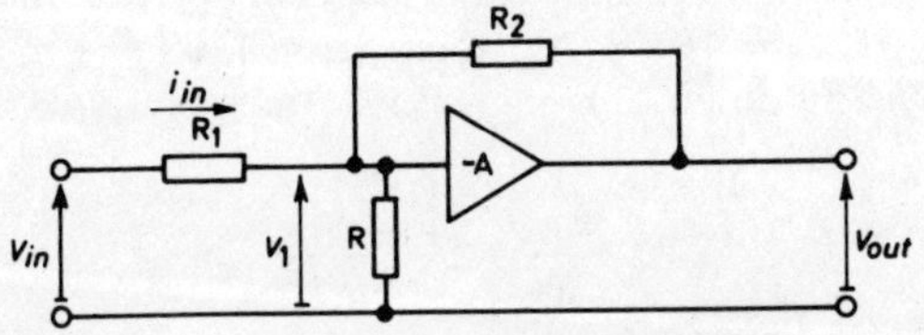

FIG. A4.1 Amplifier with shunt voltage feedback

Voltage Gain v_{out}/v_{in}

By superposition

$$v_1 = v_{out}\frac{RR_1/(R+R_1)}{R_2+[RR_1/(R+R_1)]} + v_{in}\frac{RR_2/(R+R_2)}{R_1+[RR_2/(R+R_2)]} \quad \text{...(A.49)}$$

and

$$v_1 = -\frac{v_{out}}{A} \quad \text{...(A.50)}$$

From equations (A.49) and (A.50)

$$\frac{v_{out}}{v_{in}} = -\frac{\dfrac{RR_2/(R+R_2)}{R_1+RR_2/(R+R_2)}}{\dfrac{1}{A}+\dfrac{RR_1/(R+R_1)}{R_2+RR_1/(R+R_1)}}$$

$$= -\frac{\left\{\dfrac{RR_2/(R+R_2)}{R_1+RR_2/(R+R_2)}\right\}\left[\dfrac{R_2+RR_1/(R+R_1)}{RR_1/(R+R_1)}\right]}{\left\{\dfrac{R_2+RR_1/(R+R_1)}{ARR_1/(R+R_1)}\right\}+1}$$

$$= -\frac{R_2/R_1}{1+\dfrac{R_2+RR_1/(R+R_1)}{ARR_1/(R+R_1)}} \quad \text{...(A.51)}$$

If $R \gg R_1$, this reduces to

$$\frac{v_{out}}{v_{in}} = -\frac{R_2/R_1}{1 + [(R_1 + R_2)/AR_1]} \qquad \text{...(A.52)}$$

When $AR_1/(R_1 + R_2) \gg 1$,

$$\frac{v_{out}}{v_{in}} = -\frac{R_2}{R_1} \qquad \text{...(A.53)}$$

Frequency Response

If $A = A_0/(1 + jf/f_0)$ (i.e. the 3 dB high-frequency cut-off is f_0), then from equation (A.51)

$$\frac{v_{out}}{v_{in}} = -\frac{R_2/R_1}{1 + \dfrac{(1 + jf/f_0)[R_2 + RR_1/(R + R_1)]}{A_0RR_1/(R + R_1)}}$$

This is 3 db down when the real and imaginary parts of the denominator are equal. (If $z = 1/(x + jy)$ then $|z| = 1/\sqrt{(x^2 + y^2)} = 1/[\sqrt{2}(x)]$ when $x = y$.) This occurs at $f = f'_0$, if v_{out}/v_{in} is expressed in the form

$$\frac{v_{out}}{v_{in}} = \frac{\left[\dfrac{v_{out}}{v_{in}}\right]_{LF}}{1 + jf/f_0'}.$$

Now $$\frac{f_0'}{f_0} = \frac{A_0RR_1/(R + R_1) + R_2 + RR_1/(R + R_1)}{R_2 + RR_1/(R + R_1)}$$

i.e. $$f_0' = \left\{1 + A_0\frac{RR_1/(R + R_1)}{R_2 + RR_1/(R + R_1)}\right\}f_0$$

If $R \to \infty$ and $A_0R_1/(R_1 + R_2) \gg 1$, this reduces to

$$f_0' = f_0[A_oR_1/(R_1 + R_2)]$$

A similar result is obtained for low-frequency response, i.e. improved by a factor $A_0R_1/(R_1 + R_2)$.

Input Impedance

$$Z_{in} = \left.\frac{\partial v_{in}}{\partial i_{in}}\right|_{v_{out}}$$

Now $$i_{in} = \frac{v_{in} - v_1}{R_1} = \frac{v_{in} + v_{out}/A}{R_1}$$

(from equation (A.50))

Therefore, from equation (A.51)

$$i_{in} = v_{in}\left[\frac{1}{R_1} + \frac{1}{AR_1}\left\{\frac{-R_2/R_1}{1 + \dfrac{R_2 + RR_1/(R + R_1)}{ARR_1/(R + R_1)}}\right\}\right]$$

$$= \frac{v_{in}}{R_1}\left\{\frac{A + 1 + R_2/R}{A + 1 + R_2(R + R_1)/RR_1}\right\}$$

Therefore $$Z_{in} = \left.\frac{\partial v_{in}}{\partial i_{in}}\right|_{v_{out}} = R_1 + \frac{R_2}{A + 1 + (R_2/R)} \qquad \text{...(A.54)}$$

If $A + 1 + R_2/R \gg R_2/R_1$ (i.e. $AR_1/R_2 \gg 1$),

$$Z_{in} = R_1 \qquad \text{...(A.55)}$$

Output Impedance when Amplifier Output Impedance is Z_{out}

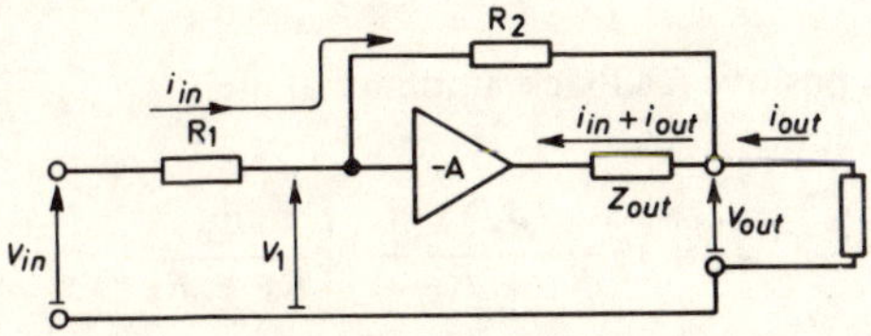

FIG. A4.2 Feedback amplifier including Z_{out}

By superposition

$$v_1 = v_{in}\frac{R_2}{R_1 + R_2} + v_{out}\frac{R_1}{R_1 + R_2} \qquad \text{...(A.56)}$$

and $$v_{out} - (i_{in} + i_{out})Z_{out} = -Av_1 \qquad \text{...(A.57)}$$

Also $$i_{in} = \frac{v_{in} - v_{out}}{R_1 + R_2} \qquad \text{...(A.58)}$$

Therefore, from equations (A.56), (A.57) and (A.58)

$$v_{out}\left[\frac{AR_1}{R_1 + R_2} + 1 + \frac{Z_{out}}{R_1 + R_2}\right] = i_{out}Z_{out} + \left.\frac{v_{in}}{R_1 + R_2}\right|(Z_{out} - AR_2)$$

Therefore output impedance =

$$\left.\frac{\partial v_{out}}{\partial i_{out}}\right|_{v_{in}} = \frac{Z_{out}}{1 + A[(R_1 + Z_{out})/(R_1 + R_2)]} \qquad \text{...(A.59)}$$

If $Z_{out} \ll R_1$ and $A\,\dfrac{R_1 + R_2}{R_1} \gg 1$, this reduces to

$$\text{Output impedance} = \frac{Z_{out}}{A[R_1/(R_1 + R_2)]} \to 0$$

Positive Feedback Within a Negative Feedback Loop

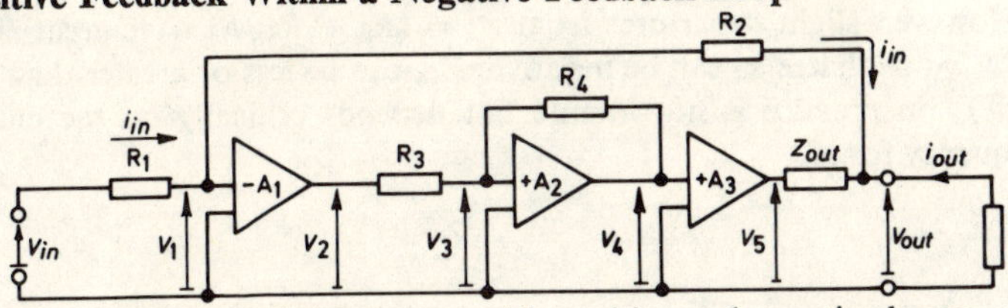

FIG. A4.3 Feedback system with positive and negative loops

Here A_1 and A_3 are the 'normal' stages of the loop amplifier. A_2 has positive feedback by R_4 and the overall gain is negative in sign.

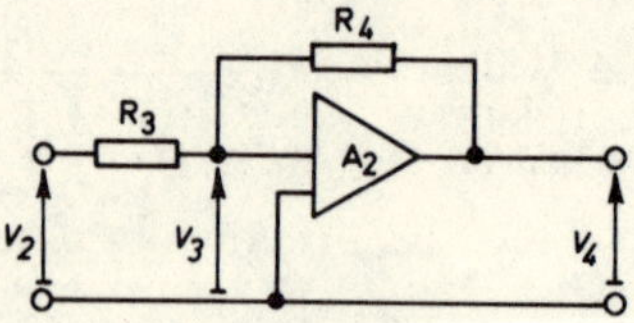

FIG. A4.4 Positive feedback loop

Consider the positive feedback amplifier alone.
By superposition

$$v_3 = v_2 \frac{R_4}{R_3 + R_4} + v_4 \frac{R_3}{R_3 + R_4} \qquad \text{...(A.60)}$$

and

$$\frac{v_4}{A_2} = v_3 \qquad \text{...(A.61)}$$

Therefore

$$\frac{v_4}{v_2} = \frac{R_4/R_3}{-1 + [(R_3 + R_4)/A_2R_3]} \qquad \text{...(A.62)}$$

Returning to the whole system,

$$\frac{v_5}{v_1} = -A_1A_3\frac{v_4}{v_2} = \frac{-A_1A_3R_4/R_3}{-1 + [(R_3 + R_4)/A_2R_3]} \qquad \text{...(A.63)}$$

The expression for v_5/v_1 may be written for $-A$ in equations (A.49)–(A.59) for a 'normal' feedback system. In each case the value of A_2 equal to $(R_3 + R_4)/R_3$ is equivalent to an infinite A.

Therefore

$$\frac{v_{out}}{v_{in}} = -\frac{R_2}{R_1}$$

$$Z_{out} = 0$$

$$Z_{in} = R_1$$

$$f_o' \rightarrow \infty$$

However, slight departures from $A_2 = (R_3 + R_4)/R_3$ have great effect on v_5/v_1, so that Z_{out} can be negative; Z_{in} can be less or greater than R_1; and f_o' in practice is not infinite but depends critically on the cut-off frequency for A.

Appendix 5

Diode pump staircase generator

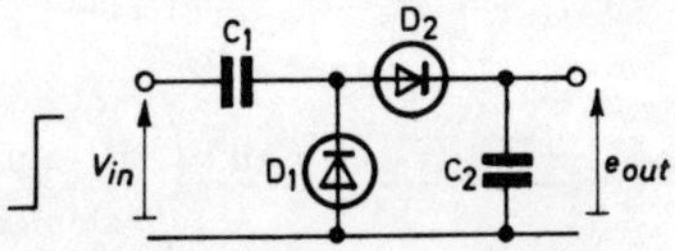

FIG. A5.1 Diode pump

After the first input pulse,

$$e_{1(out)} = \frac{C_1}{C_1 + C_2} V_{in} \qquad \text{...(A.64)}$$

If, after $(n - 1)$ pulses, $e_{(out)}$ is denoted $e_{n-1(out)}$ then the n(th) pulse increases e_{out} by $[C_1/(C_1 + C_2)]\,(V_{in} - e_{n-1(out)})$.

Therefore $$e_{n(out)} - e_{n-1(out)} = \frac{C_1}{C_1 + C_2}(V_{in} - e_{n-1(out)}) \qquad \text{...(A.65)}$$

Similarly $$e_{n-1(out)} - e_{n-2(out)} = \frac{C_1}{C_1 + C_2}(V_{in} - e_{n-2(out)}) \qquad \text{...(A.66)}$$

Equation (A.65) may be written

$$e_{n(out)} - e_{n-1(out)} \frac{C_2}{C_1 + C_2} = \frac{C_1}{C_1 + C_2} V_{in} \qquad \text{...(A.67)}$$

Equation (A.66) may be written

$$e_{n-1(out)} - e_{n-2(out)} \frac{C_2}{C_1 + C_2} = \frac{C_1}{C_1 + C_2} V_{in} \qquad \text{...(A.68)}$$

This sequence may be continued until finally, when $n = 2$,

$$e_{1(out)} - e_{0(out)} = \frac{C_1}{C_1 + C_2} V_{in} \qquad \text{...(A.69)}$$

which confirms equation (A.64) provided $e_{0out} = 0$, i.e. C_2 carries no initial charge.

By multiplying equation (A.68) by $C_2/(C_1 + C_2)$, the next equation in the sequence by $[C_2/(C_1 + C_2)]^2$ etc., and finally equation (A.69) by $[C_2/(C_1 + C_2)]^{n-1}$, the following equation for $e_{n(out)}$ is obtained by addition:

$$e_{n(out)} - e_{0(out)}\left(\frac{C_2}{C_1 + C_2}\right)^{n-1} =$$

$$V_{in}\frac{C_1}{C_1 + C_2}\left[1 + \frac{C_2}{C_1 + C_2} + \ldots + \left(\frac{C_2}{C_1 + C_2}\right)^{n-1}\right]$$

Therefore $$e_n = e_{0(out)}\left(\frac{C_2}{C_1 + C_2}\right)^{n-1} + V_{in}\frac{C_1}{C_1 + C_2}\left[\frac{1 - \left(\dfrac{C_2}{C_1 + C_2}\right)^{n}}{1 - \dfrac{C_2}{C_1 + C_2}}\right]$$

i.e. $$\underline{e_n = V_{in}\{1 - [C_2/(C_1 + C_2)]^n\}}, \quad \text{if} \quad e_{0(out)} = 0$$

Appendix 6

Low-frequency response of high-impedance bootstrap circuit

In the following analysis of a nominally unity-gain amplifier, using bootstrap feedback to obtain high input impedance, three commonly encountered non-ideal conditions are considered. These are the presence of source resistance R_s, parallel resistance R from amplifier input to earth (i.e. any un-bootstrapped component), and non-unity gain in the amplifier (i.e. imperfect bootstrapping).

For normal conditions it is shown that these have little effect on either the frequency or the gain at the response peak.

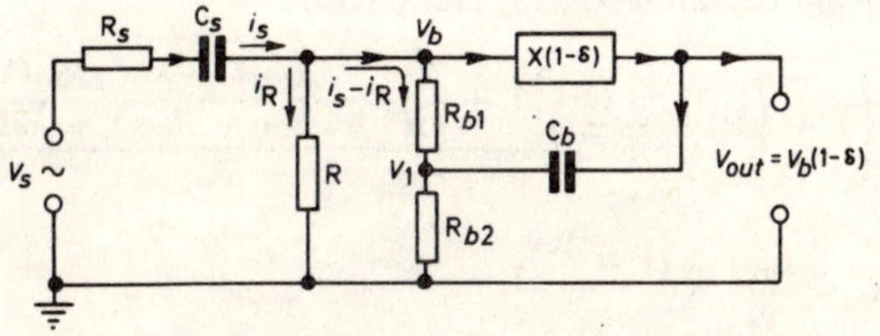

FIG. A6.1 Amplifier with bootstrap feedback

Note that it is inadmissible to state that C_s and C_b will be made so large as to be unimportant over the signal frequency band, since however large they are there is always a frequency at which a peak occurs. Even though this frequency may be outside the intended signal band, it can still cause trouble by overloading following stages after an input transient, caused by switch-on or changing the input source. In an extreme case where C_s and C_b are much larger than normal design would dictate, and where C_b/C_s is also large, an amplifier intended for audio amplification can block for several seconds after an input transient.

Analysis

The current in R_{b2}, namely v_1/R_{b2}, is the sum of two currents, one from R_{b1}, the other from C_b.

Hence,
$$\frac{v_1}{R_{b2}} = \left[\frac{v_b - v_1}{R_{b1}} + \{(1 - \delta)v_b - v_1\}j\omega C_b\right]$$

Therefore
$$v_b R_{b2}\{1 + j\omega C_b(1 - \delta)R_{b1}\} = v_1\{R_{b1} + R_{b2}(1 + j\omega C_b R_{b1})\} \qquad \text{...(A.70)}$$

Now, v_b is given by
$$v_b = Ri_R$$
and also by
$$v_b - v_1 = R_{b1}(i_s - i_R)$$
$$= R_{b1}\left(i_s - \frac{v_b}{R}\right)$$

Therefore
$$v_b\left[1 + \frac{R_{b1}}{R}\right] = v_1 + R_{b1}i_s \qquad \text{...(A.71)}$$

Therefore, from equations (A.70) and (A.71)
$$v_b\left[1 + \frac{R_{b1}}{R} - \frac{R_{b2}\{1 + j\omega C_b R_{b1}(1 - \delta)\}}{R_{b1} + R_{b2}(1 + j\omega C_b R_{b1})}\right] = i_s R_{b1} \qquad \text{...(A.72)}$$

Now,
$$v_s = i_s\left(R_s + \frac{1}{j\omega C_s}\right) + v_b \qquad \text{...(A.73)}$$

Therefore, from equations (A.72) and (A.73)
$$v_s = v_b\left\{1 + \left(R_s + \frac{1}{j\omega C_s}\right)\frac{\left[1 + \dfrac{R_{b1}}{R} - \dfrac{R_{b2}\{1 + j\omega C_b R_{b1}(1 - \delta)\}}{R_{b1} + R_{b2}(1 + j\omega C_b R_{b1})}\right]}{R_{b1}}\right\} \qquad \text{...(A.74)}$$

Therefore
$$G = \frac{v_{out}}{v_s} = \frac{(1 - \delta)v_b}{v_s}$$
$$= \frac{1 - \delta}{1 + \left(R_s + \dfrac{1}{j\omega C_s}\right)\dfrac{1 + \dfrac{R_{b1}}{R} - \dfrac{R_{b2}\{1 + j\omega C_b R_{b1}(1 - \delta)\}}{R_{b1} + R_{b2}(1 + j\omega C_b R_{b1})}}{R_{b1}}}$$
$$= \frac{1 - \delta}{1 + \left(R_s + \dfrac{1}{i\omega C_s}\right)\left[\dfrac{\dfrac{R_{b1}}{R} + \dfrac{R_{b1}\{1 + j\omega C_b R_{b2}\delta\}}{R_{b1} + R_{b2}(1 + j\omega C_b R_{b1})}}{R_{b1}}\right]}$$
$$= \frac{1 - \delta}{1 + \left(R_s + \dfrac{1}{j\omega C_s}\right)\dfrac{R_{b1} + R_{b2} + R + j\omega C_b R_{b2}\{R_{b1} + R\delta\}}{R[R_{b1} + R_{b2} + j\omega C_b R_{b1} R_{b2}]}}$$

Therefore $G = \dfrac{(1-\delta)(R_{b1} + R_{b2} + j\omega C_b R_{b1} R_{b2})}{R_{b1} + R_{b2} + R_s\left(1 + \dfrac{R_{b1} + R_{b2}}{R}\right) + \dfrac{C_b}{C_s} R_{b2}\left(\dfrac{R_{b1}}{R} + \delta\right) + j\left[\omega C_b\left(R_{b1}R_{b2} + R_s R_{b2}\left\{\dfrac{R_{b1}}{R} + \delta\right\}\right) - \dfrac{1}{\omega C_s}\left(1 + \dfrac{R_{b1} + R_{b2}}{R}\right)\right]}$

Therefore $|G| = (1 - \delta) \times$

$$\sqrt{\left\{\frac{(R_{b1} + R_{b2})^2 + (\omega C_b R_{b1} R_{b2})^2}{\left[R_{b1} + R_{b2} + R_s\left(1 + \dfrac{R_{b1} + R_{b2}}{R}\right) + \dfrac{C_b}{C_s} R_{b2}\left(\dfrac{R_{b1}}{R} + \delta\right)\right]^2 + \left[\omega C_b\left(R_{b1}R_{b2} + R_s R_{b2}\left\{\dfrac{R_{b1}}{R} + \delta\right\}\right) - \dfrac{1}{\omega C_s}\left(1 + \dfrac{R_{b1} + R_{b2}}{R}\right)\right]^2}\right\}} \quad \text{...(A.75)}$$

$|G|$ is maximum near

$$\omega C_b\left(R_{b1}R_{b2} + R_s R_{b2}\left\{\frac{R_{b1}}{R} + \delta\right\}\right) = \frac{1}{\omega C_s}\left(1 + \frac{R_{b1} + R_{b2}}{R}\right)$$

i.e. $$\omega^2 = \frac{1 + \dfrac{R_{b1} + R_{b2}}{R}}{C_s C_b R_{b2}\left[R_{b1}\left(1 + \dfrac{R_s}{R}\right) + R_s\delta\right]}$$

or $$\omega = \frac{\sqrt{1 + \dfrac{R_{b1} + R_{b2}}{R}}}{\sqrt{C_s C_b R_{b2}\left[R_{b1}\left(1 + \dfrac{R_s}{R}\right) + R_s\delta\right]}} \quad \text{...(A.76)}$$

For this value of ω, $|G|_{max.}$ is given by, from equation (A.75),

$$|G|_{max.} = \frac{(1-\delta)\sqrt{\left\{(R_{b1} + R_{b2})^2 + \dfrac{(1 + [(R_{b1} + R_{b2}/R)]C_b R_{b2} R_{b1}{}^2)}{C_s[R_{b1}(1 + R_s/R) + R_s\delta}\right\}}}{R_{b1} + R_{b2} + R_s\left(1 + \dfrac{R_{b1} + R_{b2}}{R}\right) + \dfrac{C_b}{C_s} R_{b2}\left(\dfrac{R_{b1}}{R} + \delta\right)}$$

In the typical case where $R \gg R_{b1} + R_{b2}$ (a)

$R \gg R_s$ (b)

$R_s\delta \ll R_{b1}$ (c)

and $\delta \ll 1$ (d)

equation (A.76) reduces to $\omega = \dfrac{1}{\sqrt{(C_s C_b R_{b1} R_{b2})}}$ ...(A.77) giving

$$|G|_{max.} = \frac{\sqrt{\{(R_{b1} + R_{b2})^2 + [(C_b/C_s)R_{b2}R_{b1}]\}}}{R_{b1} + R_{b2}}$$

Therefore $$|G|_{max.} = \sqrt{\left[1 + \frac{C_b}{C_s} \cdot \frac{R_{b1}R_{b2}}{(R_{b1} + R_{b2})^2}\right]} \qquad \text{...(A.78)}$$

These simplified results are combined in Fig. A6.2.

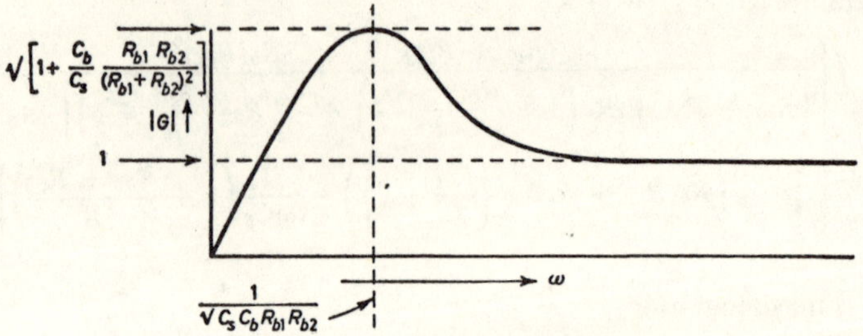

FIG. A6.2 Frequency response of bootstrap circuit

All the assumptions are valid in normal use.

(a) Implies that spurious parallel components of input impedance, which are not bootstrapped, are to be much greater than the total value of the base resistors R_{b1} and R_{b2}; this would usually be true since there would otherwise be little point in bootstrapping R_{b1}.

(b) Implies that the direct attenuation caused by source resistance and spurious parallel R is negligible.

(c) Further implies that the attenuation caused by the source resistance coupled to the bootstrapped value of R_{b1} is negligible.

(b) and (c) are always true when the bootstrap circuit is intended to give approximately unity-gain and is therefore designed so as not to load R_s appreciably.

(d) Means that the bootstrap feedback is nearly unity, a necessary condition to make the circuit effective.

The assumption made that $|G|$ is maximum when the second denominator term in equation (A.75) vanishes is valid unless C_b/C_s is very much greater than unity. In such cases the second numerator term shifts the peak frequency; this effect is usually negligible.

Inductance Analogy

The reason for the peculiar response of this circuit (Fig. A6.1) can be seen by considering the result of a step input. This step appears at the amplifier input (v_b), at the output (v_{out}), and also at v_1. Thus initially there is no change of current in R_{b1}. While the step remains at v_s and (assuming only slow discharge of C_s) at v_b, C_b begins to discharge, so that v_1 returns towards its original potential. The current in R_{b1} therefore increases exponentially and finally becomes $1/\delta$ times its original value. (This is not strictly true, since by this time C_s has discharged thus modifying the exponential.)

The above sequence shows that the current in R_{b1} begins at a low value and steadily increases, which is similar to the behaviour of an inductance. At the frequency where C_s resonates with this inductance there will therefore be a peak in the response and the peak frequency will be given by an equation of the form

$$f = \frac{1}{2\pi\sqrt{LC_s}}$$

Examination of the results of the analysis shows that the equivalent L is approximately $R_{b1}R_{b2}C_b$.

If the equivalent inductance is assumed to have internal series resistance r, then simple sine wave analysis† shows that the gain at the resonant frequency is

$$\sqrt{\left[1 + \frac{C_b}{C_s}\frac{R_{b1}R_{b2}}{r^2}\right]}$$

This is identical with the simplified result of the main analysis if $r = R_{b1} + R_{b2}$).

† *Analysis of LC_r series circuit*

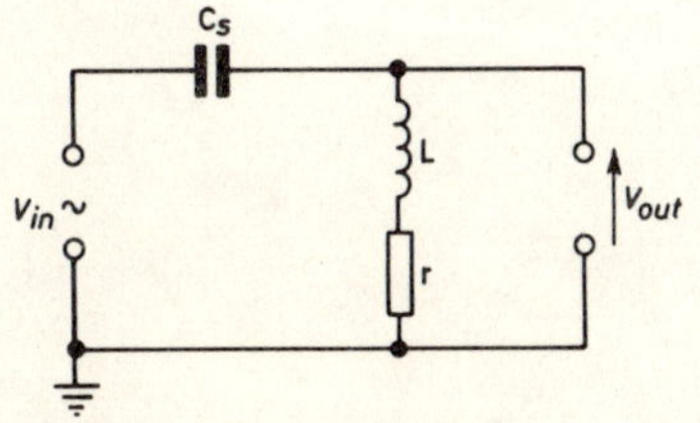

$$\frac{v_{out}}{v_{in}} = \frac{j\omega L + r}{r + j(\omega L - 1/\omega C_s)}$$

Therefore
$$\left|\frac{v_{out}}{v_{in}}\right| = \frac{\sqrt{(r^2 + \omega^2 L^2)}}{\sqrt{\{r^2 + (\omega L - 1/\omega C_s)^2\}}}$$

At resonance $\omega L = 1/(\omega C_s)$

$$\left|\frac{v_{out}}{v_{in}}\right| = \sqrt{1 + \frac{\omega^2 L^2}{r^2}} = \sqrt{1 + \frac{L}{C_s r^2}}$$

If
$$L = R_{b1}\,R_{b2}\,C_b$$

and
$$r = R_{b1} + R_{b2}$$

$$\left|\frac{v_{out}}{v_{in}}\right| = \sqrt{\left[1 + \frac{C_b}{C_s}\frac{R_{b1}\,R_{b2}}{(R_{b1} + R_{b2})^2}\right]}$$

For frequency calculations near the response peak the circuit to the right of C_s in Fig. A6.1 can therefore be represented by an inductance of $R_{b1}R_{b2}C_b$ and series resistance of $(R_{b1} + R_{b2})$ to earth, as shown in Fig. A6.3.

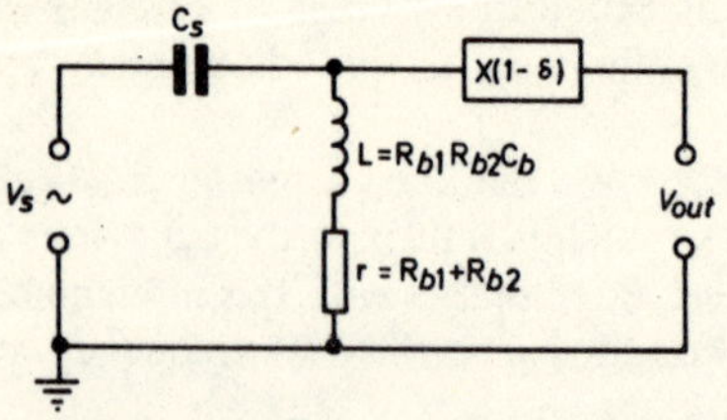

FIG. A6.3 Equivalent circuit of amplifier with bootstrap feedback (Fig. A6.1) for approximate analysis near response peak

Note that for a given low-frequency limit at which normal bootstrapping is to be effective, C_bR_{b2} will be constant, and that $(R_{b1} + R_{b2})$ will be determined by bias conditions and will also be constant. The effective L is therefore proportional to R_{b1} and the ratio R_{b1}/R_{b2} should be as small as possible consistent with R_{b1}/δ representing sufficiently high input impedance.

Appendix 7

Transistor data

For details of specific transistor types it is advisable to consult the appropriate manufacturer who is always pleased to supply up-to-date information.

Most of the symbols used for transistor parameters are defined by the manufacturer but the terms $f_0, f_\alpha, f_\beta, f_T$, and f_{hfb}, f_{hfe} often cause confusion. All are concerned with the behaviour of current gain at high frequencies.

$f_0 = f_\alpha = f_{hfb}$, often called the 'α cut-off frequency', is the frequency at which α, known as h_{fb} in h parameters, has fallen by 3 dB from its low frequency value.

$f_\beta = f_{hfe}$, often called the 'β cut-off frequency', is the frequency at which β (or h_{fe}) has fallen by 3 dB from its low frequency value.

f_T, known as the 'gain-bandwidth product' is the product of β and the frequency at which it is measured, provided this frequency is much higher than f_β. A simpler but less flexible definition of f_T is the frequency at which β has fallen to unity.

Relative Magnitude of f_α, f_β, and f_T

It is easy to calculate f_β in terms of f_α from the equations $\beta = \dfrac{\alpha}{1-\alpha}$ and $\alpha = \dfrac{\alpha_0}{1 + j(f/f_0)}$. For normal cases where $\beta \gg 1$ this gives $f_\beta = \dfrac{1}{\beta} f_\alpha$. Similarly $f_T \approx f_\alpha$.

Practical Uses of f_α, f_β, and f_T

It can be shown that the gain at high frequencies of an earthed emitter amplifier is 3 dB down compared with its low frequency gain according to the following criteria, assuming that collector capacitance is negligible:

(*a*) Emitter fully decoupled, 3 dB down when $f \approx f_\beta \approx \frac{1}{\beta} f_\alpha \approx \frac{1}{\beta} f_T$.

(*b*) Emitter undecoupled with $R_e \gg r_e$, 3 dB down when $f \approx f_\alpha \approx \beta f_\beta \approx f_T$.

For example, an amplifier using a transistor with $f_\alpha = 0{\cdot}5$ MHz and typical β of 50 would be 3 dB down at 10 kHz if fully decoupled, but 3 dB down at 0·5 MHz if an emitter load of a few kilohms were present. This is approximate and assumes that the collector load does not exceed a few kilohms. Note that there is no implication that the gain at, say, 100 kHz falls as a result of decoupling the emitter; it does in fact rise but the gain at, say, 5 kHz, rises very much more.

Use of f_T

In circuit equations f_α and f_β appear but f_T rarely occurs. The reason for its use in manufacturers' data is the relative ease with which it may be measured especially where frequencies in the GHz region are involved. Realistic comparisons between transistors are readily made and highly stable test frequencies are not required. Since β is measured at a high frequency where its law is well defined ($\beta = f_T/f$) rather than at a 3 dB point repeatability is good.

TRANSISTOR RATINGS

Most transistor ratings are easily understood provided the manufacturers' definitions are clear. It is important which collector voltage rating is stated since its permissible value depends on whether the base is open circuit, the emitter open circuit or the base emitter junction reverse biased. The symbols used for these cases are usually BV_{ceo}, BV_{cbo}, and BV_{cer}.

Power ratings are often quoted in a confusing manner. The statement that a transistor can withstand 90 W at 25°C case temperature is, by itself, useless information since the user would have to possess an infinite heat sink in an ambient temperature of 25°C to hold the case at this temperature. A statement of permissible dissipation in free air (i.e. without heat sink) is more practical but is still insufficient unless the actual air temperature coincides with that quoted on the data sheet.

To arrive at practical figures the designer needs to know firstly that the number of watts flowing through a body of thermal conductivity θ degC/W causes a temperature difference across the body of $W\theta$ degrees, and secondly that the temperature of a transistor junction must not exceed a certain figure $T_{j(max.)}$, typically 90 to 100°C for germanium and 150 to 200°C for silicon.

Taking a practical example, assume that a power transistor has a thermal conductivity from junction to case (θ_{jc}) of 1·5 degC/W and a $T_{j(max.)}$ of 90°C. This is to be bolted to a heat sink using a mica insulating washer of thermal conductivity θ_{ch} (case-to-heat sink) of 0·5 degC/W. The heat sink has a thermal conductivity θ_{ha} (heat sink to air) of 2 degC/W.

In this example, three values of θ are involved and these are simply added to give the combined θ from junction to air θ_{ja} of 1·5 + 0·5 + 2, namely 4 degC/W.

If the maximum air temperature in which the equipment has to operate is, for example, 50°C, then a drop between junction and air of $T_{j(max.)} - T_a = 90 - 50 = 40$ degC can be allowed. Since θ_{ja} is 4 degC/W, 10 W is the safe limit for transistor power dissipation. Conversely if 15 W has to be dissipated, a temperature drop of $15 \times 4 = 60$ degC will occur and the maximum safe ambient temperature is 30°C.

By using this thermodynamic equivalent of Ohm's law the required figures can be derived even if not given explicitly in the data.

Bibliography and references

Although many of the following do not deal directly with transistor circuits, they help the designer by their approach to related design problems. Littauer's work is particularly well written.

Angelo, E. J. (1965). *Electronic Circuits.* London and Philadelphia; McGraw-Hill.

Bode, H. W. (1945). *Network Analysis and Feedback Amplifier Design.* London; Van Nostrand.

Cain, W. D. (1969). *Engineering Product Design.* London; Business Books.

Cattermole, K. W. (1965). *Transistor Circuits.* London; Heywood.

Dammers, D. G., Haantje, J., Otte, J. and van Suchtelen, H. (1950). *Electronic Valves,* Vol. **4**. London; Philips Technical Library.

Deketh, J. (1950). *Electronic Valves,* Vol. **1**. London; Philips Technical Library.

Dunster, D. F. (1969). *Semiconductors for Engineers.* London; Business Books.

Hardie, A. M. (1964). *Elements of Feedback and Control.* London; Oxford University Press.

Hemingway, T. K. (1970). *Circuit Consultant's Casebook.* London; Business Books.

Heydroe, P. J. (1950). *Electronic Valves,* Vol. **7**. London; Philips Technical Library.

Jones, D. D. and Hilbourne, R. A. (1957). *Transistor A.F. Amplifiers.* London; Iliffe.

Lewis, I. A. D. and Wells, F. H. (1959). *Millimicrosecond Pulse Techniques.* Oxford; Pergamon.

Littauer, R. (1965). *Pulse Electronics.* London and Philadelphia; McGraw-Hill.

Massachusetts Institute of Technology (1943). *Radiation Laboratory Series,* Vols. **1–28**. London and Philadelphia; McGraw-Hill.

Middlebrook, R. D. (1963). *Differential Amplifiers.* New York; Wiley.

Nyquist, H. V. 'Regeneration Theory' from *Bell System Technical Journal,* Vol. **11**, Jan. 1932, pp. 126–147.

Philips, Eindhoven (1950). *Electronic Valves,* Vol. **3**. London; Philips Technical Libraries.

Philips, Eindhoven (1950). *Electronic Valves*, Vol. **3a**. London; Philips Technical Libraries.
— — (1950). *Electronic Valves*, Vol. **5**. London; Philips Technical Libraries.
— — (1950). *Electronic Valves*, Vol. **6**. London; Philips Technical Libraries.
Reintjes, J. F. and Coate, G. T. (1952). *Principles of Radar*. London and Philadelphia; McGraw-Hill.
Sturley, K. R. (1965). *Radio Receiver Design*, Vol. **1**. London; Chapman and Hall.
— (1965). *Radio Receiver Design*, Vol. **2**. London; Chapman and Hall.
Wolfendale, E. (Editor) (1958). *The Junction Transistor and Its Application*. London; Heywood.

Index